Environment and Society

Environment and Society
Advances in Research

Environment and Society

Advances in Research

Volume 2 - 2011

■ **EDITORS** ■

Paige West and Dan Brockington

Berghahn Books

New York • Oxford

Published in association with the Earth Institute of Columbia University, *Environment and Society: Advances in Research* presents articles and critical reviews reflecting the work of environmental anthropologists, geographers, scientists, human ecologists, and scholars in other environmentally oriented social sciences from all parts of the world. The field of research on environment and society is growing rapidly and becoming of ever-greater importance, not only in academia but also in policy circles and for the public at large. Climate change, the water crisis, deforestation, biodiversity loss, the looming energy crisis, nascent resource wars, environmental refugees, and environmental justice are just some of the many compelling challenges facing society today and in the future. Through this journal, we hope to stimulate advanced research and action on these and other critical issues and to encourage international communication and exchange among all relevant disciplines.

COPYRIGHT

ISSN 2150-6779 (Print) • 2150-6787 (Online)
ISBN 978-0-85745-455-3 (Paperback)

ONLINE

Environment and Society: Advances in Research is available online at www.journals.berghahnbooks.com/air-es, where you can browse tables of contents and abstracts, purchase individual articles, or recommend the journal to your library. Also visit the *Environment and Society* Web site for details including full contact information for the editors and complete submission instructions for contributors.

ADVERTISING

Environment and Society
Advances in Research

Contents

Volume 2 ■ 2011

Introduction
Environment, Society, and Food

Rebecca Feinberg, Paige West, and Dan Brockington

During the past two decades social scientists have paid an increasing amount of attention to the circulation of commodities and the effects that commodity production, distribution, and consumption have on social life (see Miller 1995). Today, social scientists are beginning to think carefully about the political ecologies of these same commodity circulations (see Bryant and Goodman 2004; Doane 2010; West 2012). We are exploring the environmental consequences of the creation, circulation, and consumption of commodities, the role of nature in shaping the commodity form, their circulation and resulting social life, and the broader political economy in which commodity circulation is found.

Food is an obvious and important entry point into the analysis of both commodity circulation and political ecology. It is important both as the commodity we consume most frequently and most intimately. Food becomes us corporeally, it changes us physically; it is a vital part of our personal, regional, and national identities, as well as the driver of, and accompaniment to, much of our social activity. Agriculture is one of the main drivers of land use change and one of the most extensive forms of land cover. Whether in the form of unrest over food prices, or protest about its industrialized production, food gets people motivated in unusual ways.

Much of the current literature on food in the social sciences fits into three categories: literature that follows food from production to distribution to consumption (the commodity chain approach), literature that focuses on the meanings attributed to foods as they circulate (the semiotics of circulation or commodity circuit approach), and literature that focuses on the lives of producers and where they fit in global political economic structures (the political economy approach). Collectively the work in these approaches provides an excellent lens with which to examine current dilemmas in social science. In particular it helps us understand different aspects of what has been called "globalization." There is now a strong contemporary literature drawing on rich histories of the analysis of agriculture, social life, and political economy from anthropology, geography, sociology, and history.

In this issue of *Environment and Society: Advances in Research* we bring together some of the insights that studying food can provide—both what they tell us with regard to food specifically, and about commodity chains, semiotics, and political economies more generally. There is a persistent theme throughout the articles; namely most take a long hard look at the transformative potential of different schemes that hope to alter the injustices and ecological violence that our food commodity chains can create. The result is not always encouraging reading but the quality of the scholarship is bracing.

Environment and Society: Advances in Research 2 (2011): 1–4 © Berghahn Books
doi:10.3167/ares.2011.020101

Cindy Isenhour's article evaluates the impact of consumer-based alternative food movements with the aim of answering the question, "Can Consumer Demand Deliver Sustainable Food?" The question is not limited to food commodity chains alone. There is a great deal of faith in the ability of consumers to change commodity chains through their purchasing decisions, in everything from clothes to energy to diamonds to tourism destinations. What sort of change they can create matters a great deal. Isenhour, by studying consumer-based movements as a category, examines the ideas at work within these initiatives and connects her central question to issues of individual agency, trends in sociopolitical culture, and inequality within the global system of production and consumption. The article outlines the promises and pitfalls of a variety of movements, demonstrating consumer-based movements' capacity to effect change, but perhaps more importantly their theoretical and practical limitations. Although consumer power is considerable, Isenhour—after an extensive and scholarly review—argues that it operates within the limits of power structures that it would be unreasonable to expect consumers to take on. Vesting too much hope in consumer power risks vesting too much responsibility on individuals to take on complex and powerful systems.

Following on from this general evaluation, Debarati Sen and Sarasij Majumder present a specific case study of consumer-driven movements. They evaluate the Fair Trade movement's success in restoring ethics to the economics of food and contesting structural inequalities within the global economy between the North and the South, with a particular focus on what the demands and rewards of Fair Trade can do to social relations in the producing communities. Their article draws upon interdisciplinary studies on Fair Trade as well as ethnographic work that illustrate the diverse outcomes of the movement. The authors argue that while Fair Trade certification has become a biopolitical regime that can act as a governing mechanism on Southern producers, constraining their lives and livelihoods, it also holds the potential for reflecting global counterpolitics. The producers themselves can, through the presence of Fair Trade, become more conscious and more active about their entitlements with the result that, in unexpected ways, Fair Trade has emancipatory potential.

Shaila Seshia Galvin examines the role of organic certification in the rapidly expanding organics sector. Drawing upon scholarship on organics and food movements, as well as fieldwork with organic smallholders in northern India, she illustrates both the diversity and complexity of organic certification and questions the extent to which it accomplishes or undermines its stated goals. Galvin concludes that the future of organic certification will be shaped by issues of trust and audit culture, quality and conventionalization, and by our understanding of the relationship between "nature" and human agency.

In "Mapping the Food Movement: Addressing Inequality and Neoliberalism," Teresa Mares and Alison Alkon investigate two main criticisms of alternative food movements, namely local food, community food security, food justice, and food sovereignty. In their comparison of these movements, they discuss the inequalities related to race and class in food production and consumption, and the neoliberal constraints of market-based solutions to systematic problems. Illuminating the fundamental differences between the first three movements and food sovereignty, the authors suggest the potential of food sovereignty to dismantle the historical and structural obstacles to more democratic and sustainable food systems.

Hannah Wittman's article takes a closer look at food sovereignty. Wittman presents a cohesive picture of the theoretical and methodological development of food sovereignty as an alternative approach to current food security thinking and practice. She explores the possibilities of a rights-based approach, grounded in agrarian citizenship and an agrarian moral economy, to reforming the dominant food regime, and reviews the demands and strategies of the food sovereignty movement, highlighting opportunities for future action and research.

Fabio Parasecoli and Aya Tasaki's concern is the role of geographical indications that are awarded to particular regional specialty products. Their concern is whether such labeling can protect agro-biodiversity and traditional knowledge, as well as to ensure the livelihood of rural communities in emerging economies. Drawing from both scholarly literature and policy-based material, the article discusses the legal, logistical, sociological, and ecological aspects of geographical indications as a form of cultural preservation, economic security, and environmental stewardship. The authors observe that the promise and potential of geographical indications for adding value to particular food and regional economies does come with significant constraints.

Shaun Ruysenaar takes on the highly charged biofuel versus food debate. He presents first the complexity of issues that lie behind the "food versus fuel" debates globally, and then explores the way the issue is being contested in South Africa, where local food insecurity, a strong corporate maize interests, and food price rises have made this a tense issue. He argues that current debates focus on either the implications for food prices or rural development, when in fact the two are interlinked. The real problem underlying these debates is that the agricultural political economy remains riven with injustice, which the debate over biofuels is not yet addressing.

Rogers-Brown, Shearer, and Harthorn offer a prescient review of an emerging topic—the public perception of nanotechnology in food and agriculture. This is a new field, but one which is likely to become more discussed in the future. Unusually for the journal, this article is distinguished by the authors' use of their own research from six deliberative workshops on nanotechnology in 2009 in the United States. Our intimate relationship to food makes nanotechnology a contested area for the application of emerging technologies, and the authors provide evidence that the concerns raised by global consumers and activist organizations about biotechnological alteration of food are material examples of unresolved social issues involving technologies and food, including access to information, government regulation, and corporate responsibility.

Finally, Joanna Mosser's article critiques the misrecognized omnipresence of the state—as both an institution and an idea—in the way we think, speak, and act in everyday life. Her particular complaint is the way in which we learn, or rather fail to learn, about the role of the state in "enviro-cidal" alliances of neoliberalism and governments, for which she draws on her own experience in the political science classroom. In order to fully understand both the development of the food and environmental movements discussed in this journal and elsewhere, Mosser demonstrates the epistemology of the state's power in shaping our concept of ourselves as largely passive individual consumers, rather than independent authors or collective actors in ecological policy and practice.

ACKNOWLEDGMENTS

Our thanks to all the authors, numerous reviewers, and colleagues who have supported the development of this issue. Dan Brockington would particularly like to thank Rob Edwards for his struggles with our current food systems, and for fighting for something better.

REFERENCES

Bryant, Raymond L., and Michael K. Goodman. 2004. "Consuming Narratives: The Political Ecology of 'Alternative' Consumption." *Transactions of the Institute of British Geographers* 29, no. 3, 344–66.

Doane, Molly. 2010. "Relationship Coffees: Structure and Agency in the Fair Trade System." Pp. 229–257 in *Fair Trade and Social Justice*, ed. Sarah Lyon and Mark Moberg. New York: New York University Press.

Miller, Daniel, ed. 1995. *Acknowledging Consumption: A Review of New Studies.* London: Routledge.

West, Paige. 2012. *From Modern Production to Imagined Primitive: The Social World of Coffee from Papua New Guinea.* Durham: Duke University Press.

I

ARTICLES

Can Consumer Demand Deliver Sustainable Food?

Recent Research in Sustainable Consumption Policy and Practice

Cindy Isenhour

ABSTRACT: From Slow Food and farmers' markets to ecolabels and fair trade an unprecedented number of consumer-based alternative food movements have risen in response to concerns about the environmental and social effects of industrialized agriculture. Some research suggests that these movements are successful in their efforts to reconnect communities, demystify global food chains, and produce sustainable foods, which are healthier for the planet and human bodies. Yet other scholars argue that the contemporary focus on consumer responsibility in policy and practice indicates much more than a process of reflexive modernization. The devolution of responsibility to consumers and the dominance of market-based solutions, these scholars argue, reflect the growing influence of neoliberal environmental governance. From this perspective these movements are naive in their assumption that consumers have the power necessary to overcome the structural barriers that inhibit significant change. These critics argue that the focus on consumer responsibility excludes those without access to consumer choice, reproduces social hierarchies, and fails to deliver the political and redistributive solutions necessary to achieve sustainability. Drawing on research across the social sciences this article surveys the existing evidence about the effectiveness of consumer-based movements in their attempts to create sustainable food systems.

KEYWORDS: citizenship, consumerism, environment, food, governance, neoliberalism, social movements, sustainability

Food has always played a central, mediating role in the relationship between humans and our environments. Anthropologists and historians, among others, have long illustrated how subsistence systems have helped to shape our societies, linking people and our views of the world to localized experiences of gathering, preparing, and consuming food. Yet today, in the context of an increasingly standardized and industrial global food system, the separation of food as a social product from the economic and natural systems that produce it has inspired significant protest (Walter 2008; Wilk 2006). Food activists have contested the conditions that alienate us—

Environment and Society: Advances in Research 2 (2011): 5–28 © Berghahn Books
doi:10.3167/ares.2011.020102

physically and emotionally—from the human and natural systems that generate our sustenance, lamented the loss of control over our food supply, and advocated a return to safe, sustainable, and healthy foods.

Certainly it can be argued that dominant forms of industrial agriculture are unsustainable in many ways, pushing human-natural systems to points that pose a fundamental threat to long-term human and environmental health. The environmental concerns associated with industrial agriculture are broad, ranging from the loss of biodiversity due to standardization, genetic engineering, and corporate consolidation (Manno 2002; Munro and Schurman 2008), to tropical deforestation linked to biofuel and cattle production (Dauvergne 2010), and the effects of agricultural chemicals in our food supply, ground waters, and living systems (Guthman 2004a). Indeed, environmental activism has focused on agriculture due to the industry's disproportionate responsibility for environmental destruction, biodiversity loss, energy use, and greenhouse gas emissions (Heinberg and Bomford 2009; Kimbrell 2002; Manno 2002).

Many concerns with the dominant food system are grounded in environmentalism but reflect a variety of other late-capitalist concerns including those associated with human rights, social justice, and unfair trade relations (Johnston 2008a). The spread of corporate agriculture and the cheap, poor-quality food it produces has threatened many small-scale farmers and the unique, localized production methods they employ (Dolan and Humphery 2000; Humphery 2010; Probyn 2011). These shifts have, over time, resulted in a significant loss of intellectual capital and have alienated many humans from the means of production. While privileged northern food activists often focus on the availability of high-quality, artisanal, and locally produced organic foods, the UN estimates that 1 billion people live with chronic hunger and malnourishment (FAO 2011). This bifurcation has significant consequences for vulnerable populations whose sustenance—due to the legacy of colonialism and unjust trade—is simultaneously dependent on the demands of northern consumers and cheap industrial food imports. These vulnerable communities bear a disproportionate burden of the environmental and social costs of industrial food production, perpetuating global inequality (Dauvergne 2010; Jaffee et. al. 2004; Wilk 2006; Wright and Middendorf 2008).

Environmental and social concerns have led to the unprecedented growth of consumer-based food activism over the last several decades. From Slow Food and boycotts of genetically modified foods to ecolabeled foods, fair trade, organics, local food, farmers' markets, community-supported agriculture, and consumer-buying cooperatives, "contests over the future of food and agriculture are thriving" (Wright and Middendorf 2008: 273).

Labeled by different disciplines and theoretical perspectives[1] as "ethical consumption" (e.g., Barnett et al. 2010; Carrier and Leutchford 2012; Harrison et al. 2005; Lewis and Potter 2011), "political consumption" (e.g., Boström et al. 2004; Micheletti 2003; Stolle et al. 2005), "sustainable consumption" (e.g., Cohen and Murphy 2001; Dolan 2002; Isenhour 2010b; Wilk 2004), and "anti-consumption" (e.g., Binkley 2008; Lee et al. 2009; Varman and Belk 2009), the diverse consumer politics formed in opposition to industrial food production are part of a larger critique of globalization and capitalism.

There seems to be a popular consensus that we consumers have a considerable amount of power to make more sustainable food choices and drive systemic change (Beck 2000). Those of us concerned about the future of our food supply are told that "it is easy being green"[2] and are encouraged by politicians, pop culture icons, best-selling novelists, and fellow citizens to take "simple steps" to consume more responsibly. Yet despite an ongoing debate about the effectiveness of these consumer-based movements (e.g., Brunori et al. 2008; Humphery 2010; Johnston 2008b; Micheletti and McFarland 2010; Wright and Middendorf 2008) and a significant amount of speculation, empirical studies on their ability to influence change are scarce and often contradictory. Focusing specifically on the ideologies of consumer responsibility, citizenship, and neoliberal environmental governance, this article critically reviews the recent literature on consumer-based alternative food movements in order to examine the promises and perils of the contemporary focus on consumer responsibility. The central question posed by this article is: Can consumer demand deliver sustainable food?

Focusing on one form of food activism (e.g., fair trade, local food, organics) would have been a more conventional, not to mention much more manageable, approach to answering this question, but I refer to all these variations as "consumer-based alternative food movements" to reflect their shared focus on consumer responsibility and empowerment. It is an admittedly precarious endeavor to group these diverse strategies into a single category, given the differences between them in values, scope, aim, and goals (Littler 2009). Alternative consumption strategies, as Humphery (2010) points out, operate in different political realms ranging from the cultural (as with culture jamming groups like AdBusters) to the communal (as with local and cooperative initiatives), the civic (as with attempt to fulfill civic duties via the purchase of "ethical" goods on the market), and lifestyle politics (via simplification, reduction, and self-provisioning). Efforts to group these myriad forms of consumer politics are further complicated by the fact that some of these consumer-based movements do not allow for the easy separation of producers and consumers as in community-based gardening initiatives. Finally, many of these movements have contrasting strategies and aims. Some, like the Slow Food movement are more clearly concerned with individual gratification and operate under the logic of consumer culture, while others—like local cooperative movements and culture jamming—are explicitly oriented in opposition to consumer logic and aim to influence political and economic reform of the structures that script for dominant modes of production and consumption.

Jo Littler is correct to argue that it is difficult to assess the political efficacy of these movements without consideration of their specific localized practices in context. Unfortunately, it is not possible to detail these difference or the contrasting values and politics that underlie each of these unique strategies here, given the challenges associated with reviewing such a broad and productive interdisciplinary literature. My aim is not to analyze these different forms of consumer activism or to compare them per se, but rather to examine the commonality between them—the emphasis on consumer responsibility, "responsibilization," or "individualization"—and how effective this consumer turn has been generating alternative food networks. I argue—following Johnston (2008b), Goodman and DuPuis (2002), and Gabriel and Lang (2005)—that while specific alternative food strategies have received considerable attention, there is a need to theorize consumer politics as they relate to governance more broadly.

The article consists of three parts. In the first section, "The Promise of Consumer Demand," I review theoretical frameworks and studies, which suggest there are reasons to be optimistic about the potential of consumer-based alternative food movements. The second section, "On the Perils of Consumer Demand," covers critiques that indicate that consumer-based food movements reproduce the very problems they claim to resist. In the third section, "Complimenting the Promise of Consumer Demand," the argument attempts to reconcile unproductive and tired distinctions between the relative power of human agency and societal structures; between consumer sovereignty and regulation. However, the argument does not rest directly in the center of these contrasting perspectives. Most of the recent literature and the current author have grown increasingly skeptical of the consumer responsibility discourse and neoliberal forms of environmental governance in the absence of a complimentary focus on citizenship, rights, and adequate political solutions. While extremely respectful of human agency and the generative power of social movements, this critical review suggests that polycentric solutions are necessary to move beyond the limited potential of consumer demand and toward those which encourage a renewed role for citizenship and the political and redistributive solutions necessary to deliver sustainable food to all global citizens.

The Promise of Consumer Demand

I imagine that nearly all readers have encountered a few of the—now ubiquitous—recommendations for sustainable eating. As Lyon (2006) points out, the past decade has been marked by an abundance of articles recommending the "small steps" that we as consumers can take to positively affect farmers, reduce our environmental impact, improve our communities, and increase personal satisfaction. In relation to food, these "small steps" nearly always include buying local, organic, and seasonal foods lower on the food chain (Dowler 2008). From Wendel Berry's *Bringing It to the Table* (2009) and Barbara Kingsolver's *Animal, Vegetable, Miracle* (2008), to Michael Pollan's *In Defense of Food* (2009), and Gary Nabham's *Coming Home to Eat* (2002), best-selling authors have echoed and amplified these calls for mindful eating.

It seems that a growing number of consumers have readily accepted this challenge. Growth in sales of "ethical" foods, including those with high animal welfare standards, organics, local produce, and fair trade certified goods have experienced significant growth in the past several decades (Dowler 2008). In 2010, 30 percent of shoppers in the UK reported buying local foods at least once a month, more than double the percentage who reported doing so in 2006 (IGD 2010). Similarly, global fair trade sales continue to outpace conventional products with 22 percent growth in 2008 alone (FLO 2011). In the US, the number of CSAs has grown from approximately 50 in 1990 to more than 2,500 (Local Harvest 2011) and the number of farmers' markets has grown from about 1,700 in 1994 to more than 6,000 in 2010 (AMS 2010). While alternative consumption movements are certainly nothing new (Furlough and Strikwerda 1999; Hilton and Daunton 2001), these recent forms are remarkable in their depth and breadth, leading many scholars to theorize their unprecedented growth and speculate about their transformative potential.

On Knowledge and the Promise of Rational Reflexivity

The most common, and perhaps optimistic, theorizations of these alternative food movements links the phenomenon to theories of "reflexive modernization" and "life politics," which suggest that individuals adapt and mold their futures as they learn about the "consequences of modernity" (Giddens 1991) and realize that the risks are quickly outpacing our capacity to manage them (Beck 1992). This perspective assumes that with this knowledge, consumers will become increasingly reflexive, altering their behaviors and lifestyles (Connolly and Prothero 2008; Halkier 1999; Wilk 2004), individualizing risk, and demanding alternatives on the free market (Adams 2001 Hobson 2002; Matti 2009). Thus food choices constitute a new arena through which one can become a political actor to influence change. Information, education, and awareness become key because reflexivity is dependent on and mediated by knowledge (Goodman and DuPuis 2002). As such many alternative food movements have focused on information dissemination and educational campaigns. As Guthman writes, "knowing where your food comes from has become one of the most prevalent idioms of the current agro-food movement" (2008: 1175).

Reflexive modernization thought is increasingly dominant in many parts of the world as is made evident by the number of states that aim to encourage consumer responsibility through educational and awareness campaigns. Examples of these campaigns are not difficult to find. Alfredson (2004), Matti (2009), and Isenhour (2010a) point to the Swedish case where several governmental agencies have supported the publication of reports and documents designed to encourage consumers to make more sustainable choices. Similarly Brunori et al. (2008) point to the UK where the Sustainable Consumption Roundtable recently produced a document titled, "I Will If You Will," which emphasized the need to encourage mainstream consumers to "shop society's way out of unsustainability."

Kate Soper (1999) suggests that through increased knowledge about the origins of our foods and the associated social and environmental risks, we are witnessing a redefinition of needs that is highly reflective of "late-modernity," within which consumers are presented with an abundance of choices in a globalized world. These choices give individuals increased power to participate in what Giddens refers to as "life politics," which "flow from processes of self-actualization in a post-traditional context, where globalizing influences intrude deeply into the reflexive project of the self, and conversely where processes of self-realization influence global strategies" (1991: 214).

On the Promise of Demystification

Through reflexivity, scholars have lauded the promise of alternative food movements to demystify commodities. Although geographic distance and the conditions of urban and productive alienation have often been perceived as a barrier to knowledge and reflexivity—and therefore a roadblock to responsible action—there is significant evidence to suggest that people are increasingly reconnecting with food by seeking knowledge about the conditions of production (Barnett et al. 2005; Dowler 2008). These efforts to understand where one's food comes from can be seen in the growing popularity of the Fair Trade movement and direct marketing schemes like CSAs and farmers' markets (Kneafsey et al. 2008). From this perspective it is essential to "defetishize" foods which, under conditions of global industrialized agriculture, take on a "mystical character" (Goss 2004; Marx [1867] 1978) and present a consideration of the labor and natural resources embodied within. Through labeling schemes and lifecycle audits, consumers gain "knowledge of chains of consequences" (Barnett et al. 2005: 24), which enables the evaluation of products based on more adequate knowledge about their destructive qualities. Fair trade advocates have lauded the potential of certification to demystify commodities. In this way, consumers of fair trade and direct marketed foods can express solidarity with producers, whether local or in the Global South (Connolly and Prothero 2008). There is also evidence to suggest that these connections have helped to open access to niche markets for struggling farmers (Shreck 2008), resulting in improved social and economic conditions for some producers (Jaffee et al. 2004: 184; Lyon 2006). This "nourishing network" (Whatmore and Thorne 1997), "moral economy" (Luetchford 2008), or "ethic of care" (Bauman 2008; Featherstone 2011) forges connections between consumers who care about socially and ecologically sustainable production and producers who are seeking dependable market access as well as fair and stable prices, effectively redistributing income to help balance global inequalities (Conroy 2001). Global justice movements, based on an increased awareness of these inequalities, are gaining momentum, reflecting a "general interest in reorienting economies away from an exclusive focus on commodification and profit maximization, and towards a more equitable and sustainable provisioning of human needs" (Johnston 2008a: 243).

On the Promise of Democratization

Ulrich Beck, whose ideas have inspired a significant and hopeful body of theory on the potential for consumer-based movements, has argued (paraphased by Sassatelli 2006: 223) "if modernity was a democracy oriented to producers, late modernity is a democracy oriented to consumers: a pragmatic and cosmopolitan democracy where the sleepy giant of the 'sovereign citizen-consumer' is becoming a counterweight to big transnational corporations."

Consumers are envisioned not only as significant contributors to the contemporary challenges presented by our food systems, but they are also viewed as empowered political actors. Armed with knowledge and the ability to "vote with their dollars" (Micheletti 2003; Shaw et al. 2006) they are seen as having the potential to challenge dominant economic and political structures and affect a new form of "globalized democracy" (Dolan 2002). Trentmann has referred to this perspective as the "new orthodoxy of the 'active consumer'" (2006: 3). According to Hobson this approach makes "perfect neoclassical sense" (2002: 107) because it does not impinge on consumer freedoms and is, as Lyon (2006) points out, based on one of the foundational principles of liberal economics—that the common good is best achieved if consumers are free to make their own rational choices without intervention or regulation (Carrier and Miller 2000). Some scholars have noted that government apathy and an erosion of public confidence in political institutions have prompted these movements, inspiring and empowering consumers and citizens to take action (Maniates 2002). This movement toward what Sassatelli (2009) refers to as "sub-politics" reflects a seemingly growing feeling that traditional forms of political participation are not working. From this perspective, the failings of the state have incited "individuals, families, communities [and] employees [to] take back to themselves the powers and responsibilities that have been acquired by states" (Rose 1999: 2–3, cited in Lewis and Potter 2011: 9). Alternative food activists thus hope that, by empowering individuals, power is relocated from the elite to the people. They blur the distinction between citizenship and consumerism, giving rise to what Micheletti (2003) refers to as "individualized collective action," through which citizens can express their values on the market and link to others with similar concerns. Alternative consumer proponents grasp the promise that a democratic movement of sovereign consumers, in aggregate, can force action and change the dominant paradigm, so that gratification and hedonism no longer need be in opposition to the common good or sustainable futures (Soper et al. 2009).

On the Promise of Inspiring Shared Responsibility

Market-based efforts thus rest on the premise that consumers can signal their values, and that these signals will be observed by peers as well as the political and industrial elite. Studies of alternative consumer intention indicate that many are inspired by the hope that they can "set a positive example for others" or that their actions will "inspire others to act" or "drive demand" for more alternative products (Isenhour 2010c). For these individuals the movement has to start somewhere and they hope their early adoption might help to push fellow citizens and policy makers to do their "fair share." The intent for many consumers is thus to drive a collective spirit of shared responsibility. In some cases, consumer activism is designed to force producers to act. According to theories of liberal economics, corporations respond to consumer demand because it is in their best interests to remain competitive, protect the value of their brands, and avoid regulatory interventions (Garsten 2009; Locke 2006). Likewise, political leaders must stay tuned into significant market trends to keep a finger on the pulse of public opinion. As Clouder and Harrison (2006) suggest, well-publicized consumer boycotts have achieved significant results, influencing powerful multinational corporations to institute more sustainable production standards and in some cases prompting government action and regulation. Munro and Schurman (2008) detail the effectiveness of a group of activists who were able to force significant regulatory constraints on biotechnology firms by engaging in a series of highly public challenges to genetically engineered foods. In 2000 several public interest groups worked together to test tacos for genetically engineered corn not yet approved by the United States Department of Agriculture (USDA). After tests revealed contamination, the USDA was forced to institute a US$20 million buyback and investigate the regulatory failure. Perhaps more important, instances like these have moved contests over the safety and sustainability of our food into the public consciousness forcing industry to respond to consumer concerns (Munro and Schurman 2008: 149). Conroy argues that in the context of a global trading system and increasing privatization and deregulation, efforts to place market pressure on firms constitute "powerful alternatives to simple invocations of corporate altruism and civic responsibility" and "may be the only alternative to the competitive downgrading of social and environmental practices by firms worldwide" (2001: 18).[3]

On the Promise of New Communities, Moral Subjectivies, and Spaces for Engagement

For Carrier and Leuchford (2011) the movement to reembed social values and relations in the economic realm is the driving process behind the "ethical consumption" movement. From this vantage point, consumers have rejected economic logics, which "reduce us to economic actors" and "strip us of other social roles" (Wright and Middendorf 2008: 278). The current system of commoditized agricultural production has indeed stripped many communities of social capital. Alternative food strategies (particularly the local food movement) can be seen as earnest attempts to rebuild community (Humphery 2010). In order to reembed food in social relations, many consumers have turned to local food movements, which act on a shared ethic of responsibility to community and environment (Brunori et al. 2008; Kloppenburg et al. 2000; Lyson et al. 1995) and have the potential to achieve democracy and social justice through the development of community-based structures for cooperative provisioning and food distribution (Johnston 2008a). Food systems embedded in community and social relations, such as farmers' markets and CSAs are thus often seen as alternatives to corporate control (Guthman 2004), successful means for strengthening community (Hinrichs 2000; Lyson et al. 1995; Wright and Middendorf 2008), and as tools for reinforcing local norms linked to stewardship (DeLind 2002). Slow Food provides another example of how imaginaries of community and place are reinserted into the process of consuming food. In opposition to industrialized, standardized and depersonalized fast food, Slow Food emphasizes foods grounded in culture and community (Kneafsy et al. 2008; Walter 2008). In these cases, consumption is about much more than the functional fulfillment of physical needs; it is also a project to fulfill political and social goals (Brunori 2007).

In resocializing food, consumers can also create alternative subjectivities (Connolly and Prothero 2008; Guthman 2008) by utilizing food to create ethical identities through sovereign consumer choice (Bauman 2008; Binkley 2008; Lyon 2006). Referred to as "moral selving" (Barnett et al. 2005), this form of identity creation allows sustainable consumers to act on their desire to do the right thing, and to reflexively construct lives consistent with their values and beliefs. Further, through the purchase of "moral products" consumers are able to demonstrate their ability to care across both spatial and temporal distance (Bryant and Goodman 2004) and identify with a community of like-minded global citizens working toward both social and ecological sustainability (whether diffuse and imagined as is often the case with ecolabeled and fair trade consumers, or embedded in local spaces as is the case with slow food, CSAs, and farmers markets). As Kate Soper has pointed out, ethical consumerism can also be about "alternative hedonism" or the reimagination of the good life via the consumption of fresh, community-produced foods with rich local histories and social meanings. The conviviality of alternative consumption can build relationships of cooperative support but also a space for enjoying the personal "sensual pleasures of consuming differently" (Soper et al. 2009).

From this perspective, the act of creating an alternative consumer identity, and the efforts required to maintain it, can facilitate entry into more significant political participation and activism within the civic realm (Barnett et al. 2005; Guthman 2008; Hartwick 2000; Sayer 2003). Indeed, many activists and scholars involved with alternative food movement feel that alternative forms of consumerism, even if they do not engage individuals more fully as citizens, serve as essential platforms for engagement, awareness, and eventual entry into more substantive forms of opposition. Scholars, for example, have observed that the longer one buys organic foods, the greater the intensity of that individual's actions and interest (Welsch and Kühling 2009); that farmers' markets help to elevate consumers' awareness of alternative agriculture and the likelihood that they will join sustainable purchasing groups (Brunori et al. 2008); and that "broader" environmental ethics frequently begin with simple strategies like recycling (Hobson 2006). Finally, writing with Hill in 2010, Hobson found that participants in urban gardening projects, despite being the subjects of governmental attempts to modify behavior, interpreted the project's goals in ways that often surpassed the intention of the intervention. Thus perhaps there is potential for an "alternative hedonism" through which consumers move from individualist projects to create pleasurable lifestyles into more meaningful and effective political projects (Soper 2007).

These optimistic perspectives on consumer-based alternative food movements have certainly been effective in mobilizing consumers who hope that increased education, reflexivity, and conscious decision making will lead to significant improvements in collective health, communities, natural environments, and social justice. Certainly these "promises" of consumer demand indicate that there is potential for these consumer-based alternative food movements. However, as the following section demonstrates, many of these promises have not yet come to fruition (Connolly and Prothero 2008).

On the Limits of Consumer Demand

Despite the rapid growth of fair trade, organic and local products make up a very small portion of food consumed by those of us living in wealthy, postindustrial urban societies. Industrial agriculture is also growing and quickly gathering market share, even in alternative markets. Fast food chains are spreading rapidly and we continue to witness the proliferation of nutrient-poor processed foods. The low price and convenience of these foods come at the cost of environmental degradation and social injustice. Meanwhile, international food prices are soaring, leaving even more families to deal with chronic hunger and malnourishment (FAO 2011). In the face of these challenges many consumers are attempting to politicize the grocery checkout line. Yet theorists have noted—and lamented—a simultaneous decline in more traditional forms of direct political action and collective organization (Allen and Kovach 2000; Guthman 2008).

On the Decline of Citizenship, Individualization, and the Neoliberal Turn

Many scholars have provided evidence to suggest that traditional forms of political participation have decreased as citizens have withdrawn from the public sector in favor of participation as consumers in the private sphere (Halkier 1999). For instance, Maniates writes, "Americans seem capable of understanding themselves almost solely as consumers who buy 'environmentally-sound' products rather than as citizens who might come together and develop political clout sufficient to alter institutional arrangements that drive pervasive consumption" (2002: 51). Critics of alternative consumer movements distinguish our roles as citizens and consumers. For them "genuine civic engagement … is seen to take place only outside the market and beyond what we buy" (Humphery 2010: 57).

The concept of the "citizen-consumer" is widely used in the academic and activist literature to describe individuals who exercise their civic duties through consumer decisions (Johnston 2008a), yet it fails to adequately conceptualize important distinctions. Gabriel and Lang point out that citizenship is largely a political concept "defining individuals standing within the state and a community, according rights and responsibilities" while our roles as consumers are defined in terms of individual economic actions with no responsibility or obligation to a community (1995: 174–175). Combining these terms into the concept of the citizen-consumer implies that there is an easy relationship between our roles as consumers—oriented toward free choice and personal utility—and notions of citizenship, which imply responsibility and a concern for the common good (Brunori et al. 2008; Johnston 2008a). Ulrich Beck (1992) refers to this as "divided citizenship," part of the liberal project within which we are supposed to use our decisions to contribute to the common good while simultaneously pursuing private interests.

However, many authors have argued that such an easy relationship between individual and collective interests does not exist; that political consumerism is more often about the fulfillment of personal interest (via more healthful products or the construction of an "ethical" identity) than about saving the environment or a concern for others (Macnaughten 2003). Indeed, several authors have argued that alternative consumer choices, like buying organic fruits and vegetables or participating in slow food movements, are driven by a concern for personal health risks or individual gratification rather than the product's impact on the environment or producer communities (Connolly and Prothero 2008; Miller 2001b). Thus Hilton (2001) argues that the concept of the citizen-consumer has absolved consumers of civic duties while simultaneously diminishing the rights of citizens in the name of free choice.

Many authors have argued that the contemporary focus on consumer responsibility at the expense of rights (Sassatelli 2006) reflects the neoliberal turn in global and state politics (Allen 2010; Guthman 2008; Hobson 2002; Lyon and Moberg 2010) and represents a significant reframing of the relationship between citizens and the state (Hobson 2002). Intimately linked to processes of market integration and liberalization, governments have increasingly dismantled regulations and institutions seen as barriers to free trade. As part of this process, responsibility for collective welfare has been devolved away from the state to the market, citizens, and other non-governmental entities (Garsten and Hasselström 2003; Stø et al. 2008). Guthman writes, "Agriculture and food sectors have been subject to some of the most intense attempts at neoliberalization—from the privatization of land and water rights, to the use of free trade agreements to dismantle national-level food safety regulations, to the protracted dismantling of food-oriented entitlement programs that exist in part to combat hunger in the US" (2008: 1171).

Indeed the state's traditional role as the regulator of both consumption and production processes has been in decline for several decades (Keiko and Ransom 2008). In this neoliberal context, Halkier writes, "it has become increasingly common to call upon so-called ordinary consumers to solve a range of societal and political problems. Environmental policies and food policies are no exception to this pattern" (2001: 205). Yet many scholars lament this "individualization" of responsibility, which would seem to frame environmental and social problems in terms of individual inadequacies rather than the product of dismantled environmental protections and diminishing social safety nets (Maniates 2002: 45).

However, as we have seen, many ethical consumers seem to accept this turn without questioning the government's devolution of responsibility and risk. According to critics, consumers have seemingly accepted (without critical examination) the government's attempts to absolve itself of responsibility by "individualizing risk" (Connolly and Prothero 2008). Johnston writes,

"While activists busy themselves with food drives, heirloom tomato sales, and garden plots, the state surreptitiously diminishes its responsibility for basic needs like food and shelter" (2008b: 97). As argued by groups like the Center for a New American Dream, it may be that consumers have essentially given up on the state as an actor able to provide basic needs like healthy foods (CNAD 2011). In this context, consumer-based alternative food movements can be seen as filling in to "accomplish what governments aren't allowed to (or don't want to or haven't got the courage to) do, and thereby release governments from domestic political pressures and nuisances" (Jacobsen and Dulsrud 2007: 475). Yet, ironically, while many food activists are essentially protesting the effects of privatization, devolution, and dismantled agricultural regulations (e.g., poor working conditions, low prices for farmers, the poorly regulated use of industrial chemicals, food inequalities), the movements in which they participate are both a reaction to and the product of the neoliberal mentality of consumer choice and individual responsibility (Allen 2010). Contemporary alternative food movements have reproduced neoliberal subjectivities and reinforced neoliberal governmentality (Guthman 2008). As such, one could argue that they are the product of "advanced liberal governance" (Lewis and Potter 2011: 17). Humphery writes, "In many respects, the new politics of consumption, particularly in the USA, is hostage to its own liberal individualism" (2010: 73).

On the Illusion of Choice

Ethical consumers certainly want to utilize their choice to influence positive change in our food systems; however, many critics argue that the emphasis on individual choice places a heavy and unrealistic burden on the shoulders of the consumer (Jacobsen and Dulsrud 2007; Isenhour 2010c; Schwartz 2005) who is "consistently helped to believe that he or she is an actor, where in fact he or she is at best a chooser" (Appadurai 2008: 59). Probyn (2011) argues that the debate about healthful and responsible eating verges on problematic moralism that labels individuals who consume cheap industrial foods as ignorant, unethical, or simply in bad taste while simultaneously saddling the same individuals with responsibility for fixing the problems associated with unsustainable food systems. This "reification of choice" mystifies the role that governmental policies and agricultural giants play in determining our choices. Illustrating the illusion of choice, Maniates (2002) argues that while consumers may be overwhelmed by choosing an automobile (e.g., size, color, make, and model), a real choice would be between a car and the ability to take public transportation, a choice for which many cannot "vote with their wallet" because it does not exist. There is a long-standing debate about the relative power of producers and consumers (e.g., Adorno and Horkheimer 2000; Baudrillard 2000; Galbraith 2000). Structuralists argue that consumers have little choice beyond packaging, but many others laud the creativity with which consumers create new meanings through an array of goods. Regardless of the stance one takes in this tired debate, Guthman (2003, 2008) and others, including proponents, question whether the consumer's personal decisions are sufficient to provide solutions on a scale adequate to address problems associated with a globally integrated food supply. If our concerns are expressed in the private realm, rather than in public discourse and through collective participation, many authors fear that they will not be sufficient to produce change on an appropriate scale. Halkier (1999) has written that if our concerns continue to be expressed in such individualistic ways, they will not become part of the collective experience, but rather "silent routines" that only send signals to retailers and producers.

Numerous studies have revealed that consumers—despite feeling a sense of responsibility and obligation to choose ethically and sustainably produced foods—are often inconsistent, ambivalent, confused, and conflicted as they make these choices (Connolly and Prothero 2008;

Halkier 2002; Isenhour 2010b; Sassatelli 2006). This literature points to a number of structural barriers, which prevent even the most aware, engaged, and committed consumers from consistently exercising their choice in favor of alternative foods. First, as previously outlined, theories of reflexive modernization suggest that if given the proper knowledge, consumers will make rational and ethical decisions on the market. Adequate information and education are certainly key to building awareness and inspiring alternative consumption; however, it seems that for informed alternative consumers, too much information is often part of the problem. In modern grocery store aisles, consumers are overwhelmed by an abundance of choice. One can choose from hundreds of breakfast cereals, countless flavors of yogurt with varying fat content, and an overwhelming array of vegetables—local and exotic, organic and conventional. Wilk remarks on the "impossibility of making informed choices in a world glutted with information and difficult contradictions between different kinds of 'good' and 'bad' products" (2010: 8). In fact, there are all sorts of contradictions and difficult decisions when it comes to buying food. Is it better to buy a conventional apple from the local farmers market, grown only a few miles away or an apple produced 1,000 miles away but without agrochemicals or underpaid labor? Further, as Brunori and colleagues (2008) point out, even the best information available cannot help consumers to make tough choices between competing priorities like price and quality, convenience and sustainability, or locals and organics. Given this "debilitating excess of choice," to use Humphery's terminology (2010), Schwartz (2005) has argued that the rhetoric of consumer freedom has actually resulted in exacerbated time poverty for already burdened citizens and rampant individualism at the expense of both sociality and freedom. This is not to say that all choice is neoliberal or bad. Certainly consumers will always want choice, regardless of the economic structure and the relationship between the market and the state. However, it is to say that there are significant barriers for consumers, even those who hope that their choices can contribute to a better world. As Humphery (2010) argues, it is necessary to compliment consumer choice with regulations that can limit the choice of products that have been shown to cause significant damage to humans and the environments upon which we depend.

It is simply unreasonable to assume that consumers, even the most committed, have the time and intellectual energy to evaluate the relative merits of all choices. Thus Jacobsen and Dulsrud argue, "non-reflective ordinary consumption will probably always comprise the larger bulk of practices we reasonably can call consumption. This severely restricts the validity of the active, ethically conscious consumer" (2007: 477). These difficulties leave many scholars deeply skeptical of the push to make consumers responsible for complicated decisions in informed and consistent ways (Belk et al. 2005). Although reflexive modernization theories continue to operate on the assumption that consumers will make rational and moral decisions if provided with adequate information, repeated efforts to spread information and awareness in many international contexts have failed to inspire significant change (Hobson 2002). Wilk therefore argues that "giving consumers more information and trusting them to make good choices is another wishful strategy with limited practical utility" (2004: 27).

Proponents of certification schemes have argued that the information provided by labels can help to ease this burden for consumers. However, critics argue that consumers often experience significant difficulties if they are presented with too little or too much information (Hoogland et al. 2007). In fact, Shreck (2008) has reported that both consumers and retailers are beginning to suffer from "label fatigue" (FAO 2000). As Carrier and Leuchford (Forthcoming 2012) point out, forms of commercial deception are fairly common. Johnston's analysis of Whole Foods Markets illustrates that corporate marketing strategies can often be misleading, even if not consciously deceptive. She notes that Whole Foods uses terms like "natural," "organic," and "sustainable" interchangeably and frequently displays photos and stories of local farmers above imported conventional produce (Johnston 2008a). Similarly, Littler has found that "greenwashing is rapidly increasing both in terms of its sheer quantity and the degree of its complex permutations" (2008: 32).

On Corporate Cooptation

Authors from Karl Marx ([1867] 1978) to Malcom Gladwell (2000) have illustrated the unique ability of the capitalist system to convert all things into commodities, even "difference, dissent, resistance and opposition" (Schor and Holt 2000: xxi). Organic foods, for example, have experienced significant growth in production. Between 2000 and 2008 global sales grew by approximately 20 percent annually (Wright and Middendorf 2008), capturing the attention of industrial agriculture and corporate retail chains (Jaffee et al. 2004). Many applaud the growing availability of organics and their movement into, for example, low-cost retail giants like Walmart. Yet many scholars and food activists have expressed concerns that the fundamental nature and values of alternative movements are lost when translated and deployed by agribusiness and retail giants (Guthman 2004). US organic standards do not prohibit the use of all potentially harmful chemicals, energy intensive factory farming, and excessive packaging, nor do they dictate that supply chains should be any different than those used by industrial agriculture (Guthman 2004; Johnston 2008a). Many long-time organic producers thus see corporate agriculture's movement into the organic sector as greenwashing and a betrayal of the movement's original intentions. Produced industrially, organics become much more about the technicalities of production than a movement toward comprehensive environmental stewardship (Vankeerberghen 2011).

Many thus fear that consumer-based ethical food movements have become little more than a niche marketing opportunity (Lyon and Moberg 2010) designed to attract the "savvy, reflective consumer-citizen … and exploit the cultural shift towards, 'affirmative purchasing' (Carrigan et al. 2004, cited in Lewis and Potter 2011: 8). Johnston writes, "From a critical perspective, ethical consumer strategies seem more like niche marketing opportunities allowing corporations to target privileged, conscientious consumers, than a substantive program for health, sustainability, and social justice at a global scale" (2008a: 240).

On Consuming Difference

Given that most consumers of alternative foods are members of the well-educated, white middle class living in wealthy postindustrial urban contexts (DuPuis and Goodman 2005; Ferrer and Fraile 2006; Raymond and Kondoh 2008), many critics have labeled alternative agro-food movements and their concern with high quality, specialty, artisanal, and local foods "elitist" and reflective of "yuppy" market segments (e.g., Donati 2005; Guthman 2003; Roseberry 1996). Certainly many alternative foods are considerably more expensive than conventional and mass-produced products. As Johnston (2008a) notes, a journalist found that a month's food supply for a small family would cost US$232.00 at Walmart and US$564.00 at Whole Foods.

Given the privileged social position of most alternative consumers, critics argue that the consumption of "ethical" foods essentially enables the consumption of social status and class-based distinction. Douglas Holt found in his study of American consumers that in the context of mass consumption individuals with high levels of cultural capital distinguished themselves by way of aesthetic consumption that was "socially scarce" (2000: 218). The emphasis on socially scarce, high quality, artisanal, and "ethically" produced foods might thus be seen as a means to distinguish oneself as having good taste and superior ethical credentials. Yet this focus on "taste" legitimizes and reproduces inequalities by failing to recognize that the ability to purchase high-quality foods is more likely dependent on income than taste. Thus Johnston argues that these elite consumption practices "create a hierarchy of moral stratification that maps onto class stratification" (2008a: 256). Thus though the state, corporations, and activists encourage consumers to signal their values via consumer choice, this strategy is not democratic. The wealthier the consumer, the more power they have to choose freely (Connolly and Prothero 2008; Dolan 2002). In contrast, though there is ample evidence to suggest that low-income consumers desire more sustainable and healthy foods (Dowler 2008), "those with the greatest needs often have the least ability to exercise individual choice, as allocations of choice are shaped by the historical demographics of inequality" (Allen 2010: 300).

Privileged consumers paint their own actions as ethical, but those who cannot afford to join a CSA or buy fair trade and organic foods at the grocery are seen as unethical, unrefined, or simply as acting in bad taste. Julie Guthman (2008), Jo Littler (2009), and Elspeth Probyn (2011) have noted the problematic moralizing that accompanies alternative food movements. Probyn writes, "ideas about what is 'good' or 'bad' are laid out in black and white terms, which often posits 'us' vs 'them'" (2011: 103). Those who buy cheap, unhealthy foods are seen as ignorant and irresponsible. Drawing on Guthman's work, Probyn goes on to write, "Rather than considering how an amalgamation of bad government policy, global retail pressures and unequal distribution of funding is creating a situation whereby people … are suffering from the medial consequences of obesity, the good food arguments individualize and damn the overweight, and more often than not the poor" (2011: 109). Many low-income families simply do not have access to sources for local food, adequate income to pay the price premiums for alternative foods, or the time and resources it takes to grow one's own food.

Similarly, Hinrichs (2003) warns that local food movements have the potential to become elitist and exclusionary by relying on privileged perspectives about what is and is not ethical. DuPuis and Goodman argue that there is an "a-political, anti-democratic and anti-reflexive bent in current food localism discourse," founded on class- and race-based hegemonic politics, which universalizes, normalizes, and elevates certain ways of eating (2005: 362). It is important to remember that local communities are also marked by internal inequalities, which can easily be reproduced by localized food systems that exclude certain members of the community. Like Daniel Miller (2001a) who has argued that we must avoid the "poverty of morality" by exposing the elitist and class-based moral agendas at the root of many forms of activism, DuPuis and Goodman argue that it is "important to pay more attention to the ways in which our possessive investments in our privilege influences how we define problems and solutions" (2005: 362).

On the Neglect of Social Justice

Some consumer-based alternative food movements are explicitly concerned with social justice (particularly fair trade), however most rarely engage with issues of equity (Dowler 2008; Goodman 2003; Jaffee et al. 2004). Patricia Allen and her colleagues (2003) found in their study of alternative food initiatives in California that their leaders rarely prioritize a concern with social justice, but rather focus their efforts on ecological sustainability. This is despite the fact that "inequities in the agro-food system abound—low wages and poor working conditions for food system workers, hunger and starvation of impoverished people and maldistribution of resources" (Allen 2010: 297). Although some alternative food movements do try to address issues of inequality, access, and hunger, Allen argues that they are not particularly effective.

> While individual choices based on an ethic of care for others (Tronto 1993) can certainly be part of working toward social justice, a concatenation of individual choices to improve social

> equity does not address the basic political economic structures, resource allocations and cultural conditions that have created inequality in the first place. At its root, inequality is a social problem that requires social analysis and solutions … [it] will require public and extra-local investment and public policies rather than devolving the functions of the welfare state onto individuals or communities. (Allen 2010: 300)

The problems that we face require attention to social justice and inequality, particularly on a global scale. More than 1.2 billion people do not have adequate access to clean drinking water (Wilk 2006) and more than a billion people live with persistent hunger (FAO 2011). Legacies of colonialism and capitalist exploitation have created a world with vast differences in power and wealth (Allen 2010), one in which the historically exploited peoples of the developing world find themselves dependent on the exportation of primary products yet, ironically, only able to afford cheap imported foods. Small-scale farmers in these contexts often cannot compete with industrial giants in a global marketplace (Lyon and Moberg 2010). Meanwhile industrial agriculture continues to externalize the environmental and social costs which are unfortunately nearly always absorbed by relatively powerless producer communities (Dauvergne 2010; Morgan et al. 2006). The stratification of the global food system is reproducing global inequalities, yet as Johnston argues, "public attention seems to be diverted elsewhere—to the health benefits of flaxseed" (2008b: 94).

Fair trade advocates have argued that through the purchase of fair trade goods consumers can ensure market access and price premiums to disadvantaged producers and thus diminish inequalities. However, as a number of scholars have shown, fair trade may not always work as consumers assume. Raynolds (2000), Jaffee et al. (2004), and Shreck (2008) have all observed that the fair trade agenda has been determined almost entirely by Northern activists and implemented via a top-down approach. Certification standards are rarely developed in cooperation with producer associations and thus often seem arbitrary and punitive (Raynolds 2000). Tad Muttersbaugh (2002) has used the term "ecological neo-colonialism" to describe the rigorous standards evaluations, controls, and surveillance that coffee producers must endure in exchange for certification. Freidberg (2004) documents how southern African farmers must endure guarded surveillance, and Lyon (2006) observes that such inspections are often carried out by foreigners with very little knowledge about agricultural production or local culture. It is no surprise then that many fair trade and certified organic farmers have come to resent inspections and ask who is certifying the North (Gereffi 2000, cited in Shreck 2008).

From this viewpoint, and despite its good intentions, attempts to alleviate global inequalities in the food chain have actually worked in some cases to reproduce unequal power relationships and the dependence of impoverished producers on the whims of privileged consumers. Rick Wilk writes, "There is no balance in the global food system. … The producers are dependent on rich consumers in a way that we are never dependent on them" (2006: 2). Further, Lyon argues that the consumption of fair trade goods "normalizes global inequities, serving to further legitimize the basic premise of the post-industrial economy—the idea that some populations should manufacture products according to the whims and desires of other populations" (2006: 388). Although many fair trade advocates and consumers are hopeful that they can "save the world" with their purchase of shade grown, fair-trade certified coffee, it seems that "the emancipator discourses of fair trade entail systems of governance" that reproduce power inequalities (Lyon and Moberg 2010: 16). Alf Hornborg, in similar fashion, argues that the idea that we can save the world is "likely a fallacy that allows the privileged North to reproduce narratives about superiority when in fact it is unequal exchange and inequality that created the conditions from which we feel we must save third world others." What is most ironic, he adds, is that the "exploited are led to believe they should be grateful to the exploiters" (2008: 1). It does seem rather naive that consumers concerned with social justice would believe that their choices, even in aggregate, could result in a redistribution of resources adequate to ensure social and environmental justice (Littler 2009).

On the Defense of Choice and the Reproduction of Consumer Culture

Most scholars agree that, despite some successes and continued growth and development, consumer-based alternative food movements have thus far not altered the basic structures of the global economy or challenged the dominance of industrial agriculture (Humphery 2010; Sassatelli 2009). Further, although alternative food movements are certainly promising in many regards, Humphery writes that we have very little information about how the "performance of virtuous personhood can actually translate into both a care for others and a remaking of social, cultural and economic life" (2010: 173). Part of the problem is that these movements have played into and, at times, even worked to reproduce the same ideologies they portend to resist. Lyon and Moberg argue that there is a "deep paradox" within consumer-based alternative food movements because they "pursue market-based solutions to the very problems developing from free markets" (2010: 7). By relying on the neoliberal logic of the market, the personalization of risk and the devolution of decision making and responsibility to individuals, these movements have inadvertently legitimized the conditions to which they are ultimately opposed. Local food movements, DuPuis and Goodman (2005) argue, legitimize the devolution of responsibility and compromise remaining state regulations designed to protect the welfare of citizens and our ecosystems.

Perhaps the root of the problem is that these movements have not adequately explored or questioned the logic of consumerism. Defined as a "set of beliefs and practices that persuade people that consumption … is, literally, at the center of a meaningful existence and that the best organized societies are those that place consumer satisfaction at the center of all major institutions" (Sklair 2001: 5), consumerism is indeed a powerful force. Central to this ideology is the notion of consumer sovereignty and choice.

Ironically, the centrality of consumer choice in alternative food movements legitimizes the very practice and effectively "strengthens the significance of consumption practice" (Dolan 2002: 177). Playing on what Johnston calls the "cornucopian ideal of nature," the logic of choice does not question consumption but rather solidifies the view of nature as boundless and human innovation as a sufficient means to overcome natural limitations (2008a: 259). Rather than encouraging individual consumers to reduce their demands on the environment and consume according to their needs but not beyond, the logic of choice neglects any discussion of the limitations implied by ecological responsibility in favor of alternative choices. Yet as Connolly and Prothero (2003), Alfredson (2004), Wilk (2010) have demonstrated, the consumption of alternative or "green" goods does not always result in environmental benefits. In his study of more than 1,000 Swedish households, Alfredson (2004) found that the adoption of a "partly green" lifestyle actually increased household carbon emissions when overall levels of consumption were not reduced. Studies like these have led a growing number of scholars to argue that alternative food movements must also bring discussions of over consumption, waste, and reduction into focus if they are to become more effective. Otherwise, alternative food movements fail to disembed themselves from "consumer culture" and thus have little hope of reforming industrial food systems (Dolan 2002: 177). Consumption—ethical or not—according to many Marxist critiques, ultimately legitimizes the existing capitalist system and its view of nature as a commodity.

Today's consumers have come to expect the freedom to choose and often resent any attempts to regulate choice. Richard Wilk writes that notions of consumer freedoms are so engrained in consumer societies that "limitation of choice or the notion of sacrifice is not likely to go over very well" (2010: 10). Indeed, studies suggest that committed ethical consumers support additional regulation to limit the availability of dangerous products (Isenhour 2010b) or limit carbon emissions (Featherstone 2011), but the majority of US consumers are not in favor of hard regulations and attribute their attitude to the potential loss of personal freedoms (Attari et al. 2009). Indeed, as Johnston argues (2008b: 102), the ideology of consumer sovereignty is so strong that it has been prioritized over collective action.

The current neoliberal emphasis on consumer choice can be seen, in many ways, as a defense of contemporary global structures of inequality. The world's most privileged defend their ability to choose, but unfortunately they often choose products or practices that are extremely harmful for the world's most vulnerable communities. These communities do not have the luxury to choose if, for example, their waters are used for the production of goods that will be shipped to Europe, their rainforests will be felled to make room for beef production destined for America, or their neighborhood will become a dumping ground for industrial wastes. Arguing that consumers should be able to choose whether or not to buy organic foods is essentially like saying that those families who cannot afford nontoxic products do not have the right to feed their children clean foods or that agricultural laborers are not entitled to safe working conditions.

The defense of unlimited consumer choice can thus be seen as a defense of power and privilege. Protecting our families from dangerous and toxic products or services associated with high-energy consumption and emissions should not be a question of consumer choice. It should be a question of sound policy and governance. Yet the contemporary focus on sustainable lifestyles and consumerism places the onus for sustainability on consumers and the "culture of consumption," as if "consumer culture" were independent of historical and contemporary political and economic structures.

All consumers are embedded in social relations and immersed in societies with social norms, which help to structure individual choices. To assume that all consumers are free in their choices, to consume as they please, and in a manner consistent with their own individual utilities and self-interests, is naive. As Dolan argues, seemingly individual acts of consumption are, in fact, macro processes at work: "Consumer practices are cultural and social practices that have historically developed and are manifestations of both local and global linkages of social interdependencies. To continually look at the consumer as the cause of the ecological problem effectively decontextualizes consumption acts from such interdependencies" (2002: 171). It is imperative that we recognize that "consumer culture" is more complex than aggregated individual choice. Our cultures are products of history and context. They rise out of, reflect, and simultaneously structure our material realities, productive systems, methods of exchange, social organizations, and political structures. To assume that consumer education and behavior modification can alter the culture of consumption without a simultaneous change in these deeply embedded structures is at best overly optimistic and at worse simply naive.

Complimenting the Promise of Consumer Demand

It is important not to undermine the sincere efforts of individual consumers to affect change. Many people around the world have invested significant emotional, psychological, and intellectual energy in their efforts to support alternative foods systems. Yet as the critiques above suggest—despite their best intentions—consumers are confronted with barriers that prevent the achievement of more significant change. Individuals can buy organic foods at the local farmers' markets, reduce meat consumption, and boycott foods produced with palm oil. But as Wright and Middendorf (2008) argue, mindful eating cannot dismantle historically embedded social structures. The achievement of wider structural changes requires collective action (Brunori et al. 2008) and a "renewed ideological debate on private vs. public responsibilities" (Jacobsen and Dulsrud 2007: 479). Princen and colleagues (2002) propose that our examinations of both consumption and environmental problems have for too long been dominated by a production perspective. Instead the authors advocate viewing these problems from a consumption perspective because consumption occurs all along the chain of production: from extraction to disposal. All along that chain there are consumers, that is individual human beings who make decisions about the consumption of resources. This perspective has important implications. By diffusing the sole emphasis of consumption study away from the point of purchase and distributing it all along the chain, we can essentially view all humans along the chain as consumers, including business owners, politicians, and individual citizens. This allows us to reimagine shared responsibility and diffuses the tired debate about the relative power of structure and agency—given that the existing political and economic structures are also populated by consumers of natural resources and human labor.

Refocusing on collective responsibilities and effective government would allow us to move away from the problematic moralization of consumer choice (Miller 2001a) and toward a more effective configuration of polycentric solutions involving consumers, citizens, producers, and the state (Wright and Middendorf 2008). It is simply not realistic or reasonable to expect consumers to carry the burden for ensuring the health and sustainability of our food systems alone. Further, historical analysis suggests that we are much more effective at forcing change when we cooperate with one another and express our values in the public realm (Gabriel and Lang 2005).

According to Allen, we "need to move beyond a discourse of choice and realize that no social advances have ever been made without a combination of social movements and legislation" (2010: 306). Without public policies and programs designed to create sustainable food systems and ensure a more equitable distribution of natural resources and environmental risks, significant long-term change is not likely. Policies are necessary to ensure that all citizens, not just the privileged, are able to eat in sustainable and healthful ways (Wilk 2004). This is certainly not to say that all our contemporary human structures must be dismantled, or that change is not possible without revolution. It is at least to argue that public participation is necessary. Manno (2002) suggests that efforts to prevent the commodification of our food networks might include increased public investment in R&D and the subsidization of sustainable agricultural methods; policies to ensure that pricing reflects the external costs of products; a shift in tax policy away from labor (because jobs are scarce and workers plentiful) and toward natural resource and energy use (inputs we need to conserve); and investments in natural capital.

However, policy solutions that require the regulation or limitation of consumer choice "are dependent on public democratic legitimacy" (Halkier 1999: 37). This, Halkier argues, is the greatest contribution of consumer-based alternative food movements—the creation of "cultural codes" that can help to enhance the potential for significant political change (1999: 37). Though alternative food movements are certainly not a panacea for the problems associated with the global dominance of industrialized agriculture and capitalist expansion, it would be simplistic to dismiss these movements without considering "the complexity of the social change process, the urgent need for innovative approaches and the possibilities for empowerment they offer" (Johnston 2008b: 98). It is difficult to declare the success or failure of consumer-based alternative food movements because, as I demonstrated, they show signs of both promise and disappointment. Further, success is relative. If we look at the alternatives created by these movements relative to the state of today's dominant food systems, we can observe significant success and improvements. According to Littler (2008: 36), consumer cooperatives can help to distribute healthy food and support alternative production methods while boycotts and buycotts pressure companies to consider the environmental and social impacts of their practices and fair trade products bring attention to issues of global inequalities and social justice. However, if we compare current alternative food movements to common aspirations for the global foods system based on rights rather than markets, health rather than convenience, stewardship rather than exploitation, and sustainability rather than growth, then it seems that these movements have not lived up to their promise. As Humphery notes, however, consumer movements are very much a "politics in the making" (2010: 157). There is significant hope and human labor invested in these movements that should not be discounted as a significant "lever for change" (Barnett and Soper 2005). The relative merits of consumer-based alternative food movements will certainly continue to be debated, but I believe the literature reviewed here suggests that it is possible to answer the central question raised by this article. Consumer demand cannot deliver sustainable foods, at least not alone and without complementary political and economic reforms to ensure that all humans are able to eat healthful foods that do not destroy the natural resources upon which we depend.

CINDY ISENHOUR is the Mellon Postdoctoral Fellow in Environmental Studies at Centre College, Danville, Kentucky. Her research interests are located at the intersection of political ecology, economic anthropology, and consumption theory. With support from the Fulbright Program, the American-Scandinavian Foundation, the Wenner-Gren Foundation for Anthropological Research, and the University of Kentucky Graduate School, Isenhour recently conducted research on sustainability policy and practice in Sweden. She has published in *American Ethnologist, Journal of Consumer Behavior, City & Society,* as well as in several edited collections.

NOTES

1. For an excellent survey of different disciplinary approaches to alternative consumerism, see Lewis and Potter's *Ethical Consumption* (2011).
2. Simplesteps.org and easybeinggreen.com are only two of the numerous websites, blogs, and twitter streams dedicated to encouraging more sustainable living and food choices.
3. It must be noted that these efforts are often enacted in the public sphere, rather than through the consumer's private role on the market. This important distinction is discussed in the concluding section of the article.

REFERENCES

Adams, William. 2001. *Green Development: Environment and Sustainability in the Third World.* New York: Routledge.

Adorno, Theodor W. and Max Horkheimer. 2000. "The Culture Industry: Enlightenment as Mass Deception" Pp. 3–9 in *The Consumer Society Reader,* ed. J. Schor & D. Holt. New York: The New Press.

Agriculture Marketing Service (AMS). 2010. "Farmers' Markets and Local Foods Marketing." http://www.ams.usda.gov/farmersmarkets (accessed 5 February 2011).

Alfredson, E.C. 2004. "Green Consumption—No Solution for Climate Change." *Energy* 29: 513–524.
Allen, Patricia. 2010. "Realizing Justice in Local Food Systems." *Cambridge Journal of Regions, Economy and Society* 3: 295–308.
Allen, Patricia, M. FitzSimmons and Michael Goodman. 2003. "Shifting Plates in the Agrifood Landscape: the Tectonics of Alternative Agrifood Initiatives in California" *Journal of Rural Studies* 19: 61–75.
Allen, Patricia and Martin Kovach. 2000. "The Capitalist Composition of Organic: The Potential of Markets in Fulfilling the Promise of Organic Agriculture." *Agriculture and Human Values* 17: 221–232.
Appadurai, Arjun. 2008. "Disjuncture and Difference in the Global Cultural Economy" Pp. 47–65 in *The Anthropology of Globalization: A Reader,* ed. J.T. Inda and R. Rosaldo. Blackwell.
Attaria, Shahzeen Z. Mary Schoena, Cliff I. Davidsona, Michael L. DeKayc, Wändi Bruine de Bruind, Robyn Dawesd and Mitchell J. Small. 2009. "Preferences for Change: Do Individuals Prefer Voluntary Actions, Soft Regulations, or Hard Regulations to Decrease Fossil Fuel Consumption?" *Ecological Economics* 68: 1701–1710.
Barnett, Clive, Paul Cloke, Nick Clarke, and Alice Malpass. 2005. "Consuming Ethics: Articulating the Subjects and Spaces of Ethical Consumption." *Antipode* 37 (1): 23–45.
———. 2010. *Globalizing Responsibility: The Political Rationalities of Ethical Consumption.* London: Wiley-Blackwell.
Barnett, Clive and Kate Soper. 2005. "Consumers: agents of change?", *Soundings: A Journal of Politics and Culture* 31: 147–160.
Bauman, Zygmunt. 2008. *Does Ethics Have a Chance in a World of Consumers?* Cambridge, MA: Harvard University Press.
Baudrillard, Jean. 2000. "The Ideological Genesis of Needs" Pp. 57–80 in *The Consumer Society Reader,* ed. J. Schor and D. Holt. New York: The New Press.
Beck, Ulrich. 1992. *Risk Society: Towards a New Modernity.* New Delhi: Sage.
———. 2000. *The Brave New World of Work.* Cambridge: Cambridge University Press.
Belk, R., T. Devinny, and G. Eckhardt, G. 2005. "Consumer Ethics across Cultures." *Consumption, Markets and Culture* 8 (3): 275–289.
Berry, Wendel. 2009. *Bringing It to the Table.* Berkeley, CA: Counterpoint Publishers.
Binkley, Sam. 2008. "Liquid Consumption: Anti-Consumerism and the Fetishized De-Fitishization of Commodities." *Cultural Studies* 22 (5): 599–623.
Boström, Magnus,A. Follesdal, M. Klintman, M. Micheletti, and B. Sorensen. 2004. "Political Consumerism: Motivations, Power and Conditions in the Nordic Countries and Elsewhere." Proceedings from the Second International Seminar on Political Consumerism, Oslo.
Brunori, Gianluca. 2007. "Local Food and Alternative Food Networks: a Communication Perspective." *Anthropology of Food* 2: 318–342.
Brunori, Gianluca, Francesca Guidi, and Adanella Rossi. 2008. "On the New Social Relations Around and Beyond Food: Analysing Consumers Role and Action." Paper presented at the Sustainable Consumption Conference. Arlon, Belgium.
Bryant, Raymond L., and Michael K. Goodman. 2004. "Consuming Narratives: The Political Ecology of Alternative Consumption." *Transcripts of the Institute for British Geographers* 29: 344–366.
Carrier, James, and Peter Leuchford. Forthcoming in 2012. *Ethical Consumption: Social Value and Economic Practice.* London: Berghahn Books.
Carrier, James, and Daniel Miller. 2000. "From Private Virtue to Public Vice." Pp. 24–47 in *Anthropological Theory Today,* Ed. Henrietta L. Moore. Malden, MA: Blackwell.
Center for a New American Dream (CNAD). 2005. "New American Dream Public Opinion Poll" Center for a New American Dream http://www.newdream.org (accessed May 5, 2006).
Clouder, Scott, and Rob Harrison. 2006. "The Effectiveness of Ethical Consumer Behavior." Pp. 89–102 in *The Ethical Consumer,* ed. Rob Harrison, Terry Newholm, and Deirdre Shaw. London: Sage.
Connolly, John, and Andrea Prothero. 2008. "Green Consumption: Life-politics, risk and contradictions." *Journal of Consumer Culture* 8 (1): 117–154.
Cohen, Maurie and Joseph Murphy. 2001. "Exploring Sustainable Consumption." London: Pergamon.

Conroy, Michael E. 2001. "Can Advocacy-Led Certification Systems Transform Global Corporate Practices? Evidence and Some Theory." Political Economy Research Institute Working Paper Series, No. 21.

Dauvergne, Peter. 2010. "The Problem of Consumption." *Global Environmental Politics* 10 (2): 1–10.

DeLind, Laura B, 2002. "Place, Work, and Civic Agriculture: Common Fields for Cultivation." *Agriculture and Human Values* 19 (3): 217–224.

Dolan, Paddy. 2002. "The Sustainability of 'Sustainable Consumption.'" *Journal of Macromarketing* 22 (2): 170–181.

Dolan, Catherine and John Humphrey. 2000. "Governance and Trade in Fresh Vegetables: The Impact of UK Supermarkets on the African Horticulture Industry." *Journal of Development Studies* 37 (2): 147–76.

Donati, Kelly. 2005. "The Pleasure of Diversity in Slow Food's Ethics of Taste." *Food, Culture and Society* 8 (2): 227–242.

Dowler, Elizabeth. 2008. "Food and Health Inequalities: The Challenge for Sustaining Just Consumption." *Local Environment* 13 (8): 759–772.

DuPuis, Melanie, and David Goodman. 2005. "Should We Go 'home' to Eat? Toward a Reflexive Politics of Localism." *Journal of Rural Studies* 21 (3): 359–371.

Food and Agricultural Organization (FAO). 2011. "One Billion Hungry." http://www.1billionhungry.org (accessed 5 February 2011).

Fairtrade Labelling Organization (FLO). 2011. "Global Growth in Fairtrade." http://www.fairtrade.net/new.0.html (accessed 5 February 2011).

Featherstone, Mike. 2001. "Foreword." Pp. xvii–xxviii in *Ethical Consumption: A Critical Introduction*, ed. T. Lewis and E. Potter. New York: Routledge.

Ferrer, Mariona, and Marta Fraile. 2006. "Exploring the Social Determinants of Political Consumerism in Western Europe." Working Papers Online Series 57/2006. http://portal.uam.es (accessed 2 December 2010).

Friedberg, Susanne. 2004. *French Beans and Food Scares: Culture and Commerce in an Anxious Age*. New York: Oxford University Press.

Furlough, Ellen, and Carl Strikwerda. 1999. *Consumers against Capitalism? Consumer Cooperation in Europe, North America, and Japan, 1840–1990*. Lanham, MD: Rowman and Littlefield.

Gabriel, Y., and T. Lang. 1995. *The Unmanageable Consumer: Contemporary Consumption and Its Fragmentations*. London: Sage.

———. 2005. "A Brief History of Consumer Activism." Pp. 39–53 in *The Ethical Consumer*, ed. R. Harrison, T. Newholm, and D. Shaw. Thousand Oaks, CA: Sage.

Galbraith, John Kenneth. 2000. "The Dependence Effect." Pp. 20–25 in *The Consumer Society Reader*. Ed. J. Schor and D. Holt. New York: The New Press.

Garsten, Christina. 2009. "Ethnography at the Interface: 'Corporate Social Responsibility' as an Anthropological Field of Enquiry." Pp. 56–68 in *Ethnographic Practice in the Present*. Ed. M. Melhuus, J. Mitchell, and H. Wulff. Oxford: Berghahn Books.

Garsten, Christina, and Anna Hasselström. 2003. "Risky Business: Discourses of Risk and (Ir)responsibility in Globalizing Markets." *Ethnos* 68 (2): 249–270.

Giddens, Anthony. 1991. *The Consequences of Modernity*. Cambridge: Polity Press.

Gladwell, Malcolm. 2000. "The Coolhunt." Pp. 360–374 In *The Consumer Society Reader*, ed. E. Schor and D. Holt. New York: The New Press.

Goodman, David, and Melanie DuPuis. 2002. "Knowing Food and Growing Food: Beyond the Production-Consumption Debate in the Sociology of Agriculture." *Sociologia Ruralis* 42 (1): 6–23.

Goss, Jon. 2004. "Geography of Consumption." *Progress in Human Geography* 28 (3): 369–380.

Guthman Julie. 2003. "Eating Risk: The Politics of Labeling Transgenic Foods." Pp. 130–151 in *Remaking the World: Genetic Engineering and Its Discontents*, eds R. Schurman and D. Takahashi-Kelso. Berkeley: University of California Press.

———. 2004a. *Agrarian Dreams: The Paradox of Organic Farming in California*. Berkeley: University of California Press.

———. 2004b. "Back to the Land: the Paradox of Organic Food Standards." *Environment and Planning A* 36 (3): 511–528.
———. 2008. "Neoliberalism and the Making of Food Politics in California." *Geoforum* 39 (3): 1171–1183.
Halkier, Bente. 1999. "Consequences of the Politicization of Consumption: The Example of Environmentally Friendly Consumption Practices." *Journal of Environmental Policy and Planning* 1: 25–41.
———. 2001. "Consuming Ambivalences: Consumer Handling of Environmentally Related Risks In Food" *Journal of Consumer Culture* 1 (2): 205-224.
Harrison, Rob, Terry Newholm, and Deirdre Shaw, Eds. 2005. *The Ethical Consumer.* Thousand Oaks, CA: Sage.
Hartwick, Elaine. 2000. "Towards a Geographical Politics of Consumption." *Environment and Planning A* 32 (7): 1177–1192.
Heinberg, Richard, and Michael Bomford. 2009. "The Food and Farming Transition: Toward a Post Carbon Food System." Post-Carbon Institute Report. http://www.postcarbon.org (accessed 1 December 2011).
Hilton, Matthew. 2001. "Consumer Politics in Post-War Britain." Pp. 241–260 in *The Politics of Consumption: Material Culture and Citizenship in Europe and America,* ed. M. Hilton and M. Daunton. London: Berg.
Hilton, Matthew, and Martin Daunton. 2001. "Material Politics: An Introduction." Pp. 1–32 in *The Politics of Consumption,* ed. M. Daunton and M. Hilton. London: Berg.
Hinrichs, Clare. 2003. "The Practice and Politics of Food System Localization." *Journal of Rural Studies* 19 (1): 33–45.
Hobson, Kersty. 2002. "Competing Discourses of Sustainable Consumption: Does the 'Rationalization of Lifestyles' Make Sense?" *Environmental Politics* 11 (2): 95–120
———. 2006. "Bins, Bulbs and Shower Timers: On the 'Techno-Ethics' of Sustainable Living." *Ethics, Place and Environment* 9 (3): 317–336.
Hobson, Kersty and Ann Hill. 2010. "Cultivating Citizen-Subjects through Collective Praxis: Organized Gardening Projects in Australia and the Philippines." Pp. 216–230 in *Ethical Consumption: A Critical Introduction,* ed. T. Lewis and E. Potter. Abingdon: Routledge.
Holt, Douglas B. 2000. "Does Cultural Capital Structure American Consumption?" Pp. 212–252 in *The Consumer Society Reader,* ed. J. Schor and D. Holt. New York: The New Press.
Hoogland, Carolien, Joop de Boer, and Jan J. Boersma. 2007. "Food and Sustainability: Do Consumers Recognize, Understand and Value On-Package Information on Production Standards?" *Appetite* 49 (1): 47–57.
Hornborg, Alf. 2007. "The Machine as Fetish: Why Space Is Conceived as Time and Wrecking the World as a National Interest." Paper presented at the American Anthropological Association Annual Meeting.
———. 2008. "Den Vite Konsumentens Börda." *Dagens Nyheter* Stockholm, 28 January.
Humphery, Kim. 2010. *Excess: Anti-Consumerism in the West.* Cambridge: Polity Press.
Isenhour, Cindy. 2010a. "Building Sustainable Societies: A Swedish Case Study on the Limits of Reflexive Modernization." *American Ethnologist* 37 (3): 511–525.
———. 2010b. "Building Sustainable Societies: Exploring Sustainability Policy and Practice in the Age of High Consumption." PhD diss, University of Kentucky.
———. 2010c. "On Conflicted Swedish Consumers, the Effort to 'Stop Shopping' & Neoliberal Environmental Governance." *Journal of Consumer Behavior,* Special Issue on Anti Consumption and Sustainability 9 (6): 454–469.
Jacobsen, Eivind, and Arne Dulsrud. 2007. "Will Consumers Save the World? The Framing of Political Consumerism." *Journal of Agricultural and Environmental Ethics* 20 (5): 469–482.
Jaffee, Daniel, Jack R. Kloppenburg, Jr., and Mario B. Monroy. 2004. "Bringing the 'Moral Charge' Home: Fair Trade within the North and within the South." *Rural Sociology* 69 (2): 169–196.
Johnston, Josee. 2008a. "The Citizen-Consumer Hybrid: Ideological Tensions and the Case of Whole Foods Market." *Theory and Society* 37 (3): 229–270.

———. 2008b. "Counterhegemony or Bourgeious Piggery? Food Politics and the Case of Foodshare.", Pp. 93–119 in *The Fight Over Food: Producers, Consumers, and Activists Challenge the Global Food System,* ed. W. Wright and G. Middendorf. University Park: Pennsylvania State University Press.
Jubas, K. 2007. "Conceptual Con/fusion in Democratic Societies: Understandings and Limitations of Consumer-Citizenship." *Journal of Consumer Culture* 7 (2): 231–254.
Keiko, Tanaka, and Elizabeth Ransom. 2008. "Citizens in the Global Agrifood System: The Cases of New Zealand and South Africa in the Global Red Meat Chain." Pp. 247–272 in *The Fight Over Food: Producers, Consumers, and Activists Challenge the Global Food System,* ed. W. Wright and G. Middendorf. University Park: Pennsylvania State University Press.
Kimbrell, Andrew. 2002. *Fatal Harvest: The Tragedy of Industrial Agriculture.* Washington, D.C.: Island Press.
Kingsolver, Barbara, Camille Kingsolver & Steven Hopp. 2007. *Animal, Vegetable, Miracle: A Year of Food Life.* New York: Harper Collins.
Kloppenburg Jr. J., S. Lezberg, K. DeMaster, G. W. Stevenson, G.W., and J. Hendrickson, 2000. "Tasting Food, Tasting Sustainability: Defining the Attributes of an Alternative Food System with Competent, Ordinary People." *Human Organization* 59 (2): 177–186.
Kneafsey, Moya, Lewis Holloway, Laura Venn, Elizabeth Dowler, Rosie Cox, and Helena Tuomainen. 2008. *Reconnecting Consumers, Producers and Food: Exploring Alternatives.* London: Berg.
Lee, M. S.W., K.V. Fernandez, and M. R. Hyman. 2009. "Anti-Consumption: An Overview and Research Agenda." *Journal of Business Research* 62 (2): 145–147.
Lewis, Tania, and Emily Potter, eds. 2011. *Ethical Consumption: A Critical Introduction.* New York: Routledge.
Littler, Jo. 2008. *Radical Consumption: Shopping for Change in Contemporary Culture.* New York: McGraw-Hill.
Local Harvest. 2011. "Community Supported Agriculture." http://www.localharvest.org/ (accessed December 1).
Luetchford, Peter. 2008. *Fair Trade and a Global Commodity: Coffee in Costa Rica.* London: Pluto Press.
Lyon, Sarah. 2006. "Migratory Imaginations: The Commodification and Contradictions of Shade Grown Coffee." *Social Anthropology* 14 (3): 377–390.
Lyon, Sarah, and Mark Moberg, eds. 2010. *Fair Trade and Social Justice: Global Ethnographies.* New York: New York University Press.
Lyson, T. A., G. W. Gillespie Jr, and D. Hilchey. 1995. "Farmers' Markets and the Local Community: Bridging the Formal and Informal Economy." *American Journal of Alternative Agriculture* 10: 108–113
Macnaughten, Phil. 2003. "Embodying the Environment in Everyday Life Practices." *The Sociological Review* 51 (1): 63–84.
Maniates, Michael. 2002. "Individualization: Plant a Tree, Buy a Bike, Save the World?" Pp. 43–66 in *Confronting Consumption,* ed. Princen, Maniates and Conca. Cambridge: MIT Press.
Manno, Jack. 2002. "Commoditization: Consumption Efficiency and an Economy of Care and Connection." Pp. 67–99 in *Confronting Consumption,* ed. Princen, Maniates and Conca. Cambridge: MIT Press.
Marx, Karl. [1867] 1978. *Capital,* vol. 1. In *The Marx-Engels Reader,* ed. R Tucker, 2nd ed. New York: W.W. Norton and Company.
Matti, Simon. 2009. "Exploring Public Policy Legitimacy: A Study of Belief-System Correspondence in Swedish Environmental Policy." PhD diss., Luleå Tekniska Universitet.
Micheletti, Michele. 2003. *Political Virtue and Shopping: Individuals, Consumerism, and Collective Action.* New York: Palgrave Macmillan.
Micheletti, Michele, and Andrew McFarland. 2010. *Creative Participation: Responsibility-Taking in the Political World.* Boulder, CO: Paradigm Publishers
Miller, Daniel. 2001a. *The Dialectics of Shopping.* Chicago: University of Chicago Press.
———. 2001b. "The Poverty of Morality." *Journal of Consumer Culture* 1 (2): 225–244.

Morgan, Kevin, Terry Marsden, and Jonathan Murdoch. 2006. *Worlds of Food: Place, Power, and Provenance in the Food Chain.* New York: Oxford University Press.

Munro, William A., and Rachel Schurman. 2008. "Sustaining Outrage: Motivating Sensibilities in the U.S. Anti-GE Movement." Pp. 145–176 in *The Fight Over Food: Producers, Consumers, and Activists Challenge the Global Food System,* ed. W. Wright and G. Middendorf. University Park: Pennsylvania State University Press.

Mutersbaugh, Tad. 2002. "The Number is the Beast: A Political Economy of Organic-Coffee Certification and Producer Unionism" *Environment and Planning A* 34 (7): 1165–84.

Nabham, Gary. 2002. *Coming Home to Eat: The Pleasures and Politics of Local Foods.* New York: W.W. Norton and Company.

Pollan, Michael. 2009. *In Defense of Food: An Eater's Manifesto.* New York: Penguin.

———. 2009. *Food Rules: An Eater's Manual.* New York: Penguin.

Pratt, Jeff. 2007. "Food Values: The Local and the Authentic." *Critique of Anthropology* 27 (3): 285–300.

Princen, Thomas, Michael Maniates, and Ken Conca, eds. 2002. *Confronting Consumption.* Cambridge: MIT Press.

Probyn, Elspeth. 2011. "Feeding the World: Toward a Messy Ethics of Eating." Pp. 103–115 in *Ethical Consumption: A Critical Introduction,* ed. Tania Lewis and Emily Potter. New York: Routledge.

Raymond, Jussaume, A., and Kazumi Kondoh. 2008. "Possibilities for Revitalizing Local Agricutlure: Evidence from Four Counties in Washington State." Pp. xxx–xxx in *The Fight Over Food: Producers, Consumers, and Activists Challenge the Global Food System,* ed. W. Wright and G. Middendorf. University Park: Pennsylvania State University Press.

Raynolds, L. 2000. "Re-embedding Global Agriculture: The International Organic and Fair Trade Movements." *Agriculture and Human Values* 17 (13): 297–309.

Roseberry, William. 1996. "The Rise of Yuppie Coffees and the Reimagination of Class in the United States." *American Anthropologist* 98 (4): 762–775.

Sassatelli, Roberta. 2006. "Virtue, Responsibility and Consumer Choice: Framing Critical Consumerism." Pp. 219–250 in *Consuming Cultures, Global Perspectives: Historical Trajectories, Transnational Exchanges,* ed. John Brewer and Frank Trentmann. Oxford: Berg.

———. 2009. "Representing Consumers: Contesting Claims and Agendas." Pp. 25–42 in *Better than Shopping. The Politics and Pleasures of Consuming Differently,* ed. K. Soper, M. Ryle, and L. Thomas. London: Palgrave.

Sayer, A., 2003. "De(commodifcation), Consumer Culture, and Moral Economy." *Environment and Planning D: Society and Space* 21 (3), 341–357.

Schor, Juliet and Douglas Holt. 2000. The Consumer Society Reader. New York: The New Press.

Schwartz, Barry. 2005. *The Paradox of Choice: Why Less Is More.* New York: Harper Collins.

Shaw, D., Newholm, T., and Dickinson, R. 2006. "Consumption as Voting: An Exploration of Consumer Empowerment." *European Journal of Marketing,* Special Issue: Consumer Empowerment 40 (9–10): 1049–1067.

Shreck, Aimee. 2008. "Resistance, Resdistribution, and Power in the Fair Trade Banana Initiative." Pp. 121–144 in *The Fight Over Food: Producers, Consumers, and Activists Challenge the Global Food System,* ed. W. Wright and G. Middendorfl. University Park: Pennsylvania State University Press.

Sklair, Leslie. 2001. *Globalization: Capitalism and its Alternatives.* Oxford: Blackwell.

Soper, Kate. 1999. "Other Pleasures: The Attractions of Post-Consumerism." Pp. 114–132 in *Necessary and Unecessary Utopias: socialist Register,* ed. L. Panitch and C. Leys. New York: Monthly Review Press.

———. 2007. "Re-Thinking the 'Good Life': The Citizenship Dimension of Consumer Disaffection with Consumerism." *Journal of Consumer Culture* 7 (2): 205–229.

Soper, Kate, Martin Ryle, and Lyn Thomas. 2009. The Politics and Pleasures of Consuming Differently. London: Palgrave Macmillan.

Stolle, D., M. Hooghe, and M. Micheletti. 2005. "Politics in the Supermarket: Political Consumerism as a Form of Political Participation." *International Political Science Review* 26 (3): 245–269.

Stø, Eivind, Harald Throne-Holst, Pål Strandbakken, and Gunnar Vittersø. 2008. "Review: A Multi-dimensional Approach to the Study of Consumption in Modern Societies and the Potential for Radical Sustainable Changes." Pp. 234–254 in *System Innovation for Sustainability: Perspectives*, ed. Tucker. Sheffield: Greenleaf.

Trentmann, Frank. 2001. "Bread, Milk and Democracy: Consumption and Citizenship in Twentieth-Century Britian." Pp. 129–164 in *The Politics of Consumption: Material Culture and Citizenship in Europe and America*, ed. M. Hilton and M. Daunton. London: Berg.

———. 2006. *The Making of the Consumer: Knowledge, Power & Identity in the Modern World.* London: Berg.

Tronto, J. 1993. *Moral Boundaries: A Political Argument for an Ethic of Care.* New York: Routledge.

Vankeerberghen, Audrey. Forthcoming in 2012. "Today, One Can Farm Organic without Living Organic: Belgian Farmers and Recent Changes in Organic Farming." Pp. 146–172 in *Ethical Consumption: Social Value and Economic Practice*, ed. J. Carrier and P. Leuchford. London: Berghahn Books.

Varman, Rohit, and Russell W. Belk. 2009. "Nationalism and Ideology in an Anti-Consumption Movement." *Journal of Consumer Research* 36 (4): 686–700.

Walter, Lynn. 2008. "Slow Food and Home Cooking: Toward a Relational Aesthetic of Food and Relational Ethic of Home." Paper presented at the 2nd Sustainable Consumption Conference in Arlon, Belgium.

Welsch, Heinz, and Jan Kühling. 2009. "Determinants of Pro-environmental Consumption: The Role of Reference Groups and Routine Behavior." *Ecological Economics* 69 (1): 166–176.

Wilk, Richard. 2004. "Questionable Assumptions about Sustainable Consumption." Pp. 17–22 in *The Ecological Economics of Consumption*, ed. L. Reisch and I. Røpke. Cheltenham, UK: Edward Elgar.

———. 2006. *Home Cooking in the Global Village: Carribean Food from Buccaneers to Ecotourists.* Oxford: Berg.

———. 2010. "Consumption Embedded in Culture and Language: Implications for Finding Sustainability." *Sustainability: Science, Practice and Policy* 6 (2): 38–48.

Wright, Wynne, and Gerad Middendorf. 2008. "Introduction: Fighting Over Food- Change in the Agrifood System." Pp. 1–26 in *The Fight Over Food: Producers, Consumers, and Activists Challenge the Global Food System*, ed. W. Wright and G. Middendorf. University Park: Pennsylvania State University Press.

Fair Trade and Fair Trade Certification of Food and Agricultural Commodities

Promises, Pitfalls, and Possibilities

Debarati Sen and Sarasij Majumder

ABSTRACT: The global circulation of food and agricultural commodities is increasingly influenced by the ethical choices of Western consumers and activists who want to see a socially and environmentally sustainable trade regime in place. These desires have culminated in the formation of an elaborate system of rules, which govern the physical and social conditions of food production and circulation, reflected in transnational ethical regimes such as fair trade. Fair trade operates through certifying producer communities with sustainable production methods and socially just production relationships. By examining interdisciplinary academic engagements with fair trade, we argue that fair trade certification is a transnational bio-political regime; although, it holds the potential for reflecting global counterpolitics. By reviewing the literature on the emergence and history of fair trade certification, agro-food chains, case studies on certified producer communities and the certification process, this article shows that fair trade certification is a new governing mechanism to discipline farmers and producers in the Global South by drawing them into globalized market relationships. However, recent studies suggest that fair trade also leaves open the potential for creative iterations of the fair trade idea in producer communities to give voice to their situated struggles for justice. Thus, fair trade constitutes a contested moral terrain that mediates between the visions of justice harbored by producers and activists in the Global South and reflexive practices of the Western consumers. To map these critical developments around fair trade and fair trade certification, close ethnographic attention to the material and symbolic life of certification is vital.

KEYWORDS: agro-food networks, creative iterations of fair trade, ethical trade, fair trade certification, organic certification, politics of certification

Fair trade and fair trade certification of food and agricultural commodities restore ethics and morality to economic practices of trading, investing, buying, and selling. Economic actions remain unmoored from ethical considerations of how they affect people's lives and livelihoods. Fair trade as an alternative model aims to provide an ethical anchor by establishing partnership-based dialogue between producers or farmers and ethical consumers (Murray and Raynolds

Environment and Society: Advances in Research 2 (2011): 29–47 © Berghahn Books
doi:10.3167/ares.2011.020103

2007). The dialogue depends on certification of producers and their products for ethical and ecologically sound production practices (West 2010). It is also a process of meaning making, where an "economy of semiology" (Dolan 2008) constructs an emotional bridge between cosmopolitan or Western consumers and the Third World or Southern producers.

Cosmopolitan consumers attempt to reinscribe ethics into their shopping practices through dialogue, partnership, and equal exchange with producer-farmers; however, to what extent are these relational ethics actualized? Does the "economy of semiology" also operate as a new technical rationality with a moral twist? Does fair trade contest the unequal market based exchanges between Northern markets and Southern producers, or does it reinforce existing asymmetrical market relations that constrain and discipline food producers and small farmers in new and subtle ways?

To address the above questions, we locate fair trade and fair trade certification in the context of economic globalization. Recent economic globalization is guided by neoliberal principles, that is doctrines and policies that accord the market rather than the state the main role in satisfying basic human needs (Edelman and Haugerud 2005). Fair trade as a nonstatist (Fridell 2007) way of regulating the global trade in agricultural commodities tends to fill the gap left by the rollback of the state. Thus fair trade practices and state practices become comparable in many ways. Provisioning of basic needs and dispensing of social and economic justice—the activities of a public institution, the state—is now partly taken up by private and transnational organizations administering fair trade.

This partial privatization of state actions through fair trade also signals the emergence of private transnational regimes of governance (Busch and Bain 2004; Mutersbaugh 2005; Vandergeest 2007). The logic of these transnational governing practices is similar to the ones of the state that rule over a national society. Michel Foucault (1990: 13) identifies the logic of the modern state's ruling practices as biopolitical, since administrative and political practices of the state derive their legitimacy by fostering self-disciplining of its citizens by regulating "bio"—the ordinary and existential life of individuals or citizens. These individuals and citizens constitute a national society where individuals and groups depend on each other and impersonal bureaucratic rules for their basic existence. Thus biopolitics operates by regulating society through the self-regulation of individual citizens.

In the era of globalization, mutual dependence extends beyond the borders of nations. Hence, transnational governance mechanisms like fair trade have emerged to govern trade relationships that traverse continents. The fair trade model and its practice of certifying and labeling food and agricultural commodities are means of nurturing ecological and equitable life on a transnational scale. Fair trade practices also operate as a regulating, disciplining, and governing mechanism that is both restrictive and productive. It entails technical interventions in producer communities by monitoring and inspecting farms and farming techniques, checking the conditions of on-farm labor relations, and tracking the use of fair trade premium by farming communities to approve fair trade labeling of agricultural products. In this sense, fair trade fosters individual and collective self-regulation among producer communities and is a transnational biopolitical regime.

Moreover, the effects of fair trade as a biopolitical regime are similar to the effects of bureaucratic and governing practices of the state in instructive ways. A contradiction between a bureaucratically inclusive national society and exclusions that curtail rights of minorities characterizes state-centric biopolitics (Ong and Collier 2005: 15). This incessant clash between inclusive ideals and actual facts of exclusion is also the emerging trait of fair trade, which promises equality and justice for marginalized producers. Incidents of exclusion of marginalized farmers, who find meeting fair trade standards costly, are very common as corporations are appropriating the fair trade model or making use of certification practices. Thus the fair trade label, whose primary purpose is to convey on-the-ground situations in producer and farmer communities, is displaced to give a false sense of moral choice to consumers buying labeled products. Consequently, the goal of defetishizing the commodity is defeated by further fetishizing the imagined moral relationship that the consumer strikes with the producer. This has lead Mark Moberg and Sarah Lyon (2010: 8) to see fair trade as "shaped advantage" that enables "a limited number of producers to enter the global market under more favorable terms, utilizing enhanced institutional capacity and marketing skills to tap into a growing niche market."

However, the resemblance of state-centric biopolitics and fair trade extends beyond simply being a mechanism of domination. Disenfranchisement and exclusion notwithstanding, the idea of inclusiveness of state practices nurture expectations among citizens. These expectations make people aware of their rights and give rise to counterpolitics. Similarly, fair trade and its certification practices cultivate certain kinds of expectations in producer communities, even among the excluded groups (Lyon 2008; Sen 2009). These expectations serve as a launching pad for critical counterpolitics that tend to challenge and contest national (Majumder 2010) and transnational political economic relationships undergirding fair trade, conventional trade, state practices, as well as various kinds of inequalities within the producer communities, such as class and gender inequalities, that keep the poorest or the most disadvantaged groups from accessing the benefits of fair trade or trade in general (Renard and Pérez-Grovas 2007; Sen 2009). This aspect of fair trade is relatively unexamined and can be highlighted through ethnographic attention to the effects of fair trade certification in producer communities. Fair trade has set in motion a process that helps global market relations to penetrate the remotest corners of the world for resources, but it also creates potentials and possibilities for a global counterpolitics.

In this review article[1] we show that the trajectory of fair trade is best understood as an interaction between fair trade as a social movement with commitment to ethical and environmental issues, and fair trade as a transnational biopolitical regime. The trajectory is marked by shifts, displacement of goals, reassertion of activist values, and uneven impacts on the ground. We argue that fair trade blurs the distinction between market-based exchanges of agricultural commodities that extract resources from the Global South, and the global counterpolitics that challenge inequities of market-based exchange. This blurring, we contend, has created new possibilities for producers to articulate their situated demands for social justice, but it is also a new system of disciplining producers.

To map out the contested terrain of fair trade and fair trade certification of food and agricultural commodities, we divided the article into four sections and a conclusion. The article begins with a brief history of the emergence of fair trade and fair trade certification, followed by a review of the theoretical literature on agro-food chains. The third section presents case studies on the diverging impacts of agricultural commodity certification. The next section presents findings from the first author's ethnographic research in India (Sen 2009) to uphold how creative iterations of the fair trade idea are used to contest the disciplining of certification. We conclude with thoughts about the pitfalls and possibilities of this emerging trade initiative.

A Brief History of Fair Trade and Fair Trade Certification of Agricultural Commodities

The Fair Trade movement emerged out of the progressive solidarity movements, like the cooperative movement of the 1970s (Moore 2004), to question the harmful excesses of global free trade. Its roots lay in the post–World War II era of agricultural Fordism, which entailed use of state-subsidized and productivity-enhancing technological inputs to produce for the mass market. Under this kind of production regime, fair trade was meant to protect marginalized Southern farmers from price fluctuations of conventional exports such as coffee and sugar (Raynolds and Wilkinson 2007). Thus, fair trade came to represent a direct moral connection between Western consumers and the farmer-producers in the South. It reflected the growing desires of Western consumers for safe and organically produced food products. Consequently, the target market for agro-food products under the fair trade system shifted from mass-market to the up-market quality conscious consumers. The shift required independent monitoring and certification of producers with little state involvement (Busch and Bain 2004).

The direct marketing links that activist consumers supported after 1970 (Murray and Raynolds 2007: 7) have gradually taken more concrete shape in pushing for ethical trade initiatives (ETIs) at the global scale, replacing small-scale solidarity buying that marked the initial phase of the development of alternative trade organizations. Further, the dominance of neoliberal policies in global trade, the collapse of the Bretton Woods, the creation of the World Trade Organization, the proliferation of structural adjustment programs (Fridell 2007: 30), and continued poverty of Third World workers and farmers raised concerns over the plight of the marginalized agricultural producers. As usual, Western consumer citizens began voting with their dollars to change business (Jaffee 2007).

These concerns from consumer groups influenced the formation of market-based regulation and certification schemes that have affected the global traffic in agricultural commodities including coffee, tea, bananas, shrimp, cocoa, and wine. The initial efforts came from organizations that fought against poverty and hunger, like Christian Aid, World Development Movement, and Oxfam. Tradecraft in the UK and Ten Thousand Villages in the US first pushed for ethical labels in handicrafts. Starting with the Max Havelaar label in coffee in the 1980s, ethical labeling was extended to agricultural commodities generating new standards of production and trade. Over the years food labeling initiatives have proliferated, such as International Federation of Organic Agriculture Movements, the Rainforest Alliance, Fair Trade Labeling Organizations International (FLO) (Hatanaka 2010), EurepGap/GlobalGap (Friedmann and McNair 2008), Forest Stewardship Council Standards (Klooster 2010), and Aquaculture Certification Council (Vandergeest 2007). Collectively they constitute Transnational Alternative Agrifood Networks (TAAFNs), which prioritize social justice and sustainability within trade initiatives. Formed in 1997, FLO emerged as an umbrella organization that generates some basic rules of fair trade food certification and disseminates information on whether producer communities are receiving benefits of fair trade while engaging in sustainable social and ecological production practices. Based in Germany, FLO operates through national initiatives like Transfair USA and Fair Trade Foundation UK.

These food certification initiatives may have different and contradictory standards; however they concur on some basic values of social justice and sustainable production methods. There is a difference between simple organic certification and fair trade certification, although many producer communities might be compliant and certified under both standards (Sen 2009). Organic certification focuses on the quality of the product being certified, but fair trade certification involves product quality enhancement as well as linking marginalized agricultural producers in the Global South—removed from profitable circuits of global trade—with ethical consumer groups and retailers (Getz and Shreck 2006: 496). Tad Mutersbaugh (2005) further emphasizes that the fair trade label for agricultural products is crucial, as the certified label helps mark the difference in fair trade agricultural products that otherwise taste and smell the same as conventionally produced agro-food products. Nongovernmental institutions run this certification process, and their regulatory frameworks reflect the perceived desires and expectations of Northern consumers, retailers, and activists (Raynolds et al. 2007). Since 2000, large corporations like Starbucks, which were the critical targets of the early progressive movements, have begun participating in the certification processes, sometimes generating their own codes like the Café Practices Program (MacDonald 2007). The other characteristic is that the producer organizations' participation is voluntary, but these organizations are lured by the possibility of reaching niche markets in the West. In many cases, the lack of economic opportunities provided by the local state and issues of state accountability increase the moral reach of these privatized ethical regimes.

Under the broad rubric comprising economic, social, and environmental development, the FLO (n.d.) outlines general objectives of fair trade on its website:

- ensure that producers receive prices that cover their average costs of sustainable production;
- provide an additional fair trade premium which can be invested in projects that enhance social, economic and environmental development;
- enable pre-financing for producers who require it;
- facilitate long-term trading partnerships and enable greater producer control over the trading process;
- set clear minimum and progressive criteria to ensure that the conditions of production and trade of all fair trade certified products are socially and economically fair and environmentally responsible.

Through these standards, FLO tries to guarantee that the bargaining power of food producers in the Global South is protected. In addition to providing guaranteed prices and a social premium, FLO ensures that producers are represented by democratically elected bargaining institutions, like producer-cooperatives (for farmers) and unions (for hired workers). It also tries to enhance production and marketing skills for producer families beyond the production of the certified commodity by creating opportunities for nonfarm employment.

To guarantee producer compliance, in the past decade FLO and other TAAFNs (Getz and Shreck 2006; Hatanaka 2010) have devised a system of third-party certification (TPC) that aims to increase "partnership and traceability" (Friedmann and McNair 2008; Raynolds 2009: 1083). FLO subsidiaries send independent inspectors to visit producer organizations each year to check for compliance with these goals. They conduct periodic trainings in producer communities, making sure that producers understand the goals of the larger fair trade initiatives. FLO verification and auditing rules for certification are based on the International Standards Organization 65 quality standards for organic agriculture. Recently, many certification initiatives have begun an internal control system, where key representatives of producer communities, like Peasant Inspectors or Community Technical Officers (CTOs), work to ensure quality production (Mutersbaugh 2008).

Despite good intentions behind creating the fair trade label, what remains unclear sometimes is what exactly is being tested under the broad umbrella of quality. The label assures consumers of a variety of standards and values that they might want to see fulfilled within the fair trade system, which often leads to confusion (Moore 2004). Peter Vandergeest (2007: 1154) distinguishes between "experience based quality certification" (taste) and "credence based" (nonmaterial, ethical). The challenges of certifying for extrinsic ethical qualities, such as fair labor use in production, are difficult and attained through what Tad Mutersbaugh calls "just in space" regulation of labor practices. Mutersbaugh distinguishes this form of product quality maintenance from the Fordist system of production, where the final product's quality was tested at the end of the production process. In this new system of product quality maintenance, monitoring of quality begins at the site of procurement of raw materials. For example, in the case of fair trade tea production in Darjeeling, the monitoring would start with an inspection of whether workers (in plantations) and farmers (in tea producing households) were being treated with respect. When the fair trade label is found on a package of tea it means that the product, the people producing the product, and those involved in certifying it have been checked for quality (Mutersbaugh 2005).

In addition to transnational certification processes being getting centralized through FLO, there has been a push for regional certification and labeling initiatives. The impetus for designing and promoting these so-called local initiatives stands as a critique of the perceived top-down directives of transnational certification regimes. Harriet Friedmann and Amber McNair (2008) describe these as alternative regulatory schemes, which give more freedom to local producers to innovate and maintain their regional distinction, defying fair trade's biopolitical aspect. For instance, the Slow Food Foundation for Biodiversity created Presidia—an initiative of local producer groups, which have more power to maintain quality through their own efforts to innovate, instead of following the narrow standards of quality defined in global value chains. This enables a greater respect for the producers' own methods, instead of the mainstreaming of organics (Guthman 2004). Similarly in Mexico there is the Coyotè Royal regional label for agriproducts aimed to protect indigenous varieties and local methods of farming. The Local Food Product campaign in Canada resulted in a productive partnership of local growers to promote farm products from the area that were being wiped out by larger corporations promoting organic or sustainable agriculture. Some scholars see these new initiatives as new regional institutional alternatives with concrete local effects, compared to the more militant antiglobalization movements (Friedmann and McNair 2008: 427).

Indigenous cultivation methods are being celebrated through these regional labeling initiatives, but other scholars propose the need for adjusting certification rules to reflect the contested regional histories, because much of the promise of labels is to enlighten consumers about the potential ecological and social impacts of sustainable production. In post-apartheid South Africa, wineries in the Western Cape region have seen changes in ownership and management composition as a result of the reforms made by the African National Congress. Initiatives like Black Economic Empowerment have resulted in white winery owners seeking out partnership with black owners. However, land distribution and some changes in farm ownership have done little to change the existing unequal worker relations in the farms (Moseley 2008). In 2004, FLO developed some country-specific criteria to reflect the specificity of fair trade certification to address place-specific inequities. Wineries would now have to make sure that there would be worker co-ownership of 25 percent of the business to ensure more power for black partners.

To assess further whether fair trade and fair trade certification of food and agricultural commodities is engendering any structural changes in the power imbalances within global commodity networks, it is important to look at the theoretical frames used to analyze such networks and ascertain their adequacy.

Fair Trade, Fair Trade Certification, and Agro-Food Networks: An Evaluation of Existing Frameworks

Scholars trying to understand the relationship between fair trade and conventional agro-food chains can broadly be divided into two camps, fair trade skeptics and fair trade enthusiasts, both influenced directly or indirectly by the commodity chain approach. We briefly discuss the commodity chain approach here to clearly distinguish between these two groups.

From the perspective of commodity chain approach, global trade in agricultural commodities looks like a chain of relationships through which value of the goods is appropriated. The value appropriation depends on distribution of power in the commodity chain. At one end of the chain figures the farmer or the producer and at the other end the Northern supermarkets and retailers. Globalization, commodity chain theorists contend, gives unprecedented power to Northern retailers because marketing of agricultural products become crucial for realizing the value of the commodities. Thus, commodity chains become buyer-driven rather than producer-driven, that is global corporations and retailers appropriate the value of agricultural products at the expense of marginalized producers (Gereffi et al. 1994).

Following this line of analysis, fair trade skeptics find little difference between the workings of fair trade–influenced commodity chains and conventional commodity chains. They see fair trade as further entrenching the market system (Guthman 2007; Moberg and Lyon 2010; Wilson 2010) through a "new system of private governance" (Gereffi et al. 2001: 57). Conversely, fair trade enthusiasts claim that fair trade is an alternative globalization process that empowers farmers in the South (Bacon 2005; Taylor 2005).

A middle ground between these two approaches is required to appreciate the effects of fair trade on producer communities and wider power inequalities. Seeing fair trade as a moral terrain of struggle between various kinds of groups, such as big corporations, activists, marginalized producers, big plantations, wealthy farmers, and women farmers is a possible middle ground. Borrowing a metaphor from Daniel Jaffe (2007), fair trade can be seen as a "dance with the devils." The metaphor suggests the contradictory nature of fair trade, which primarily wanted to counter the big corporation-dominated agro-food chains. But expansion of fair trade means bringing the conventional agro-food chains into its ambit and potentially compromising its goals. However, the "dance" also signals a process and performance whose outcome is unpredictable and open-ended with radical possibilities.

Conceptualizing fair trade as a terrain of struggle requires complicating the commodity chain analysis. Laura Raynolds and John Wilkinson (2007) argue that the image of commodity networks helps conceptualize fair trade in nondeterministic ways in contrast with the global commodity or value chain imagery that creates a dichotomy between buyer-driven versus producer-driven commodity chains and thereby loses sight of power struggles within them. Commodity chain analysts have also recognized the importance of the term network over chain (Gerreffi et al. 2005). Networks, Gerreffi and his colleagues argue, constitute one of the three modes of organizing relationships between firms or in the case of agricultural commodities, among farmers, retailers, and consumers. The other two modes are the market and the hierarchical bureaucratic control. In the case of market-based transactions, producers, traders, and consumers can change their trading partners easily; under hierarchical bureaucratic control, switching partners is almost impossible. Networks are an intermediate form of conducting trading partnerships or transactions among entities located across borders. Networks are also contingent on a constant interplay between economic and symbolic aspects of life. Flow of power is much more complex in a network than in a bureaucratic and hierarchical setting where actions and interactions are completely codified.

Gerreffi et al. (2005) have developed a typology of commodity and value chain governance based on the complexity of transactions, ability to codify transactions and capabilities of suppliers: modular case, relational mode, and the captive mode. In the modular case the production chain is segmented—that is commodities move across many entities—allowing relative supplier or producer autonomy. In the relational mode, solidarity and cooperation among the producers, suppliers, and retailers stabilize power across the network. In the captive mode, producers are strictly controlled by demands of retailers with strict contracts (see Ponte 2002). The relational mode best describes fair trade–based transactions, where retailers, certifying institutions, and producers come together to uphold sustainability in trade. Fair trade–based transactions may also shift toward captive mode or modular form depending on the degree of power asymmetry between Northern retailers and Southern producers.

This typology locates fair trade within a broader context of agro-food trade, but it can also be used to show how power operates in an indirect fashion often coupled with morality. Although in the relational mode trust is a key coordinating element, power and disciplining often work through such relationships of trust as producers in many cases are eager to comply with fair trade certification rules for reaching niche markets. When producer communities are fair trade certified, a relationship of trust ensues between monitoring institutions like FLO and representatives of producer organizations like cooperative governing body members. Producer-community insiders are responsible for ensuring community compliance with fair trade or organic standards (Mutersbaugh 2008). Here, the biopolitical aspect of fair trade becomes salient, as power does not work through coercion or constraints, but through communities self-regulating their production priorities themselves.

Moreover, the relationship of trust at times masks the internal battles over resources and representations in producer communities. Thus in fair trade certified Darjeeling's tea cooperatives (Sen 2009) and Guatemalan coffee cooperatives (Lyon 2008), women producers find it increasingly difficult to claim a share of the monetary benefits coming from fair trade certification. Male dominance of key cooperative affairs persists in the face of fair trade certification, which belies the promise of the larger movement aimed at empowering the weakest among the producers. There are several studies of fair trade that point out that broader and deeper historical inequalities are unaddressed through certification (Busch and Bain 2004; Dolan 2008; Shreck 2005). Nonetheless there are also instances where fair trade certification acts as the primary vehicle of redistribution of resources. Fair trade premium brings money for development (Sen 2009; Valkila and Nygren 2009) helps producer communities enter niche markets, and reduces small farmer vulnerability (Bacon 2005). Fair trade networks serve as an excellent tool for redistributing wealth from Northern consumers to Southern producers. For instance, after the first three years of its banana program, FLO assessed that annually two million dollars was being transferred to banana producers in eleven registered producer associations (Shreck 2005: 223).

Thus fair trade complicates the workings of power within commodity networks. It does not just work in terms of captive relationships between producers, retailers or buyers, but much more subtly. Fair trade draws producers from remote areas into trade networks through an allure of the possibility of entering niche markets. Certification practices give rise to community-based monitoring and changes community dynamics, thereby making it similar to operation of biopower through self-disciplining as described by Foucault (1990). Partial successes or "shaped advantage" (Moberg and Lyon 2010) keep the allure of trade alive even in the minds of groups less affected by the redistributive benefits of fair trade. As a transnational initiative fair trade has given rise to a range of possibilities where one finds a genuine commitment to social justice and environmental issues at one extreme and token responses at the other.

Seeing fair trade as a gamut of occurrences helps us discern the complex interaction between mainstream trade regimes theorized under commodity chain analysis and alternative trade regimes that grew out of social movements. It is because of their interaction that neoliberal tendencies of controlling the global food economy through depoliticized and transnational market relations have been pushed to address the social, political, and environmental issues, which are otherwise considered externalities. Thus fair trade blurs the distinction between market-based exchanges of agricultural commodities that extract resources from the Global South, and the global counterpolitics or social movements that challenge inequities of market-based exchange. For Laura Raynolds and John Wilkinson, "the terrain of social movements has now become the terrain of competing market growth strategies. The cost of the neoliberal dream of transforming market economies into market societies has become an unimaginable expansion of market responsibility and accountability" (2007: 42). Although expansion of market responsibility may remain limited, Raynolds and Wilkinson certainly point toward the blurring between market-based exchanges of agricultural commodities and the global counter-politics.

Value Contradictions, Resource Disparities, and Conflicts over Representation in Fair Trade Certification

One can categorize the tensions within fair trade into three broad headings. First, value contradictions, that is contradictory demands of market and profit-maximization principles, on the one hand, and social justice and ethical priorities on the other hand. Second, resource and access inequality among the Southern producers that affect certification. Third, representational problems, that is exclusion of Southern farmers from the fair trade labeling and certifying organizations. Struggles, critiques, claims, and counterclaims are constitutive of the fair trade system because it seeks to defetishize the commodity by further fetishizing it. It tries to use brand certification, labeling, and other market strategies to contest fetishism on which global market relations are based.[2]

Value Contradictions

Recent case studies on fair trade report a gradual waning of commitment to social justice goals in fair trade practices. The seeds of this shift reside in fair trade's success, its acceptance in mainstream trade networks (Jaffee and Howard 2010). Fair trade monitoring organizations, such as FLO and Transfair USA, have partly given into the demands of mainstream trade networks. One can see a goal displacement in the monitoring organization from more activist-oriented goals to more technical or commercial goals. FLO affiliates measure their success more in quantitative and statistical terms, which narrowly pursue increasing fair trade sale volumes as an end in itself (Raynolds and Murray 2007: 225). In doing so they are driven more by market demands than by the development needs of producer-communities thereby upholding the dominant economic model of trade. They tend to rely on dependable sourcing rather than exploring and connecting the absolutely marginalized. This trend is pronounced in Transfair USA's certification of corporate produced bananas for the US market (Raynolds 2007).

In their study of supermarket branding of fair trade in the UK, Barrientos and Smith (2007) discuss a scenario where the key elements of fair trade, such as partnership, are lost when supermarkets introduce their own brand of fair trade goods. The authors demonstrate that the ability to brand fair trade goods, such as cocoa from Ghana and fresh fruits from South Africa, give the supermarkets enormous power to dominate the trade network. The trading practices of the supermarkets tend to resemble the practices of conventional agro-food chains where buyers call the shots. The supermarkets with FLO-certified brands pass on the risk that they face in serving the consumers to the producers. Supermarkets require year-round supply, competitive prices, and just-in-time production. The producers are disciplined by these requirements. In the case of South African fresh fruits, the concerned UK supermarket with its own fair trade brand also switched from one producer to another to maximize its profits and meet the consumer demands. Although the supermarkets that adopted fair trade gave more farm gate prices than conventional traders, some farmers questioned the benefits of fair trade for them because following fair trade conditionalities also doubles the cost of production. Thus fair trade becomes the arena of dispute over prices, requirements, and costs of production. The recent extension of the fair trade label to plantation-produced crops and beverages have escalated concerns over inequities within the system. Plantations have material and symbolic resources to meet certification standards without much hassle. This is one aspect of value compromise with the extension of fair trade to certify more products. What plantations do once they are fair trade certified is alarming and points to further value contradictions within the fair trade certification system (Sen 2009). For instance, tea plantations in India decrease their vulnerability in the global tea trade as they have more resources (money, personnel, and contacts) to make direct connections with retailers and independent buyers interested in selling fair trade tea to the West. In this way plantations protect themselves from the volatile local auctions where prices for tea export are set. It is impossible to track how these "fair trade" gains are actually translating into worker empowerment (Sen 2009). The way in which benefits of fair trade can accrue to laborers may require standards specific to a country region or a place. In South Africa, which has a history of racism and white domination over farm ownership, simple implementation of FLO standards is inadequate (Kruger and Du Toit 2007; Moseley 2008). For similar reasons, in Latin America, fair trade fails to reduce inequities but indirectly radicalizes the demands of the marginalized producers. Latin American and Caribbean Coordination of Fair Trade Small Producers asserts that structural changes within plantation type producer organizations can only occur when fruit plantations become worker-owned collectives (Raynolds and Murray 2007: 228).

Although plantation certification perpetuates the vulnerability of workers, the process of retrocertification of tea, prevalent before 2009, posed challenges for small tea producers in Kenya (Dolan 2008: 308–309). Unlike other agricultural products, tea was not labeled fair trade at its moment of entry into the commodity chain. At the Mombasa auction, tea was sold to brokers as black tea. How much of it was to be sold under the fair trade label was determined when Western retailers let importers know how much fair trade tea they wish to purchase. Tea factories in Kenya were thus never fully sure about the amount of fair trade premium they were going to receive in the end, which negatively affected their social development plans with the premium and they also lost interest they could have earned on premium money over time. These case studies demonstrate that despite its inclusive ideals, mainstreaming of fair trade and extension of the label to large producers and retailers, compromises the social justice goals.

Resource Inequality

Fair trade seeks to directly connect producers in the Global South and consumers in the Global North. However, reaching the marginalized poor farmers and wage laborers in the South by cutting the middlemen off is easier said than done. Success, acceptance, and popularity of the fair trade business model based on branding and labeling tends to take a toll on distributing benefits of the global trade in agricultural commodities to the most disadvantaged groups. Some argue that the recent expansion of fair trade shows its effectiveness to transform wage labor and small farmer enterprises and broaden the distribution of benefits; others contend that this process is promoting the integration of fundamentally unjust operations into the fair trade network. For example, in the case of bananas (Raynolds 2007), exporting is so complicated that harmful intermediaries are required to get quality fruit to the market. Entering the fair trade network also requires significant collective capacities and connections in which the more economically savvy producers and producer collectives succeed. Recent studies show that market trends are discouraging small farmers from joining fair trade networks. Increasing quality and volume requirements allow larger and more commercially oriented farmers to take advantage of the system. For example in Brazil's orange juice sector, stringent quality standards, blend requirements, and the demand for large continuous supplies have sharply increased the share of large capital-intensive enterprises in fair trade exports (Wilkinson and Mascarehnas 2007). The same is the case in the coffee sector in Mexico and the quinoa sector in Bolivia. In Bolivia, export incentives given to large quinoa-growers through fair trade often makes quinoa, a staple diet for the local population, expensive in the domestic market (Cáseres et al. 2007).

Moreover, acquiring organic and fair trade certification is difficult for small and marginal farmers. First, there are material barriers (like high certification fees) to getting quality or organic certification, buying or preparing organic manure, and sustaining organic production. Often the shift to organic cultivation is funded by fair trade premium, but the premium money is not always adequate for small and marginal farmers to keep up with the organic and fair trade standards. Tad Mutersbaugh demonstrates that smaller producer organizations are most burdened by certification as fair trade premium price covers the costs of sustainable production, but certification costs are paid "as poll tax, such that all producers, regardless of farm size, contribute equal amounts" (2002: 1170) to certification.

Second, the level of literacy among farming communities determines their ability to follow certification guidelines (Mutersbaugh 2008). Thus participation in or accessing the fair trade network depends on material and symbolic resources unevenly distributed among rural producers. Although landholding size is an important determinant of this unevenness, political and social connections also create differences within the small and marginal farmer groups. Thus social movements in Brazil and other Latin American countries question the appropriateness of fair trade for small and marginal farmers' development priorities. Many scholars have underscored how the process of gaining fair trade and organic certification can create inequality in resource access among producers. Even when vegetable and herb farmers in del Cabo Mexico (Getz and Shreck 2006: 494) decided to form a cooperative to pool in for meeting the high certification costs, problems of cooperation between members resulted in closing the cooperative to new members. This created a situation where existing cooperative members with membership numbers became objects of envy for farmers who could no longer enter the coop. The membership numbers, which were proof of compliance with fair trade organic certification, raised the symbolic and material resources of certain farmers. These producer hierarchies created further cleavages in farming communities. Certified farmers would now have to give up long standing sourcing relationships held with noncertified farmers to maintain certification, undermining localized systems of exchange and reciprocity. Farmers with membership numbers also engaged in unethical practices and emerged as a new class of middlemen who gradually moved away from direct farming. They engaged more in rent-seeking practices by sourcing their produce from nonmembers. Fair trade certification is also creating a class of local experts on whom producer communities depend on for getting through the certification process (Mutersbaugh 2008), but such dependence creates harmful externalities. As Christy Getz and Aimee Shreck note,

> cooperative's agronomists at times abused their discretionary control over the certification process and production quotas that determined how much product the cooperative would purchase from each farmer. This abuse exacerbated income inequality among members themselves. This "politicization" of certification allowed some members the ability to have significantly more land certified than others ... One of the cooperative's six technical advisors

> was in charge of certification, and he developed a reputation for only certifying extra land for those farmers he liked. (2006: 494–495)

As these case studies demonstrate, local resource differentials among producers not only affect who can get certified, but resource differentials result from the process of certification itself, making it possible for certification to perpetuate exclusionary practices. We discuss the exclusionary tendencies in certification next.

Representation and Power

Scholars researching fair trade and its empowerment potential are raising serious questions about representation to assess the effectiveness of fair trade. Fair trade places importance on the inclusion and adequate representation of marginalized groups in producer organizations. For instance, women's empowerment is one of FLO's key impact areas. According to FLO, "Important investments can be made in women's income generating activities that are not related to the farm, thereby strengthening their income, business experience, and position in the family" (cited in Lyon 2008: 259). Sarah Lyon, in her study of gender equity within fair trade certified coffee producing organizations alerts us to serious shortcomings of certification to mitigate gender-based inequities in three areas: first, women cannot democratically participate in voting and other procedures of the coffee cooperative; second, male cooperative members are not very supportive of women's nonfarm income generating activities; third, there is a general lack of support for women coffee producers. Local patriarchal norms and the burden of household chores prevent women from effective participation in the cooperative's affairs. Further, certification guidelines do not engage with the issue of labor appropriation within households. Lyon proposes a participatory certification process, where certifiers would work with ordinary cooperative members to identify the interests and desires of all groups within the cooperative. Participatory certification would also ensure that certifiers find place specific solutions to promote gender equity.

The question of representing the interests of marginalized producers gets even more complicated when we consider the mainstreaming of fair trade, which has resulted in sourcing from producers whose conditions of operation are not conducive to representing the interests of women workers. Stephanie Barrientos, Catherine Dolan, and Anne Tallontire (2003) underscore the growing informalization and feminization of the labor process in African horticulture and the fresh produce business. Although buyers such as supermarkets and retailers are increasingly forced to comply with a plethora of codes and their costs, they try to minimize the burden by transferring some of the costs to producers who remain in the informal economy. Being in the informal sector, women workers balance both production and family maintenance at their own cost. Even though they are supplying to organizations abiding by sound labor codes, supporting maternity leave and collective bargaining, it does not matter for women workers in the informal economy as they do not have an organization to represent them.

The exclusion of targeted beneficiaries in setting standards and making certification decisions is another major area of concern. Peter Vandergeest (2007) advocates taking the approach of Community Based Natural Resource Management (CBNRM) to give local communities more power in setting certification standards. Based on his study of how Environmental Regulatory Networks (ERN) affects organic shrimp farming certification. He claims that "when communities are mentioned in the standards, it is almost always in relation to the so-called social standards typically included in certification templates. These standards often mandate varying combinations of consultation, sharing of benefits, labor rights, and respecting property and/or human rights. But they do not allow for including local communities in the process of setting, monitoring, or enforcing technical standards" (Vandergeest 2007: 1155).

In Thailand, community-based collective action and local government initiatives kept the negative environmental impacts of shrimp farming under control. The work of environmental regulation performed here reflects that of CBNRM. In recent years, concern among many Western buyers and retailers over the environmental and social impacts of shrimp farming have increased monitoring of shrimp farms by transnational regulatory bodies, most notably the Aquaculture Certification Council. These two regulatory imperatives do not necessarily work together. The environmental standards of certification are put in a technical language in which local communities have no say. Vandergeest (2007) proposes that no matter how difficult the task, local communities should be involved in setting some of these standards because transnational ethical initiatives are after all meant for protecting the interests of local communities.

As mentioned earlier, the fair trade certification system also alters its certification regimen as seen in the case of South African wineries, where country-specific criteria for certification emerged to address historic inequities (Moseley 2008). In the next section we illustrate that despite the failure of its intentions, fair trade—in an indirect way—has the potential to raise the awareness of marginalized producers about their value in producing sustainable organic commodities for the global market. Marginalized women farmers become conscious of their entitlements through their engagements with and critique of fair trade certification.

Counterpolitics and Creative Iterations of Fair Trade

While the existing literature on fair trade tries to determine the movement's outcome vis-à-vis its stated intentions and promise, a new and exciting line of enquiry is emerging. The latter places emphasis on the meaning of fair trade and certification in people's everyday lives; how the subjective social experience of fair trade certification is lived and mobilized for situated struggles for empowerment. Such nuanced analysis of the "ethical cultural politics" (Mutersbaugh and Lyon 2010: 29) of certification processes is possible through intense ethnographic fieldwork in fair trade certified producer communities to understand how fair trade articulates with the subjective aspirations for justice nurtured by its targeted beneficiaries. Catherine Dolan's (2008) qualitative study of fair trade tea in Kenya locates fair trade within the broader field of development and examines how the "key tenets of the Fair Trade system—empowerment, transparency, partnership, and democratic participation—are realized among tea producers in Kiegoi." She further suggests that the attainment of ideals of fair trade "remains at best inchoate and at worse absent from the experience of many actors engaged in Kenyan Fair Trade tea, where the trajectory of ethical outcomes is mediated by an array of conflicting interests, both within and beyond the commodity chain" (2008: 306).

Dolan's claim is powerful and is substantiated by her findings in which tea farmers make no qualitative distinction between this new potentially empowering form of trade and earlier forms of development assistance carried out by missionaries and development practitioners. Tea farmers in Kiegoi understand fair trade to be a more permanent form of development assistance compared to projects that have dried up. They do not see themselves as empowered partners in this new sustainable trade regime as they cannot make a meaningful distinction between the top-down workings of fair trade. Dolan brings out the complexity of how fair trade's meaning gets situated within place-based understandings of development. In Mexico, certification of organic coffee depends on the hard labor of Community Technical Officers (CTOs)who take on the arduous job of checking for quality. CTOs use this service work to derive pride from organic certification work, because organic farming is seen as part of their indigenous tradition of voluntary work called *cargo* (Mutersbaugh 2008). The meaning of fair trade or organic certification is powerfully shaped by factors outside the commodity chain, like colonial and postcolonial legacies of development assistance, or localized symbolic economies of work.

The social life and meaning making around fair trade is a powerful site for examining the potential for empowerment. Though Dolan's formulation of the absence and inchoateness of fair trade's ideals from producer's lived realities is accurate, ethnographic findings from Darjeeling, India (Sen 2009) extends this line of research by examining the gendered resource battles within Fair Trade and organic certified tea cooperatives.

The context of these gendered resource battles was the shift to fair trade-organic tea cultivation in Darjeeling in the 1990s. Women tea farmers and their organic methods of cultivation came to the limelight when big plantations were forced to adopt organic tea production in the 1990s, giving up green revolution technologies popularized by the postcolonial Indian state in the 1960s. Such conversion was not easy. The switch to organic methods and abandoning of chemical fertilizers resulted in a 30 percent decline in the total tea production for each plantation. To counter the productivity loss resulting from organic and fair trade certification, plantations increasingly relied on contractual agreements with tea farmers who were categorized by the local state as producers of "illegal tea," majority of them are women.

By giving prominence to women producers of illegal tea, the advent of fair trade overturned a colonial legacy that had deep sociopolitical implications. Legally, tea could only be produced in large-scale plantations with processing factories. This law remained in effect under the postcolonial regime. As a result, poor village women grew tea leaves illegally in the existing plantation fringes, which were their backyards, and sold them at local markets at dirt-cheap prices. Men went to work in the towns and cities; women who managed the household became tea farmers by default. They grew "illegal" tea with homemade/organic manure to supplement cash income.

Seeing the economic potential of fair trade alliances for small producers, an NGO helped organize some of the region's small farmers into cooperatives so that they could better bargain with plantations. In the process of forming the cooperatives, male members of tea farming households gained prominence, even when the actual work of tea production was done by women. Women tea farmers were realizing their worth in the organic tea production process but they were also feeling a sense of loss, as male community members and local middlemen controlled the profits from tea sales. This realization of self-worth led them to wage a battle both within their households and in the male-dominated tea farmer's cooperative, by creatively appropriating the ideals of fair trade and coining their own terminology—*Swachcha Vyapar*. In Nepali, *swachcha* means fair, transparent, clean, pure, and *vyapar* means business. By using the phrase *Swachcha Vyapar,* women implied trade in tea without the involvement of middlemen. Therefore, *Swachcha Vyapar* was a creative and powerful iteration of the fair trade idea that indexed a gendered awareness about a fair trade network dominated by big plantations, male-dominated tea cooperatives, and male traders.

Women farmers juxtaposed fair trade with *Swachcha Vyapar* to contest the inadequacies of fair trade certification. Such contestations often took the shape of withholding their participation in the fair trade certified and male-dominated tea farmers' cooperative in their community by strengthening a separate Women's Wing of the cooperative. Through the Women's Wing, women tea farmers wanted to claim a separate share of the cooperatives fair trade premium for their own economic projects, like returning microcredit loans. However, the male heads of the cooperative had refused such requests. Fair trade inspectors, who did not know much about intracommunity tussles over premium, believed that including women tea farmers in the operations of the tea cooperative was a positive step toward attaining gender equality, but women tea farmers did not consider this directive to be empowering. Women's Wing members, however, refused to give in to the inspector's suggestion, since they feared being completely dominated by the male cooperative members and losing a share of the premium. Members of Women's Wing knew that defying the inspector could result in the loss of fair trade certification for the cooperative, resulting in monetary losses, yet they went ahead with their decision. As many Women's Wing members mentioned, defying the inspector was a way to ensure real fair trade—*Swachcha Vyapar*—because certification did not guarantee them assistance for their income generation projects.

Through these contestations over inclusion and exclusion, the fair trade idea was infused with new situated moralities that signal how fair trade is made to work, despite the formal failure of its intended outcomes of maintaining gender equality. Without the advent of fair trade, there would be no mobilization for *Swachcha Vyapar* among women tea farmers. Ironically, fair trade and *Swachcha Vyapar* had an oppositional relationship in Darjeeling's villages as the fair trade certified co-operative was male-dominated.

The incongruity in fair trade ideals and practice and resultant counterpolitics demonstrates similar contradictions that Ong and Collier (2005: 15) point out with respect to biopolitics. The case of fair trade tea certification in Darjeeling and also the cases that Moberg and Lyon (2010) describe as "shaped advantages" are proof of tensions that underlie any bio-political regime of governance. Such inconsistencies, as we have shown, also provide communities and marginalized actors the much needed cultural rhetoric and political handle to question existing power structures both local and transnational that tends to control every detail of their productive and social life.

Conclusion

In this review article on fair trade and fair trade certification of food and agricultural commodities we have synthesized different aspects of the existing interdisciplinary literature. Fair trade is a recent entry in the history of global trade in food and agricultural commodities, yet it holds the potential for shaping the fundamental values of global trade. A movement, which began with activist aims of correcting inequalities of market-based global trade (Guthman 2007), and promoting quality food production across the world, has chosen to operate within market parameters. Working within the dominant paradigm of free market trade in agricultural commodities, fair trade has redirected a small amount of profit from global trade back to producer groups that serve as critical development assistance for their communities in the absence of state assisted development (Sen 2009). In that respect, fair trade certification has fulfilled some of the founding objectives of the movement to make trade become a tool for empowerment, not impoverishment, despite various shortcomings in application. Such material gains from fair trade for producer communities are important, however, the movement has also reached a critical moment where its social justice goals stand to be compromised as powerful corporations, plantations, and supermarkets—which have contributed to inequities in global trade—can acquire a fair trade label.

The expansion of fair trade through certifying more agro-food products is both a reason for celebration and a cause for concern. To be fair in its global operations, fair trade has had to centralize rule making. To satisfy quality requirements of its patrons in the West, it devised certification compliance guidelines in which producers from the Global South have very little say. Fair trade certification of food and agricultural commodities is ultimately based on tests for compliance and not so much on partnerships; the latter would entail more intense localized negotiations over what constitutes fair trade. It is for this reason we have claimed fair trade's operations as biopolitical, as it wants to manage producers individually and collectively under the broad goal of maintaining quality and equality. However, in the numerous reiterations, interpretations, and appropriations of fair trade ideals by marginalized farmers, the movement opens up new dialogues in producer communities. Such critical dialogues constantly rejuvenate the search for justice, equity, and sustainability, making fair trade and fair trade certification speak to the place-based justice needs of producers.

ACKNOWLEDGMENTS

The authors thank the editors of this special issue for their guidance and encouragement. Debarati Sen's ethnographic research in India was funded by a Wenner Gren Foundation Individual Research Grant (no. 7495) and a NSF DDIG Grant (no. 0612860). She also thanks Aaron Theodore Samuel for his wonderful research assistance and Rachel Watkins for support. Sarasij Majumder thanks the American Institute of Indian Studies and NSF DDIG (no. 0612845) for their generous support of his research in rural West Bengal. The authors are grateful to the four anonymous reviewers whose appreciation of their work and pointed criticisms have vastly improved this article. Any remaining errors are the authors'.

DEBARATI SEN is assistant professor of International Conflict Management and Anthropology at Kennesaw State University, Kennesaw, Georgia. She is a cultural and feminist anthropologist who researches the impacts of alternative trade regimes on women's community activism. Over the past several years she has conducted research in Eastern India, specifically the Darjeeling district in West Bengal, on the effects of fair trade-organic tea production. Her research and teaching endeavors embody a strong commitment to issues of social justice from a feminist perspective.

SARASIJ MAJUMDER is assistant professor of Asian Studies and Anthropology at Kennesaw State University, Kennesaw, Georgia. He is a cultural and political anthropologist whose research focus is on neoliberalism and agrarian transition in South Asia. His current research looks at the politics of farmer identity formation and activism, and food security in Eastern India. He is also interested in the changing meanings of land under postcolonial development regimes in India. At Kennesaw State he teaches courses on Capitalism and Culture in Asia, Understanding Asia, Linguistic Anthropology, and Cultural Anthropology.

NOTES

1. The literature review in this article is drawn from the most cited articles on fair trade and fair trade certification in the past fifteen years.
2. Convention theory emphasizes how divergent norms are negotiated within and between commodity networks. During the Fordist period, agroindustries would assert the superiority of industrial/market conventions. This is first kind of fetishizing where agricultural products became like industrial commodities. However, growing mistrust for such commodities among consumers pushed corporations to highlight civic/ domestic conventions. This is defetishizing by further fetishizing.

REFERENCES

Bacon, Christopher. 2005. "Confronting the Coffee Crisis: Can Fair Trade, Organic, and Specialty Coffees Reduce Small-Scale Farmer Vulnerability in Northern Nicaragua?" *World Development* 33 (3): 497–511.

Barrientos, Stephanie, Catherine Dolan, and Anne Tallontire. 2003. "A Gendered Value Chain Approach to Codes of Conduct in African Horticulture." *World Development* 31 (9): 1511–1526.

Barrientos, Stephanie, and Sally Smith. 2007. "Mainstreaming Fair Trade in the Global Food Production Networks: Own Brand Food and Chocolate in UK Supermarkets." Pp. 103–123 in *Fair Trade: The Challenges of Transforming Globalization,* ed. L. T. Raynolds, D. T. Murray, and J. Wilkinson. New York: Routledge.

Busch, Lawrence, and Carmen Bain, 2004. "New! Improved? The Transformation of the Global Agrifood System." *Rural Sociology* 69 (3): 321–346.

Cáseres, Zina, Aurélie Carimentrand, and John Wilkinson. 2007. "Fair Trade and Quinoa from the Southern Bolivian Altiplano." Pp. 180–198 in *Fair Trade: The Challenges of Transforming Globalization,* ed. L. T. Raynolds, D. T. Murray, and J. Wilkinson. New York: Routledge.

Dolan, Catherine S. 2008. "In the Mists of Development: Fairtrade in Kenyan Tea Fields." *Globalizations* 5 (2): 305–318.

Edelman, Marc, and Angelique Haugerud. 2005. "Introduction: The Anthropology of Development and Globalization." Pp. 1–74 in *The Anthropology of Development and Globalization,* ed. M. Edelman and A. Haugerud. London: Blackwell.

FLO. n.d. "Aims of Fair Trade Standards." http://www.fairtrade.net/aims_of_fairtrade_standards.html (accessed 1 December 2010).

Foucault, Michel. 1990. *The History of Sexuality,* vol. 1. New York: Vintage Books.

Fridell, Gavin. 2007. *Fair Trade Coffee: The Prospects and Pitfalls of Market-Driven Social Justice.* Toronto: University of Toronto Press.

Friedmann, Harriet, and Amber McNair. 2008. "Whose Rule Rules? Contested Projects to Certify 'Local Production for Distant Consumers.'" *Journal of Agrarian Change* 8 (2&3): 408–434.

Gereffi, Gary, Ronie Garcia-Johnson, and Erika Sasser. 2001. "The NGO-Industrial Complex." *Foreign Policy* 125: 56–65.

Gereffi, Gary, John Humphrey, and Timothy Sturgeon. 2005. "The Governance of Global Value Chains." *Review of International Political Economy* 12 (1): 78–104.

Gereffi, Gary, Miguel Korzeniewicz, and Roberto P. Korzeniewicz, eds. 1994. "Introduction: Global Commodity Chains." Pp. 1–14 in *Commodity Chains and Global Capitalism.* London: Praeger.

Getz, Christy, and Aimee Shreck. 2006. "What Organic and Fair Trade Labels Do not Tell Us: Towards a Place-Based Understanding of Certification." *International Journal of Consumer Studies* 30 (5): 490–501.

Guthman, Julie. 2004. *Agrarian Dreams: The Paradox of Organic Farming in California.* Berkeley: University of California Press.

———. 2007. "The Polanyian Way? Voluntary Food Labels as Neoliberal Governance." *Antipode* 39: 456–478.

Hatanaka, M. 2010. "Certification, Partnership, and Morality in an Organic Shrimp Network: Rethinking Transnational Alternative Agrofood Networks." *World Development* 38 (5): 706–716.

Jaffee, Daniel. 2007. *Brewing Justice: Fair Trade Coffee, Sustainability, and Survival.* Berkeley: University of California Press.

Jaffee, Daniel, and Philip Howard. 2010. "Corporate Cooptation of Organic Fair Trade Standards." *Agriculture and Human Values* 27 (4): 387–399.

Klooster, Dan. 2010. "Standardizing Sustainable Development? The Forest Stewardship Council's Plantation Policy Review Process as Neoliberal Environmental Governance." *Geoforum* 41 (1): 117–129.

Kruger, Sandra, and Andries Du Toit. 2007. "Reconstructing Fairness: Fair Trade Conventions and Worker Empowerment in South African Horticulture." Pp. 200–219 in *Fair Trade: The Challenges of Transforming Globalization,* ed. L. T. Raynolds, D. T. Murray, and J. Wilkinson. New York: Routledge.

Lyon, Sarah. 2008. "We Want to Be Equal to Them: Fair-Trade Coffee Certification and Gender Equity within Organizations." *Human Organization* 67 (3): 258–268.

Macdonald, Kate. 2007. "Globalizing Justice within Coffee Supply Chains? Fair Trade, Starbucks and the Transformation of Supply Chain Governance." *Third World Quarterly* 28 (4): 793–812.

Majumder, Sarasij. 2010. "'Peasants' Against the Nano? Neoliberal Industrialization and Land Question in Marxist-Ruled West Bengal, India." PhD diss., Rutgers University.

Moberg, Mark, and Sarah Lyon, eds. 2010. "What's Fair? The Paradox of Seeking Justice through Markets." Pp. 1–27 in *Fair Trade and Social Justice: Global Ethnographies.* New York: New York University Press.

Moore, Geoff. 2004. "The Fair Trade Movement: Parameters, Issues and Future Research." *Journal of Business Ethics* 53 (1) : 73–86.

Moseley, William G. 2008. "Fair Trade Wine: South Africa's Post-Apartheid Vineyards and the Global Economy." *Globalizations* 5 (2): 291–304.

Murray, Douglas T., and Laura T. Raynolds. 2007. "Globalization and its Antonomies: Negotiating a Fair Trade Movement." Pp. 3–14 in *Fair Trade: The Challenges of Transforming Globalization,* ed. L. T. Raynolds, D. T. Murray, and J. Wilkinson. New York: Routledge.

Mutersbaugh, Tad. 2002. "The Number Is the Beast: A Political Economy of Organic-Coffee Certification and Producer Unionism." *Environment and Planning A* 34 (7): 1165–1184.

———. 2005. "Just-in-Space: Certified Rural Products, Labor of Quality, and Regulatory Spaces." *Journal of Rural Studies* 21 (4): 389–402.

———. 2008. "Serve and Certify: Paradoxes of Service Work in Organic Coffee Certification." Pp. 261–288 in *Confronting the Coffee Crisis: Fair trade, Sustainable Livelihoods and Ecosystems in Mexico and Central America,* ed. C. Bacon, E.V. Méndez, and S. Gliessman. Boston: MIT Press.

Mutersbagh, Tad, and Sarah Lyon. 2010. "Transparency and Democracy in Certified Ethical Commodity Networks." *Geoforum* 41: 27–32.

Ong, Aihwa, and Stephen. J. Collier, eds. 2005 "Global Assemblages, Anthropological Problems." Pp. 3–21 in *Global Assemblages: Technology, Politics, and Ethics as Anthropological Problems.* Oxford: Blackwell.

Ponte, Stefano. 2002. "The 'Latte Revolution'? Regulation, Markets and Consumption in a Global Coffee Chain." *World Development* 30 (7): 1099–1122.

Raynolds, Laura T. 2007. "Fair Trade Bananas: Broadening the Movement and Market in the United States." Pp. 63–82 in *Fair Trade: The Challenges of Transforming Globalization,* ed. L. T. Raynolds, D. T. Murray, and J. Wilkinson. New York: Routledge.

———. 2009. "Mainstreaming Fair Trade Coffee: From Partnership to Traceability." *World Development* 37 (6): 1083–1093.

Raynolds, Laura T., and Douglas T. Murray. 2007. "Fair Trade: Contemporary Challenges and Future Prospects." Pp. 223–234 in *Fair Trade: The Challenges of Transforming Globalization,* ed. L. T. Raynolds, D. T. Murray, and J. Wilkinson. New York: Routledge.

Raynolds, Laura T., Douglas T. Murray, and Andrew Heller. 2007. "Regulating Sustainability in the Coffee Sector: A Comparative Analysis of Third-Party Environmental and Social Certification Initiatives." *Agriculture and Human Values* 24 (2): 147–163.

Raynolds, Laura T., and John Wilkinson. 2007. "Fair Trade in the Agriculture and Food Sector: Analytical Dimensions." Pp. 33–47 in *Fair Trade: The Challenges of Transforming Globalization,* ed. L. T. Raynolds, D. T. Murray, and J. Wilkinson. New York: Routledge.

Renard, Marie-Christine, and Victor Pérez-Grovas. 2007. "Fair Trade Coffee in Mexico: At the Center of Debates." Pp. 138–156 in *Fair Trade: The Challenges of Transforming Globalization,* ed. L. T. Raynolds, D. T. Murray, and J. Wilkinson. New York: Routledge.

Sen, Debarati. 2009. "From Illegal to Organic: Fair Trade-Organic Tea Production and Women's Political Futures in Darjeeling, India." PhD diss., Rutgers University.

Shreck, Aimee. 2005. "Resistance, Redistribution, and Power in the Fair Trade Banana Initiative." *Agriculture and Human Values* 22 (1): 17–29.

Taylor, Peter Leigh. 2005. "In the Market But Not of It: Fair Trade Coffee and Forest Stewardship Council Certification as Market-Based Social Change." *World Development* 33 (1): 129–147.

Valkila, Joni, and Nygren, Anja. 2009. "Impacts of Fair Trade Certification on Coffee Farmers, Cooperatives, and Laborers in Nicaragua." *Agriculture and Human Values* 27 (3): 321–333.

Vandergeest, Peter. 2007. "Certification and Communities: Alternatives for Regulating the Environmental and Social Impacts of Shrimp Farming." *World Development* 35 (7): 1152–1171.

Wilkinson, John, and Gilberto Mascarenhas. 2007. "Southern Social Movements and Fair Trade." Pp. 52–62 in *Fair Trade: The Challenges of Transforming Globalization,* ed. L. T. Raynolds, D. T. Murray, and J. Wilkinson. New York: Routledge.

Wilson, Bradley. R. 2010. "Indebted to Fair Trade? Coffee and Crisis in Nicaragua." *Geoforum* 41 (2010): 84–92.

West, Paige. 2010. "Making the Market: Specialty Coffee, Generational Pitches, and Papua New Guinea." *Antipode* 42 (3): 690–718.

Nature's Market? A Review of Organic Certification

Shaila Seshia Galvin

ABSTRACT: As organic food becomes more widely available, great faith is placed on the seal or logo that certifies organic status. This article treats the mark of certification as a starting rather than an end point, critically reviewing literature from diverse national and regional contexts. Exploring questions concerning the extent to which organic certification assists or undermines the goal of ecological sustainability, abets the advance of large-scale agricultural capital, and supports the livelihood of smallholder farmers, the article considers the theoretical foundations, methodologies and modes of inquiry that have guided studies of organic agriculture and certification. It brings this research into conversation with literatures on audit cultures, quality, and with ongoing nature-culture debates. Through critical review of the literature and the author's extensive fieldwork with organic smallholders in northern India, the article suggests possible directions in which the literature may be expanded and advanced.

KEYWORDS: audit cultures, capitalism, globalization, nature, organic agriculture, organic certification, organic standards

Numbers tell a story. In 2008, 35 million hectares of land across some 150 countries were under certified organic cultivation, with worldwide sales of organic food and drink amounting to US$50.9 billion (Willer and Kilcher 2010). This represents a quadrupling in the area of certified land since 1998, and more than a doubling of sales since 2000 (Willer et al. 2008: 16). The story told by these numbers is unambiguously one of growth and expansion. Its unfolding is evident in, among other things, the rise of farmers' markets and direct marketing schemes (Brown 2002; Kirwan 2004), the appearance of organic product lines in major food retailers (Browne et al. 2000; Pollan 2006a), the rapid expansion of a supermarket chain dedicated entirely to natural and organic foods (Martin 2007), and the increasing scope for imports to European and American markets of organic products from countries in the Global South (Barrett et al. 2002; Barrett et al. 2004; FAO et al. 2001). For a consuming public located largely in the Global North, in an age of heightened awareness and anxiety pertaining to health and food safety issues, the seal attesting to the certified status of an ever-wider range of carefully branded organic products is one of the principal icons of this story of growth.[1] Yet, this story also unfolds in ways currently less visible to such publics – for example, in bureaucratic-legal efforts undertaken nationally and internationally to establish and harmonize standards for organic certification and, as I discovered during twenty months of fieldwork, in the everyday efforts of cultivators on India's Himalayan frontier to adhere to practices enabling them to obtain certification that

Environment and Society: Advances in Research 2 (2011): 48–67 © Berghahn Books
doi:10.3167/ares.2011.020104

would confer their status as organic not only within India, but also in the United States and the European Union.

Numbers, then, tell only part of a story, and the purpose of this article is not to explain the rapid surge in organics, which has been undertaken elsewhere (Allen and Kovach 2000; Guthman 1998, 2004a; Raynolds 2000). Rather, reviewing scholarship and thinking that has emerged largely within the past fifteen years, this article inquires into the role of organic certification in this expansion, and examines the consequences it has for producers. To this end, the article considers how organic certification has been understood as a social, political, and cultural process and, reviewing both established debates and emerging scholarship, identifies areas in which research may be further developed. I suggest that, to date, approaches informed to varying degrees by political economy have dominated the literature on organic certification over the past decade. I argue that there is additional room for studies that refine conceptual thinking about organic certification by broadening out both the theoretical and empirical approaches to its study.

The article is organized into three sections. The first section considers the ways in which traditions of political economy, especially within agro-food studies, have informed research on certified organic agriculture to date. The second section directs attention to a distinctive set of concerns emerging from literature focused on the Global South, elaborating on and extending these ideas through a case study based on my own research in Uttarakhand. The third section outlines three areas in which the literature on certified organic agriculture may be expanded: the relation of certification to the wider phenomena of audit cultures; its engagement with the "quality turn" in agro-food studies; and the conceptualization of nature-culture relations.

The Political Economy of Certified Organic Agriculture

Work on organic certification spans, blurs, and sometimes divides disciplinary and subdisciplinary boundaries such as geography, anthropology, sociology, economics, and their hybrids (e.g., economic sociology, cultural geography, environmental studies). Despite this diversity, a substantial body of research locates its theoretical and analytic origins in agro-food studies and is informed to varying degrees by political economy (Goodman and Redclift 1991; Goodman and Watts 1997). Much work in this area maintains a broad commitment to discerning the social, political, and economic conditions of production, informed by concepts of commodity fetishism, and the related argument that capitalism obscures relations of production. Such approaches have shaped the much broader tradition of commodity systems analysis and, as several scholars have argued (Goodman and DuPuis 2002: 6–9; see also Goodman 1999, 2003), proved formative in the development of research in agro-food studies.

Within the broad ambit of political economy, scholarship that focuses on organic agriculture traces several different threads. Drawing attention to the constitution of agrarian capital and relations of power and inequality within the organics sector, a key debate animating much path-breaking scholarship in this area turned on the question of whether organic agriculture would become "conventionalized" along the lines of industrialized modes of agricultural production often associated with large agribusiness, and examined the mechanisms of conventionalization (Buck et al. 1997; Campbell and Liepins 2001; Coombes and Campbell 1998; Guthman 2002, 2004a, 2004b, 2004c). It is in the context of this larger debate on conventionalization that the regulation of organic agriculture through the articulation of national and international standards and certification emerged as distinct line of inquiry (Campbell and Liepins 2001; Jaffee and Howard 2010; Michelsen 2001; Vos 2000).

Conventionalization and Its Discontents

The conventionalization thesis, as it has come to be known, was first articulated by Buck, Getz and Guthman (1997: 4) in an article based on "exploratory" research in California in which they argued that "agribusiness is finding ways to industrialize organic production." Situating the growth of organic agriculture in California within a long-standing inquiry into the "agrarian question," the authors argued that organic agriculture in northern California would, in theory, seem to lend sustenance to Kautsky's ([1899] 1988) famous contention that the biological processes on which agriculture is based, and the uncertainties inherent in agricultural production, posed obstacles to the incursions of capital. Thus, organic agriculture founded on agroecological precepts that supported diverse cropping practices, low mechanization, and the use of on-farm inputs (Altieri and Nicholls 2004; Gliessman 2004; Pimentel et al. 2005) would arguably resist capitalization. Instead, however, the authors found that organic agriculture increasingly resembled its conventional counterpart—a process they famously termed conventionalization.

A landmark study of the development of organic agriculture in California by Guthman (2004a) developed the conventionalization thesis by exploring how, in the late years of the twentieth century and early years of the twenty-first, Californian organic agriculture departed from its movement-based origins to increasingly take the form of large agribusiness. The ideals of the organic movement in California originate in what Guthman terms the "new agrarianism," a form of agrarian populism that emerged in the mid-twentieth century to critique both the state and corporate power in an increasingly industrialized agriculture and food sector, and valorized the image of the yeoman farmer and the family-owned small scale farm (Guthman 2004a: 10–12). While this ideal was posited in direct opposition to the industrial character and history of Californian agriculture, Guthman observes a certain path dependence and argues that ultimately "organic production was layered onto an already existing landscape of agricultural industrialization" (2004a: 172). The kernel of the conventionalization thesis is also expressed in the work of Michael Pollan (2006a, 2006b), who asks whether the ecological principles on which organic agriculture is based can be reconciled with industrial modes of farming and queries whether, for agribusiness and food retailers, "organic is not a revolution so much as a market niche" (Pollan 2001). The paradox that both Guthman and Pollan point to stems from the distinction between what is alleged to be the purer, and more authentic, ideals associated with organic—stemming from its association in the Californian context with counterculture and back-to-the-land movements—and its more recent industrial form.

Although the conventionalization argument provides a useful framework for considering political and economic processes underlying the rapid growth of organic agriculture, its early expression by Buck, Getz, and Guthman (1997) generated debate among scholars who resisted conceptualizing conventionalization as a general, universal tendency inherent in the logic of agrarian capital, and instead saw it as highly context dependent (Michelsen 2001). Coombes and Campbell (1998) were among the first to articulate this position, observing that in New Zealand organic agriculture is bifurcated into a highly capitalized, export-oriented sector and a small-scale sector catering to the domestic market. They further argue that the flexibility of small-scale producers in New Zealand and their ability to tailor production to low-volume niche markets will ensure their viability even in the face of expanded presence by agribusiness. They conclude that "the evolution of organic production is inherently specific to regional conditions and national regulation" (Coombes and Campbell 1998: 130).

Guthman (1998: 149) acknowledges California's "exceptionalism" and demonstrates the ways in which the specificities of its agrarian history shape the conventionalizing tendencies of California's organics sector. Within this history, and among a larger set of factors, she identifies the "imperative of agricultural intensification" (Guthman 2004c: 312) achieved through technical efforts such as the development of irrigation systems, fast-growing varietals, and synthetic inputs. This intensification is a key mechanism driving the capitalization of land value, a pressure to which even smaller organic farmers are not immune. Thus, Guthman argues, in California "the cost of land makes some degree of conventionalization hard to resist" and maintains that smaller producers will find it increasingly difficult to compete with large agribusiness (2004c: 312). In this sense, though Guthman is sensitive to the diversity of national and regional contexts in which organic agriculture is practiced, she maintains that the interaction between agricultural intensification and capitalization of land values are crucial to the logic of conventionalization.

Codifying Agriculture, Conventionalizing Standards

To what extent is the development and formalization of standards and regimes of certification related to processes of conventionalization? How might we parse the conceptual and practical links among standards, certification, and regulation in the organics sector? Early studies of organic certification, especially in the US context, emerged in relation to the conventionalization thesis. Guthman (1998) sees what she terms the "codification" of California organic agriculture as intimately connected with the sector's larger conventionalizing tendencies. National standards promulgated in the name of organic agriculture have expanded sites for the accumulation of capital, particularly by agribusiness engaged in interstate and international commerce. Thus, Guthman (1998: 150) argues, "organic regulation makes organic agriculture safe for capitalism."

This research emerged in the US in the context of the controversial formulation of the National Organic Program Proposed Rule in 1997, which sought to implement the Organic Food Production Act of 1990. Vos (2000) examines closely the debate generated by the release of this rule, noting a wide gulf between small-scale organic producers and the emerging United States Department of Agriculture (USDA) treatment of organic agriculture. He characterizes their differences as a clash of two inimical worldviews (Vos 2000: 245). The proposed USDA rule permitted the inclusion of genetically modified organisms, sewage sludge, and irradiation in the definition of organic agriculture, and was thus an affront to organic producers who described organic agriculture as a "way of life" based on a holistic view of agriculture and nature. Vos (2000: 247) suggests that the USDA involvement in the formation of organic standards was suspect from the outset, given the department's historic involvement in enabling the rise of industrialized modes of farming. Its promotion of a narrow definition of organic agriculture that relies simply on the substitution of prohibited inputs for allowable ones, moreover, leaves unexamined the social conditions of agricultural production (see Allen and Kovach 2000). Implicit in the latter claim is the issue of corporate cooptation, an issue that Jaffee and Howard (2010) examine explicitly in a recent article exploring the subject in relation to the regulation of organic agriculture and fair trade. Here, it is perhaps significant that while Guthman (1998), in some respects contrary to Vos (2000), found little evidence of widespread corporate cooptation in the formulation of the 1997 USDA rule, Jaffee and Howard (2010) argue that the approval by the US Congress of the Organic Trade Association Rider in 2005 was the outcome of corporate lobbying and reflective of a regulatory sector increasingly "captured" by corporate actors.[2] Johnston et al. (2009) also pursue this analysis by examining corporate cooptation of discourses related to place, locality, and the family farm historically produced within the organic movement. Beyond highlighting the extent to which organic regulation has created a new, lucrative arena for the accumulation of agrarian capital, these studies together suggest the need for further exploration of the dynamic interaction of sectoral growth, regulation, and processes of cooptation or conventionalization.

Scholars writing of the US context link the elaboration of standards and certification requirements with the industrialization of organic agriculture, but those in other regional settings contest this association and thus extend the contours of the conventionalization debate. Concurring with Guthman (1998) about the significance of regulation, Michelsen's study (2001: 62) of the Danish experience suggests that standards are "at the core" of organic farming. Far from being inimical to the essence of organic agriculture, Michelsen contends that standards are a means to achieve the values of agricultural and environmental sustainability. Yet he distinguishes between what he terms self-regulation, or voluntary regulation, practiced by farmers' associations and the subsequent development of formalized, legal regulation and acknowledges that the latter cannot necessarily achieve the broader social, environmental, and ethical goals of the former (Michelsen 2001: 66–73). Campbell and Liepins (2001) challenge the conventionalizing role attributed to organic standards and certification, noting that small-scale producers in New Zealand have increasingly opted out of a certification system that caters to larger producers by ensuring compliance with international frameworks and standards. In this manner, they suggest, producers who opt-out need not be subject to the potentially conventionalizing consequences of certification and they point to efforts by the New Zealand government to support the development of a certification system tailored specifically for small, domestic producers (Campbell and Liepins 2001: 35).[3] Thus, while Guthman contends that the "standards-oriented regulation" (1998, 2004b) carries powerful, conventionalizing implications for Californian agriculture, Michelsen (2001) and Campbell and Liepins (2001) do not see this as an inherent property of standards themselves.

Cumulatively, these cases highlight the significance of national and regional specificities in the form that standards take, the processes through which they come into being, and the consequences that they have in diverse agrarian contexts. The question, thus, is not as simple as whether standards and certification are conventionalizing. Rather these studies point to the need to attend to important ways in which factors such as distinctive agrarian histories, the structure of the agricultural sector and the significance of cross-border (interstate or international) trade interact with systems of organic regulation to shape the growth and development of the sector.

They further highlight the need to distinguish among diverse forms of standards and systems of certification. In recent scholarship (Clarke et al. 2008: 220; Getz and Shreck 2006; Philpott et al. 2007), the terms are tightly linked and the elaboration of standards almost always implies a certification scheme—usually third-party—to enforce them. Yet earlier literature suggests that this was not always, and is not necessarily, the case. Here, discussion of standards is not automatically connected to certification. Thus, despite their differences, Coombes and Campbell (1998), Guthman (1998), and Michelsen (2001) find in each of their diverse national and regional contexts that small associations of organic producers often held up standards of production based on voluntary compliance, trust, and personal relations, rather than the practices of monitoring and surveillance that characterize independent third-party certification regimes. Moreover, a recent study conducted in Norway reveals that organic farmers ceasing certified organic production are, in a significant number of cases, continuing to adhere to organic practices (Flaten et al. 2010). This suggests a further need to inquire into the work of certification as distinct from, but related to, the wider issues of organic practice and standards, and calls for closer scrutiny of the conditions under which certification became a necessary component of frameworks codifying organic standards.

Early work on standards and certification in organic agriculture demonstrates a concern for the way that standards came to be progressively codified through national regulation during the 1980s and 1990s. Several studies published in the first years of the millennium compared organic and fair trade labeling, often noting a distinction between the two in terms of their ability to achieve goals of environmental sustainability and/or improve social conditions of production. An early study of UK organizations, ranging from supermarket retailers to smaller alternative trade organizations, finds that the notion of ethical trading typically used by the latter type of organization to be more encompassing of the conditions of work, environmental issues, and animal welfare than the organic label preferred by larger retailers, which was usually more narrowly focused on the technical facets of production (Browne et al. 2000). In the Californian context, surveys and interviews with farmers reveal an ambivalence in relation to the question of whether organic standards should include social sustainability goals (Shreck et al. 2006). Deepening Guthman's paradox of organic agriculture in California, these authors find that larger producers were better able to provide living wages, paid holidays, sick days and insurance for farm workers than the small-scale, family-owned farms which often could not afford such benefits for their own members.[4] In a recent study comparing certified organic and certified fair trade standards in the US, Jaffee and Howard (2010: 396) identify the "locus of standard setting" as especially important in determining the extent to which larger social or environmental ideals will be honored, and in this regard find movement or civil society-based regulatory bodies such as the Fairtrade Labelling Organization (FLO) preferable to the state-based system of regulation that governs organic agriculture in the US. This literature therefore indicates that the ability of organic agriculture to live up to its movement-based ideals of ecological and, in some cases, social sustainability, is conditioned not only by the standards through which it comes to be defined but also by the institutional location of standard-setting. Yet, while comparisons of organic and fair trade certification often conclude that the latter is better able to achieve its aims than the former, recent scholarship somewhat complicates this picture. In their contribution to this issue, Sen and Majumder argue that even within systems of fair trade certification, the ecological and social ideals of fair trade are blurred, and bound up, with a Foucauldian biopolitics of compliance, discipline, and self-regulation.

Producing Organic in a Standardizing World

The scope of the literature on organic regulation has expanded to consider the internationalization of standards and their harmonization across national regulatory regimes. Such literature brings different sets of actors into the foreground, among them the International Federation of Organic Agriculture Movements (IFOAM), the Codex Alimentarius Commission of the Food and Agricultural Organization (FAO), the World Health Organization (WHO), and the International Standards Organization (ISO). Although focused on different empirical contexts, in some respects this literature adopts a stance that is compatible with the conventionalization thesis. Along these lines Raynolds, in a study on the internationalization of regulatory frameworks, argues that "despite the organic movement's historical commitment to domestic and civic values, rooted in personal trust, local knowledge, ecological diversity, and social justice, organic certification appears to reassert industrial and commercial quality conventions, based on efficiency, standardization, bureaucratization and price competitiveness" (2004: 731). In an analysis akin to that of Guthman (2004a, 2004c), DeLind (2000), and others (e.g., DuPuis and Gillon 2009), Raynolds juxtaposes the industrializing consequences of international frameworks with trust and personal relationships. As she does so, however, the distinct meanings of domestic are elided in significant ways as the term is used as both a referent for the nation-state as in "domestic production" or "domestic commodities" and, following conventions theory (Boltanski and Thévenot 2006) as a gloss for trust and place-attachment.

Mutersbaugh (2005a) to some extent corroborates, but also departs from, Raynolds's analysis of the internationalization and harmonization of standards in agro-food networks generally, and their potentially homogenizing and simplifying consequences. In particular, he distinguishes multilateral institutions such as the ISO and World Trade Organization (WTO) from nongovernmental or movement-based organizations such as IFOAM and FLO. The former, he argues, emphasizes auditing practice, excludes environmental and labor criteria, and threatens to render obsolete smaller, independent labeling schemes (Mutersbaugh 2005a: 2038). To some extent, however, the distinction between movement-based and multilateral frameworks for standards and certification is increasingly blurred. Thus, for example, the 2005 version of the IFOAM Accreditation Criteria[5] conforms with ISO standards relating to certification (IFOAM 2011a). Attune to the ways in which movement-based and multilateral standard-setting organizations are networked and interconnected, and the possible conventionalizing consequences that arise therefrom, Mutersbaugh (2005a: 2045–2048) nonetheless points to a potential counterweight in single-label schemes that combine standards for organic, fair trade, and ecologically sustainable production. In this way he does not see conventionalization as the inevitable outcome of codified standards and certification requirements. Given the proliferation of different initiatives to regulate agro-food quality at transnational, national, and regional levels, Mutersbaugh argues that the sheer plurality of initiatives mean that "harmonization does not yield homogenous standards" (2005a: 2034).

As processes of harmonization and standardization work to bring equivalence, if not uniformity, to organic regulation across national diverse contexts, a distinct branch of inquiry has developed to examine how processes of standards formation play out in the Global South (Mutersbaugh 2002b, 2005a; Raynolds 2002, 2008; Tovar et al. 2005). The consequences of organic regulation were debated in the early years of the millennium within the context of comparisons between organic agriculture and fair or ethical trade. Rice (2001) argues that organic and fair trade organizations engaged in coffee production, and comprised mainly of smallholders, have successfully carved a niche in a sector otherwise dominated by more industrial interests. Bacon (2005) reports on a survey of over two hundred coffee growers in Nicaragua, and finds that those engaged in fair trade or organic production experience reduced livelihood vulnerability compared to their conventional counterparts, due to higher price premiums, lower input costs, and crop diversification. These findings align with the econometric analysis of Bolwig et al. (2009), which indicates that participation in a scheme for certified organic contract farming of coffee resulted in net revenue increases for Ugandan farmers compared to their uncertified counterparts, a finding that Sen and Majumder (this issue) also note in their broad review of fair trade certification.

Although these authors see the potential for organic agriculture to assist smallholders, others are less sanguine and argue more specifically that organic certification may be inimical to their interests. Raynolds (2000) adopts a Polanyian ([1944] 1957) lens, arguing that fair trade schemes, more than those for organic production, hold the potential to directly challenge the "disembedding" tendencies of conventional agriculture by directly tackling the question of social equity. In subsequent scholarship, Raynolds roots organic certification in larger relations of structural economic inequality, arguing that "onerous and expensive organic certification requirements create significant barriers to entry for poor Southern producers and encourage the concentration of organic production and price premiums in the hands of large corporate producers" (2004: 738). A subsequent commodity network study of the organics sector in the Dominican

Republic elaborates on this contention, demonstrating that stringent quality parameters regarding uniformity and appearance make it more difficult for smaller producers to compete with larger buyers and retailers (Raynolds 2008). Other studies also draw attention to the challenges experienced by smallholder farmers who undertake certified organic production, pointing similarly to the complexities of regulation, the requirement of extensive documentation, and the additional demands on time and labor (Gonzalez and Nigh 2005; Hatanaka 2010; Tovar et al. 2005). Cumulatively, this research suggests that the promise held by certified organic agriculture is ambivalent at best and suggests the need to better understand the conditions under which it may yield the greatest benefits for smallholders.

In assessing the potential consequences of organic certification for producers, scholars working in diverse regions have turned their attention to the institutional forms and avenues along which certified organic agriculture is enacted. These encompass systems of group certification that enable smallholders—organized into producer cooperatives, federations, or other forms of association—to comply with organic standards and requirements of third-party organic certification required for international trade (IFOAM 2011b). Mutersbaugh et al. (2005: 382) note that producer cooperatives have proved crucial in providing channels for technical assistance needed to set up systems of standards and certification compliant with the requirements for transnational trade. Raynolds (2004, 2008: 179–180), though otherwise skeptical of the possibilities for organic certification to benefit smallholders, notes that such organization may provide an important counterweight to an otherwise marked tendency for certification systems to favor large producers. Goldberger (2008) highlights the role played by NGOs as "boundary organizations" that link donors, research institutions, the Kenyan state, and farmers. Nelson et al. (2010: 227) examine the growth of what they term "beyond organic" initiatives in Mexico, exploring the development and operations of a Participatory Guarantee System that relies on a system of peer review rather than third-party certification. The plurality of institutional forms that exist to coordinate and certify organic production for smallholders is evident even in the necessarily limited range of studies described briefly here, making apparent that the rich institutional life of certified organic agriculture offers considerable scope for further empirical research.

Becoming Organic in Uttarakhand, India

In the northern Indian state of Uttarakhand where I conducted twenty months of fieldwork between 2006 and 2008, the practice of organic farming engages with a number of issues relating to the regulation of organic agriculture discussed above. Scholars such as Raynolds (2004) and Mutersbaugh (2005a, 2005b) view the expansion of certified organic agriculture as a process led by multilateral institutions of global governance, the private sector, or movement-based organizations, with the state role typically limited to the codification of standards. However, in Uttarakhand it is the state government that has catalyzed the development of the sector.[6] It has done so against the grain of a more mainstream approach to agricultural development in India, focused on green and "gene" revolution technologies. For this recently created state on India's Himalayan frontier, organic agriculture is part of a discursive repertoire that places broad emphasis on eco-development around which the state seeks to carve a distinctive regional identity within the larger mosaic of the Indian union.

Yet the promotion of organic agriculture in Uttarakhand carries institutional as well as discursive entailments. Shortly after its formation in 2000, the state government of Uttarakhand, through the leadership of its then Forest and Rural Development Commissioner, moved swiftly to create the bureaucratic infrastructure and capacity to undertake training, coordinate group certification, and facilitate marketing through a semi-autonomous organic commodity board, and also to provide third-party certification by establishing India's first state-run organic certification agency. These institutional innovations followed closely on India's establishment in 2001 of its own organic standards through the National Programme on Organic Production (NPOP). Testifying to the workings of transnational efforts at harmonization, India's NPOP system adheres to IFOAM norms while agreements established with the European Commission and the USDA enable organic products produced and certified in India to be sold as organic in the US and European markets.

Creating a viable system of certified organic production among smallholder cultivators scattered across often geographically challenging terrain, which adheres to and links up with national and transnational criteria for organic production, poses a number of logistical challenges. In this regard organic producers' groups, or farmers' federations, which connect cultivators from different villages within a geographically contiguous area, play a critical role as units of certification. They represent the condition of possibility for the development of commercial certified organic agriculture among smallholder farmers, who see them as a mechanism for extending commercial ties beyond local and regional markets and into India's ever-expanding metropolises, growing urban centers and, increasingly, also abroad. These associations act as units of group certification through an internal control system mandated under IFOAM norms as well as NPOP. This system enables organic certification to be held collectively by an association of farmers rather than individually, a structure that is intended to spread the costs of certification and to smooth the otherwise difficult logistics of individually certifying large numbers of smallholders. The associations also act as units that contract with large retailers to supply basmati rice and wheat and, as part of contracting arrangements, channel the supply of inputs and technical assistance.

Beyond these roles, however, these associations prove themselves to be dynamic, internally diverse, and often deeply political entities. In Uttarakhand's fertile Doon Valley these associations, called farmers' federations, each consist of several hundred farmers who cultivate organic basmati rice and wheat, and whose average holdings are less than two hectares. Here, the federations that were initially established as one component of a World Bank project, are now coordinated by the organic commodity board and with its assistance have entered into contract agreements with a large Indian rice retailer to produce organic basmati rice for export. During the course of ethnographic fieldwork, I found that some federations operated smoothly and maintained good relations with the retailer, the state government, and their farmer members, while in others the farmer members failed to supply contracted quantities and at times alleged that federation office holders misappropriated funds. The uneven efficacy of such associations has, to some extent, been noted by others who find that though membership confers certain benefits—such as premium prices—these associations also may also face challenges such as problems with contract compliance and access to credit (Rice 2001: 53–61), rent seeking, the effacement of existing systems of trade, barter and exchange, and the production of new inequalities within rural communities (Getz and Shreck 2006).

Findings from my own fieldwork demonstrate that the codification and certification of organic agriculture draws together diverse sets of actors, including but not limited to cultivators, bureaucrats, inspectors, extension workers, and corporate retailers, as it redefines the parameters of agricultural production. The incorporation of smallholder farmers in Uttarakhand's Doon Valley into harmonized frameworks of organic production might appear to align nicely with perspectives that use the rubric of neoliberal governmentality to understand the proliferation of standards, regulation, and third-party certification (Agrawal 2005; Hughes 2001). The neoliberal dimensions of Uttarakhand's promotion of organic agriculture notwithstanding, everyday practices of organic agriculture and its certification that unfold in the fields and offices in which certified status is established prove themselves to be arenas for negotiation, contestation, and compromise that complicate a simple reading of certification-as-governmentality. For example, buffer zones that establish spatial separation between organic and conventional plots have been codified in harmonized IFOAM-compliant national frameworks. Though these zones may appear as a top-down technique of managerial control within which there is little room for maneuver, this proved not to be the case in my research. Instead, during certification inspections I observed, queries surrounding the maintenance of buffer zones invited animated discussions among farmers, fieldworkers, and certification inspectors about field elevation and microclimatic variables such as wind pattern, and demonstrated an understanding and delimitation of such zones in sophisticated, nonspatial ways. Moreover, during such farm visits and inspections fieldworkers and inspectors routinely impressed on me that, to fulfill their roles, *visvās* (the meaning of which is connoted by the English expression of trust, or faith) was as central to certified organic agriculture as the qualities of inspectability and traceability that one more typically expects of it. The scope for negotiation and accommodation within systems of certification, and the significance of trust and personal relationships in them, urge the expansion of the study of organic agriculture, standards, and certification beyond the frames of political economy and governmentality.

Expanding the Study of Organic Standards and Certification

To date, understandings of the growth and globalization of the organics sector have been enabled largely through conceptual frameworks that draw attention to the operation of power and the constitution of inequalities at different levels within it, and to the structural conditions that shape production opportunities and constraints (Higgins et al. 2008). This literature also sows seeds that may move future scholarship beyond these foundations.

Cultures of Trust and Audit

Trust emerges as a concern in literature on organic standards and certification in, typically, one of three ways. First, some scholars contend that trust is a central element of the boundary work accomplished by organic agriculture as it establishes itself as different from conventional agriculture (see, e.g., DuPuis and Gillon 2009; Gieryn 1999; Hetherington 2005). The incorporation of trust as an element of this boundary-work acquires particular salience in light of heightened public anxiety over food scares such as E. coli (Dunn 2007), bovine spongiform encephalopathy, or salmonella, as well as growing mistrust about particular aspects of conventional food production (DuPuis 2000; Guthman 2004a). A second, related strand is developed by Guthman (2003), who contends that the channeling of information through food labeling or certification is one means through which reflexive taste, which relies critically on the production of knowledge and information, is cultivated among consumers.[7] The dynamic among consumer demand, reflexivity, and sustainable food is explored thoroughly by Isenhour in her contribution to this issue. A third commonly expressed perspective sees certification as replacing or substituting for more personalized relations of trust. This is implicit, for example, in work by Campbell and Liepins (2001: 27–28) who describe organic agriculture in New Zealand prior to the shift to formalized standards as an "on trust" system, in which the integrity of the term organic was based on knowledge of the grower rather than a seal testifying to the conditions of production. DeLind (2000) invokes trust in the US context to argue that the advent of national standards for organic certification will act as a poor substitute for relations of trust formed locally through farmers' markets, community-supported agriculture (CSA), and other direct marketing schemes, while other scholars include organic production within a larger concept of civic agriculture (e.g., Lyson 2004).

Despite the frequency with which it appears in the literature, the term itself is perhaps understudied in literature on organic agriculture and certification and it is typically invoked in descriptive rather than analytic terms. This, then, suggests room for further research to specify and refine the conceptualization of trust in relation to organic certification. To build such an inquiry, one might look to broader scholarship on trust and modernity. Giddens (1990: 26), for example, argues that trust is reconfigured through modernity as its locus shifts from individuals to abstract or expert institutions. Arguably, the expansion of direct marketing schemes, CSAs, and farmers' markets react against such systems of expertise as they work to shorten supply chains and shift the locus of trust back into more personal relations and encounters. The growth of such arrangements troubles Giddens's easy association of modernity with ever-greater levels of abstraction, expertise, and time-space distanciation. These forms of alternative trade are decidedly not pre- or anti-modern, but express a different vision of modernity from that captured by Giddens, one brought into being through a dialectic engagement with capitalized modes of industrialized agriculture.

The shift Giddens posits, however, maintains contemporary relevance and is exemplified in what Strathern (2000) describes as the rise of audit cultures, a phenomenon associated in the UK with the shift from "the welfare to the regulatory state" (Power 1997: 52) and globally with what some have argued is the expansion of "neoliberal governmentality" (Shore and Wright 2000: 61). In one sense, literature on audit cultures appears to lend support to Giddens's analysis of expert systems—audit is a case, par excellence, of an abstract and often opaque system of expertise. Yet this literature also complicates and extends Giddens's arguments in possibly useful ways. Strathern (2000: 7) highlights the paradox that audits are premised on mistrust, but demand that trust be placed in their procedures and conclusions. On this point Power (1997: 13) writes that "audits themselves are necessarily trusting practices," that indeterminacy and lack of information are a crucial feature of audit work (see also Pentland 1993). This observation has, thus far, been unrecognized in the literature on organic standards and certification, which tends to juxtapose both conventional agriculture and conventionalized organic systems with what are often represented as the more authentic forms of organic agriculture based on personal relations, direct marketing, and relations of trust and solidarity. Such assessments pay little heed to the ways in which trust operates within certification systems, or within other systems of agriculture and food production. A recent exception to this general tendency is found in a study by Campbell, Murcott, and MacKenzie (2011), which compares the regulation and auditing practices for kosher and halal foods within, respectively, Jewish communities in New York and Maghrebi Muslims in Aquitaine. These authors usefully caution against viewing audit culture as an icon of contemporary neoliberalism, arguing for greater attentiveness to a wider set of issues concerning the constitution of trust in food (Campbell et al. 2011: 78).

This lacuna raises questions that may warrant more careful empirical and theoretical study: How is trust constituted differently under different regimes of agricultural production? How is it generated through processes of organic certification? Addressing such questions may require methodological and theoretical approaches different from those most commonly used in the certification literature. Scholarship on audit has drawn attention to its performative or dramaturgical dimensions (Harper 2000; Munro 1999), while others, inspired by Douglas (1986), examine its "rituals of verification" (Power 1997). Several scholars note the ways in which seemingly procedural work is shot through with more emotive motivations to alleviate anxiety and produce comfort (Pentland 1993; Power 1997), leading Pentland (1993: 610) to ask "what is the role of affect in the formation of audit opinion?" Close attention to the constitution of affect and the performative, ritualistic aspect of certification,[8] as an instance of audit practice, promises to yield further insight into the configuration of trust in systems of certified organic agriculture.

The Quality Turn in Agro-Food Studies and Moral Economies of Certification

The progressive codification of organic agriculture is often regarded in the literature as a technical phenomenon (Allen and Kovach 2000; Guthman 1998; Murdoch et al. 2000; Raynolds 2000), turning on allowable inputs, requirements of documentation, conversion years, and the size of buffer zones. Yet the varying qualities and associations called forth by the term organic itself warrants further critical study. In this sense, organic agriculture is very much a part of what has been described as the "quality turn" in agro-food studies (Goodman 2000, 2003; Murdoch et al. 2000). Quality, Murdoch and colleagues (2000: 108) contend, refers to both "more local" and "more natural." The turn to quality has sparked a different set of theoretical commitments, including conventions theory (Boltanski and Thévenot 2006), actor network theory and science and technology studies (Callon et al. 2002; Haraway 1991; Latour 1993; Latour and Woolgar 1986), and the sociology of embededness (Granovetter 1985; Polanyi [1944] 1957). These perspectives draw attention to the ways in which particular qualities are negotiated, constructed, and institutionalized in agro-food networks (Busch and Tanaka 1996; Konefal and Hatanaka 2011; Raynolds 2008), and emerge from a broader contention that economic actions are enmeshed in larger webs of social relations.

The distinct question of the moral economy of standards is taken up explicitly by Busch and Tanaka (1996) and Busch (2000). Drawing on science and technology studies, and the call for greater symmetry of the human and nonhuman, these scholars argue that standards are never simply measures of things, but are also measures of people. Standards promote uniformity and work against the heterogeneity of people and things in ways that abet the formation of capitalist markets (Busch and Tanaka 1996). Standards also constitute a moral economy as they create parameters of who/what is good and who/what is bad (Busch 2000). In an insight that is arguably deserving of more attention within the literature on organic certification, Busch and Tanaka (1996: 23) write "[p]eople care about tests of things *because* they are also tests of people."

To date, literature on organic agriculture often links ethics with scale, where local, small-scale producers are endowed with moral qualities of "civicness" (DuPuis and Gillon 2009) lacking in industrialized or conventionalized operations. Drawing the ethical lines of agrarian practice in this manner has obscured recognition of the way in which standards and certification schemes impose their own moral parameters on those who participate in them. My own fieldwork in Uttarakhand testified to this, as compliance with organic standards was one metric informally used by officials in the public and private sectors to distinguish "progressive" and "backward" farmers. In Mexico, Mutersbaugh examines how participation in alternative trade organizations and regimes of organic certification reconfigures the moral economy of coffee producer groups in Oaxaca (Mutersbaugh 2002a, 2002b). Attention to the moral economy of standards and certification thus draws attention to the work of certification through and on, not only things being certified but the people who produce them. Building on this, future scholarship may seek to further specify how organic certification inaugurates or reconfigures moral economies of production.

The Quality of Nature

Perhaps the most ubiquitous quality deployed in the marketing of organic products and reproduced in scholarly literature is that of nature, and the natural. Nature figures in the literature in varying ways—as an object that can be appropriated by capital (Guthman 1998), as a subject with an agency independent of human efforts to control it (Murdoch et al. 2000: 110; Dunn 2007) or with which "rapprochement" is sought (Vos 2000: 246). In that latter respect Vos (2000: 245–246) argues that it is through agriculture that "human beings encounter and most intimately connect with nature." Yet even as Vos (2000: 252) maintains that organic agriculture dismantles the modernist nature-culture dichotomy, in much of the literature on organic certification nature is conceptualized as external to human activity, and therefore outside the realm of social, cultural, political, and economic processes. The literature on organic agriculture, and on organic certification more specifically, is thus susceptible to a more general critique leveled at agro-food studies that it reproduces the nature-culture dichotomy (Goodman 1999: 17).

Given the frequency with which nature is invoked in relation to discussions of organic agriculture, it is curious that more attention has not been paid to a well-known observation that "the idea of nature contains, though often unnoticed, an extraordinary amount of human history"(Williams 1980; see also Cronon 1996). Such a line of inquiry encourages the development of more historically refined understandings of the conditions under which particular ideas about nature became enfolded within the conceptualization of organic agriculture. Alternatively, work within and inspired by the domain of science and technology studies further troubles such dichotomies, calling attention to the proliferation of hybrids and advocating a more symmetrical study of the human and nonhuman (Busch 2000; Busch and Tanaka 1996; Goodman 1999; Kirksey and Helmreich 2010; Latour 1993; Morgan et al. 2006). These diverse approaches challenge, in different ways, the dualism often established between nature and culture or society. Their insights, arguably, hold promise for further refinement of the understanding and conceptualization of nature within scholarship on organic agriculture.

These debates exist at a theoretical level, but the conceptualization of nature and "the natural" constitutes a practical problem when it comes to delimiting the boundaries of certified organic agriculture. This is discussed explicitly by Mansfield (2004), in a study of the US debate on whether aquatic animals should be eligible for organic certification. She notes that those who supported the inclusion of wild-caught fish within organic certification parameters did so precisely based on their wildness and naturalness, qualities that are often invoked as central to the meaning of organic. Yet their inclusion was vigorously opposed by many in the organic movement, who argued that it would further erode already weak organic standards proposed by the USDA. Among the principal arguments levied against their inclusion was the fact that USDA standards provided for process rather than product certification (Deaton and Hoehn 2005). Thus, the quality of organicness was to be determined not by characteristics inherent in the final product, but in the practices used to produce it. Along these lines, Mansfield observes that those opposing the inclusion of aquatic fish often invoked the importance of deliberate practices of management and control in organic cultivation, and the involvement of "the farmer as an active agent" (2004: 221).

Mansfield's (2004) study of the problematic status of aquatic animals in relation to organic certification illuminates the complex relations among nature, culture, and human agency that are enfolded into notions of organic. Like the case of wild-caught fish in the US, in Uttarakhand, farmers whose practices are often described as "organic by default" because of a lack of access to synthetic inputs do not automatically qualify for certified organic status. The importance of this distinction was brought into sharp relief by a senior civil servant in the Uttarakhand state government, who explained to me one day that to achieve certification "many precautions, many steps … have to be taken." Certified organic agriculture, he emphasized, can never be established by default; rather, it demands a demonstrated compliance with organic standards, and is therefore, always, organic "by design." These relations invite further study, calling for critical interrogation of the conceptual basis of the organic paradoxes and ironies noted by Guthman and Pollan, which hinge on the way in which the contemporary practice of organic agriculture has deviated from what Sir Albert Howard (1940) once termed "nature's farming." They open avenues for future inquiry into the ways in which notions of organic are constituted by, and in turn work to constitute, historically situated ideas of nature, and for further developing the theorization around the enduring question of the relation between nature, human agency, and culture.

Conclusions

This article set out to consider the ways in which organic certification has been studied as a social, cultural, and political process. Perspectives drawing on political economy have proved to be an important lens through which to understand the emergence of, and expansion and change in, the organics sector. In their early years, studies in this vein engaged the question of whether organic agriculture represented a true alternative to dominant modes of agro-industrial production, or whether it too would be conventionalized. In Northern, industrialized economies, debates have centered on whether organic agriculture and the formation of organic standards and certification are inevitably and inherently conventionalizing. Although conventionalization appears to characterize the Californian organic sector with a large internal market and a history of heavily capitalized, industrial agriculture (Guthman 1998, 2004a), the same does not hold true in New Zealand (Coombes and Campbell 1998) where the sector is bifurcated into industrialized/export-oriented and alternative/local sectors that have come to be regulated through different regimes of certification. In both contexts, however, small-scale farmers are challenged by the high costs and stringent requirements of certification. A similar challenge is faced by smallholder farmers in the Global South, where the organic sector is largely export-oriented and where internal markets for certified organic products are more limited relative to their Northern counterparts. A growing number of scholars argue that although organic agriculture may offer the promise of price premiums for smallholder farmers, such producers are increasingly unable to conform to stringent quality parameters that accompany organic status (Nigh 1997; Raynolds 2008; Tovar et al. 2005).

The roots of such scholarship in concerns informed by political economy have facilitated effective and insightful engagement with questions addressing the workings of structural power and inequality within this dynamic agricultural sector. Yet as the sector itself grows, as processes of harmonization unfold globally, as new participants emerge and schemes for organic production proliferate, different sets of questions, concerns, and problems arise. Thus, this article also sketches the contours of an expanded scholarship on organic agriculture and certification that brings existing literature into conversation with broader empirical and theoretical concerns. In this regard, questions of trust have been discussed, on the one hand, as a consequence of certification and food labeling (Guthman 2003) and, on the other hand, as a casualty of it (DeLind 2000). In a slightly different vein, my own research in Uttarakhand highlights the way in which the conduct of certification may be understood as an affective enterprise that hinges as much on faith, belief, and trust as it does on the technical aspects of audit procedure demanded by regulatory frameworks. Efforts to more clearly parse the constitution and meanings of trust in these different contexts are warranted, and may benefit from closer engagement with existing scholarship on audit. Emerging scholarship on certified organic agriculture and "beyond organic" initiatives engages fruitfully with literature on quality, conventions theory, and science and technology studies. These studies facilitate analysis of the mechanisms by which food quality is established through systems such as organic certification, and also invite study of the ways in which such systems inaugurate and reconfigure moral economies of production. Finally, this article suggests that scholarship on organic agriculture and certification may benefit from more refined conceptual and empirical attention to the enfolding of historically constituted ideas about nature within the meaning of organic agriculture. It suggests further that the relation between nature and human agency within certified organic agriculture merits more extensive theorization and empirical study.

As this article demonstrates, the contemporary certified organics sector encompasses a multiplicity of actors ranging from, among others, corporate agribusiness and international certification agencies to CSAs and smallholder farmers. The very dynamism and diversity of the organics sector thus necessitates these same qualities in the scholarship that is produced about it. Attune to the diversity produced through distinctive agrarian histories and national or regional contexts, scope continues to exist both for scholarship that analyzes broad processes such as conventionalization or harmonization at work in organic regulation, and for attention to the everyday practices, negotiations, and compromises through which the term certified organic is brought into being.

ACKNOWLEDGMENTS

I thank Michael R. Dove, Karen Hébert, K. Sivaramakrishnan, Chris Galvin and participants in the Dove Doctoral Lab, Environmental Anthropology Collective, Agrarian Studies Graduate Student Workshop, and South Asian Studies Graduate Student Colloquium, for their comments and suggestions. Detailed, constructive feedback from three anonymous reviewers proved invaluable in the revision of this article for publication. Fieldwork discussed in this article was funded by the Wenner-Gren Foundation, the Social Sciences and Humanities Research Council of Canada, the MacMillan Center for International and Area Studies at Yale University, and the Yale Program in Agrarian Studies. I am grateful to all those individuals in Uttarakhand, whose generosity and insight made my dissertation fieldwork possible. Errors and omissions in this work are mine alone.

SHAILA SESHIA GALVIN is a doctoral candidate in Anthropology and Forestry and Environmental Studies at Yale University. Her research interests include political ecology, the anthropology of environment and development, and political anthropology. Her dissertation, tentatively titled "State of Nature: Agriculture, Development, and the Making of Organic Uttarakhand," examines the conjunction of a commercially oriented development strategy with the formation of regional identity through the promotion and practice of organic agriculture in the northern Indian state of Uttarakhand in the central Himalaya.

NOTES

1. Ninety-seven percent of global revenues from the sale of organic food and drink are generated in North America and Europe. See Helga Willer, Minou Yussefi, and Neil Sorensen (2008).
2. Jaffee and Howard (2010: 388) to some degree extend the conventionalization thesis by linking it directly with the sociological concept of cooptation. They argue that while markets for organic agriculture and fair trade seek ostensibly to "reclaim ethical values" they have emerged as "new spaces for

extracting profits" and new sites of accumulation. In this context, regulatory capture is an important mechanism for achieving corporate cooptation, and one of its results is a weakening of standards.

3. In the years since the publication of their article, as Campbell and Liepens (2001) allude to in their closing remarks, an agency dedicated to organic provisioning in the domestic market has been established with the support of the Soil and Health Association and a newly elected national government. The agency, OrganicFarmNZ operates a system of certification based on peer review, which distinguishes it from the third-party certification regimes that are often a requirement for participation in international trade and export markets.
4. Benefits and insurance are, arguably, but one element of work conditions as Shreck et al. (2006) acknowledge.
5. Revealing the ways in which systems of accreditation and certification are nested, these define the requirements that certification agencies must meet to become accredited.
6. For a discussion of state involvement in the organic sector, see also Thiers (2002).
7. At the same time as she advances the notion of "reflexive tastes," Guthman (2004a) questions the extent to which it is a meaningful term to describe the consumption habits of organic consumers. The idea of reflexivity, of course, also raises questions about how information is produced through certification, something that Guthman does not consider in her article.
8. Hetherington (2005), for example, characterizes the conversion that farmers must undergo to be certified as organic as an example of a "rite of passage."

REFERENCES

Agrawal, Arun. 2005. *Environmentality: Technologies of Government and the Making of Subjects, New Ecologies for the Twenty-First Century.* Durham: Duke University Press.

Allen, Patricia, and Martin Kovach. 2000. "The Capitalist Composition of Organic: The Potential of Markets in Fulfilling the Promise of organic Agriculture." *Agriculture and Human Values* 17 (3): 221–232.

Altieri, Miguel, and Clara I. Nicholls. 2004. "An Agroecological Basis for Designing Diversified Cropping Systems in the Tropics." Pp. 81–103 in *New Dimensions in Agroecology,* ed. David R. Clements and A. Shrestha. Binghamton: Food Products Press.

Bacon, Christopher. 2005. "Confronting the Coffee Crisis: Can Fair Trade, Organic, and Specialty Coffees Reduce Small-Scale Farmer Vulnerability in Northern Nicaragua?" *World Development* 33 (3): 497–511.

Barrett, H. R., A. W. Browne, P. J. C. Harris, and K. Cadoret. 2002. "Organic Certification and the UK Market: Organic Imports from Developing Countries." *Food Policy* 27 (4): 301–318.

Barrett, Hazel R., Angela W. Browne, and Brian W. Ilbery. 2004. "From Farm to Supermarket: the Trade in Fresh Horticultural Produce from Sub-Saharan Africa to the United Kingdom." Pp. 19–38 in *Geographies of Commodity Chains,* ed. A. Hughes and S. Reimer. London: Routledge.

Boltanski, Luc, and Laurent Thévenot. 2006. *On Justification: Economies of Worth.* Princeton: Princeton University Press.

Bolwig, Simon, Peter Gibbon, and Sam Jones. 2009. "The Economics of Smallholder Organic Contract Farming in Tropical Africa." *World Development* 37 (6): 1094–1104.

Brown, Allison. 2002. "Farmers' Market Research, 1940–2000: An Inventory and Review." *American Journal of Alternative Agriculture* 17: 167–176.

Browne, A. W., P. J. C. Harris, A. H. Hofny-Collins, N. Pasiecznik, and R. R. Wallace. 2000. "Organic Production and Ethical trade: Definition, Practice and Links." *Food Policy* 25 (1): 69–89.

Buck, Daniel, Christina Getz, and Julie Guthman. 1997. "From Farm to Table: The Organic Vegetable Commodity Chain of Northern California." *Sociologia Ruralis* 37 (1): 3–20.

Busch, Lawrence. 2000. "The Moral Economy of Grades and Standards." *Journal of Rural Studies* 16 (3): 273–283.

Busch, Lawrence, and Keiko Tanaka. 1996. "Rites of Passage: Constructing Quality in a Commodity Subsector." *Science, Technology, & Human Values* 21 (1): 3–27.

Callon, Michel, Cécile Méadel, and Vololona Rabeharisoa. 2002. "The Economy of Qualities." *Economy and Society* 31 (2): 194–217.

Campbell, Hugh, and Ruth Liepins. 2001. "Naming Organics: Understanding Organics Standards in New Zealand as a Discursive Field." *Sociologia Ruralis* 41 (1): 21–39.

Campbell, Hugh, Anne Murcott, and Angela MacKenzie. 2011. "*Kosher* in New York City, *Halal* in Aquitaine: Challenging the Relationship between Neoliberalism and Food Auditing." *Agriculture and Human Values* 28 (1): 67–79.

Clarke, Nick, Paul Cloke, Clive Barnett, and Alice Malpass. 2008. "The Spaces and Ethics of Organic Food." *Journal of Rural Studies* 24(3): 219–230.

Coombes, Brad, and Hugh Campbell. 1998. "Dependent Reproduction of Alternative Modes of Agriculture: Organic Farming in New Zealand." *Sociologia Ruralis* 38 (2): 127–145.

Cronon, William, ed. 1996. *Uncommon Ground : Rethinking the Human Place in Nature.* New York: W.W. Norton.

Deaton, B. James, and John P. Hoehn. 2005. "The Social Construction of Production Externalities in Contemporary Agriculture: Process versus Product Standards as the Basis for Defining 'Organic.'" *Agriculture and Human Values* 22 (1): 31–38.

DeLind, L. B. 2000. "Transforming Organic Agriculture into Industrial Organic Products: Reconsidering National Organic Standards." *Human Organization* 59 (2): 198–208.

Douglas, Mary. 1986. *How Institutions Think, The Frank W. Abrams Lectures.* Syracuse, New York: Syracuse University Press.

Dunn, Elizabeth. 2007. "*Escherichia coli*, Corporate Discipline and the Failure of the Sewer State." *Space and Polity* 11 (1): 35–53.

DuPuis, E. Melanie. 2000. "Not in My Body: BGH and the Rise of Organic Milk." *Agriculture and Human Values* 17 (3): 285–295.

DuPuis, E. Melanie, and Sean Gillon. 2009. "Alternative Modes of Governance: Organic as Civic Engagement." *Agriculture and Human Values* 26 (1): 43–56.

FAO, ITC, and CTA. 2001. *World Markets for Organic Fruit and Vegetables: Opportunities for Developing Countries in the Production and Export of Organic Horticultural Products.* Rome: Food and Agriculture Organization of the United Nations.

Flaten, Ola, Gudbrand Lien, Matthias Koesling, and Anne-Kristin Løes. 2010. "Norwegian farmers Ceasing Certified Organic Production: Characteristics and Reasons." *Journal of Environmental Management* 91 (12): 2717–2726.

Getz, Christy, and Aimee Shreck. 2006. "What Organic and Fair Trade Labels Do Not Tell Us: Towards a Place-Based Understanding of Certification." *International Journal of Consumer Studies* 30 (5): 490–501.

Giddens, Anthony. 1990. *The Consequences of Modernity.* Stanford: Stanford University Press.

Gieryn, Thomas F. 1999. *Cultural Boundaries of Science: Credibility on the Line.* Chicago and London: University of Chicago Press.

Gliessman, Stephen R. 2004. "Integrating Agroecological Processes into Cropping Systems Research." Pp. 61–80 in *New Dimensions in Agroecology,* ed. D. R. Clements and A. Shrestha. Binghamton: Food Products Press.

Goldberger, Jessica. 2008. "Non-Governmental Organizations, Strategic Bridge Building, and the 'Scientization'" of Organic Agriculture in Kenya." *Agriculture and Human Values* 25 (2): 271–289.

Gonzalez, A. A., and R. Nigh. 2005. "Smallholder Participation and Certification of Organic Farm Products in Mexico." *Journal of Rural Studies* 21 (4): 449–460.

Goodman, David. 1999. "Agro-Food Studies in the 'Age of Ecology': Nature, Corporeality, Bio-Politics." *Sociologia Ruralis* 39 (1): 17–38.

———. 2000. "Organic and Conventional Agriculture: Materializing Discourse and Agro-Ecological Managerialism." *Agriculture and Human Values* 17 (3): 215–219.

———. 2003. "The Quality 'Turn' and alternative Food Practices: Reflections and Agenda." *Journal of Rural Studies* 19 (1): 1–7.

Goodman, David, and E. Melanie DuPuis. 2002. "Knowing Food and Growing Food: Beyond the Production-Consumption Debate in the Sociology of Agriculture." *Sociologia Ruralis* 42 (1): 5–22.

Goodman, David, and M. R. Redclift. 1991. *Refashioning Nature : Food, Ecology, and Culture.* London and New York: Routledge.

Goodman, David, and Michael Watts. 1997. *Globalising Food : Agrarian Questions and Global Restructuring.* London and New York: Routledge.

Granovetter, Mark. 1985. "Economic Action and Social Structure: The Problem of Embededness." *American Journal of Sociology* 91 (3): 481–510.

Guthman, Julie. 1998. "Regulating Meaning, Appropriating Nature: The Codification of California Organic Agriculture." *Antipode* 30 (2): 135–154.

———. 2002. "Commodified Meanings, Meaningful Commodities: Rethinking Production-Consumption Links through the Organic System of Provision." *Sociologia Ruralis* 42 (4): 295–311.

———. 2003. "Fast Food/Organic Food: Reflexive Tastes and the Making of 'Yuppie Chow.'" *Social & Cultural Geography* 4 (1): 45–58.

———. 2004a. *Agrarian Dreams: The Paradox of Organic Farming in California.* Berkeley: University of California Press.

———. 2004b. "Back to the Land: The Paradox of Organic Food Standards." *Environment and Planning A* 36 (3): 511–528.

———. 2004c. "The Trouble with 'Organic Lite' in California: A Rejoinder to the 'Conventionalisation' Debate." *Sociologia Ruralis* 44 (3): 301–316.

Haraway, Donna Jeanne. 1991. *Simians, Cyborgs, and women : The Reinvention of Nature.* London: Free Association Books.

Harper, Richard. 2000. "The Social Organization of the IMF's Mission Work: An Examination of International Auditing." Pp. 21–53 in *Audit Cultures: Anthropological Studies in Accountability, Ethics and the Academy,* ed. M. Strathern. London and New York: Routledge.

Hatanaka, Maki. 2010. "Certification, Partnership, and Morality in an Organic Shrimp Network: Rethinking Transnational Alternative Agrifood Networks." *World Development* 38 (5): 706–716.

Hetherington, Kregg. 2005. *Cultivating Utopia: Organic Farmers in a Conventional Landscape.* Halifax: Fernwood.

Higgins, Vaughan, Jacqui Dibden, and Chris Cocklin. 2008. "Building Alternative Agri-Food Networks: Certification, Embededness and Agri-Environmental Governance." *Journal of Rural Studies* 24 (1): 15–27.

Howard, Sir Albert. 1940. *An Agricultural Testament.* Oxford: Oxford University Press.

Hughes, Alex. 2001. "Global Commodity Networks, Ethical Trade and Governmentality: Organizing Business Responsibility in the Kenyan Cut Flower Industry." *Transactions of the Institute of British Geographers* 26 (4): 390–406.

IFOAM. 2011a. "The IFOAM Norms." http://www.ifoam.org/about_ifoam/standards/norms.html (accessed 12 August 2011).

———. 2011b. "Internal Control Systems (ICSs) for Group Certification." http://www.ifoam.org/about_ifoam/standards/ics.html (accessed August 14 2011).

Jaffee, Daniel, and Philip H. Howard. 2010. "Corporate Cooptation of Organic and Fair Trade Standards." *Agriculture and Human Values* 27 (4): 387–399.

Johnston, Josée, Andrew Biro, and Norah MacKendrick. 2009. "Lost in the Supermarket: The Corporate-Organic Foodscape and the Struggle for Food Democracy." *Antipode* 41 (3): 509–532.

Kautsky, Karl. [1899] 1988. *The Agrarian Question.* London: Zwan.

Kirksey, S. Eben, and Stefan Helmreich. 2010. "The Emergence of Multispecies Ethnography." *Cultural Anthropology* 25 (4): 545–576.

Kirwan, James. 2004. "Alternative Strategies in the UK Agro-Food System: Interrogating the Alterity of Farmers' Markets." *Sociologia Ruralis* 44 (4): 395–415.

Konefal, Jason, and Maki Hatanaka. 2011. "Enacting Third-Party Certification: A Case Study of Science and Politics in organic Shrimp Certification." *Journal of Rural Studies* 27 (2): 125–133.

Latour, Bruno. 1993. *We Have Never Been Modern.* Trans. C. Porter. Cambridge: Harvard University Press.
Latour, Bruno, and Steve Woolgar. 1986. *Laboratory life : The Construction of Scientific Facts.* Princeton: Princeton University Press.
Lyson, Thomas A. 2004. *Civic Agriculture : Reconnecting Farm, Food, and Community, Civil Society.* Hanover, NH and London: University Press of New England for Tufts University Press.
Mansfield, B. 2004. "Organic Views of Nature: The Debate over Organic Certification for Aquatic Animals." *Sociologia Ruralis* 44 (2): 216–232.
Martin, Andrew. 2007. "Whole Foods Makes Offer for a Smaller Rival." *New York Times,* 22 February.
Michelsen, Johannes. 2001. "Organic Farming in Regulatory Perspective. The Danish Case." *Sociologia Ruralis* 41 (1): 62–84.
Morgan, Kevin, Terry Marsden, and Jonathan Murdoch. 2006. *Worlds of Food: Place, Power, and Provenance in the Food Chain.* Oxford: Oxford University Press.
Munro, Rolland. 1999. "The Cultural Performance of Control." *Organization Studies* 20 (4): 619–640.
Murdoch, Jonathan, Terry Marsden, and Jo Banks. 2000. "Quality, Nature, and Embeddedness: Some Theoretical Considerations in the Context of the Food Sector." *Economic Geography* 76 (2): 107–125.
Mutersbaugh, Tad. 2002a. "Ethical Trade and Certified Organic Coffee: Implications of Rules-Based Agricultural Product Certification for Mexican Producer Households and Villages." *Transnational Law and Contemporary Problems* 89 (1): 89–107.
———. 2002b. "The Number Is the Beast: A Political Economy of Organic-Coffee Certification and Producer Unionism." *Environment and Planning A* 34 (7): 1165–1184.
———. 2005a. "Fighting Standards with Standards: Harmonization, Rents, and Social Accountability in Agrofood Networks." *Environment and Planning A* 37 (11): 2033–2051.
———. 2005b. "Just-in-Space: Certified Rural Products, Labor of Quality, and Regulatory Spaces." *Journal of Rural Studies* 21 (4): 389–402.
Mutersbaugh, Tad, Daniel Klooster, Marie-Christine Renard, and Peter Taylor. 2005. "Certifying Rural Spaces: Quality-Certified Products and Rural Governance." *Journal of Rural Studies* 21 (4): 381–388.
Nelson, Erin, Laura Gómez Tovar, Rita Schwentesius Rindermann, and Manuel Gómez Cruz. 2010. "Participatory Organic Certification in Mexico: An Alternative Approach to Maintaining the Integrity of the Organic Label." *Agriculture and Human Values* 27 (2): 227–237.
Nigh, R. 1997. "Organic Agriculture and Globalization. A Maya Associative Corporation in Chiapas, Mexico." *Human Organization* 56 (4): 427–436.
Pentland, Brian T. 1993. "Getting Comfortable with the Numbers: Auditing and the Micro-Production of Macro-Order." *Accounting, Organizations and Society* 18 (7–8): 605–620.
Philpott, Stacy M., Peter Bichier, Robert Rice, and Russell Greenberg. 2007. "Field-Testing Ecological and Economic Benefits of Coffee Certification Programs." *Conservation Biology* 21 (4): 975–985.
Pimentel, David, Paul Hepperly, James Hanson, David Douds, and Rita Seidel. 2005. "Environmental, Energetic and Economic Comparisons of Organic and Conventional Farming Systems." *BioScience* 55 (7): 573–582.
Polanyi, Karl. [1944] 1957. *The Great Transformation.* Boston: Beacon Press.
Pollan, Michael. 2001. "Naturally: How Organic Became a Marketing Niche and a Multibillion-Dollar Industry." *New York Times Magazine,* 13 May.
———. 2006a. "Mass Natural." *New York Times Magazine,* 4 June.
———. 2006b. *The Omnivore's Dilemma: A Natural History of Four Meals.* New York and London: Penguin.
Power, Michael. 1997. *The Audit Society: Rituals of Verification.* Oxford: Oxford University Press.
Raynolds, Laura T. 2000. "Re-Embedding Global Agriculture: The International Organic and Fair Trade Movements." *Agriculture and Human Values* 17 (3): 297–309.
———. 2002. "Consumer/Producer Links in Fair Trade Coffee Networks." *Sociologia Ruralis* 42 (4): 404–424.
———. 2004. "The Globalization of Organic Agro-Food Networks." *World Development* 32 (5): 725–743.

———. 2008. "The Organic Agro-Export Boom in the Dominican Republic: Maintaining Tradition or Fostering Transformation?" *Latin American Research Review* 43 (1): 161–184.
Rice, Robert A. 2001. "Noble Goals and Challenging Terrain: Organic and Fair Trade Coffee Movements in the Global Marketplace." *Journal of Agricultural and Environmental Ethics* 14 (1): 39–66.
Shore, Cris, and Susan Wright. 2000. "Coercive Accountability: The Rise of Audit Culture in Higher Education." Pp. 57–89 in *Audit Cultures: Anthropological Studies in Accountability, Ethics and the Academy,* ed. M. Strathern. London and New York: Routledge.
Shreck, Aimee, Christy Getz, and Gail Feenstra. 2006. "Social Sustainability, Farm Labor, and Organic Agriculture: Findings from an Exploratory Analysis." *Agriculture and Human Values* 23 (4): 439–449.
Strathern, Marilyn. 2000. *Audit Cultures: Anthropological Studies in Accountability, Ethics, and the Academy.* New York: Routledge.
Thiers, Paul. 2002. "From Grassroots Movement to State-Coordinated Market Strategy: The Transformation of Organic Agriculture in China." *Environment and Planning C-Government and Policy* 20 (3): 357–373.
Tovar, Laura Gómez, Lauren Martin, Manuel Angel Gómez Cruz, and Tad Mutersbaugh. 2005. "Certified Organic Agriculture in Mexico: Market Connections and Certification Practices in Large and Small Producers." *Journal of Rural Studies* 21 (4): 461–474.
Vos, Timothy. 2000. "Visions of the Middle Landscape: Organic Farming and the Politics of Nature." *Agriculture and Human Values* 17 (3): 245–256.
Willer, Helga, and Lukas Kilcher, eds. 2010. *The World of Organic Agriculture: Statistics and Emerging Trends 2010.* Frick and Bonn: Research Institute of Organic Agriculture and International Federation of Organic Agriculture Movements.
Willer, Helga, Minou Yussefi, and Neil Sorensen. 2008. *The World of Organic Agriculture : Statistics and Emerging Trends 2008.* London: Earthscan.
Williams, Raymond. 1980. *Problems of Materialism and Culture.* London: Verso.

Mapping the Food Movement

Addressing Inequality and Neoliberalism

Teresa Marie Mares and Alison Hope Alkon

ABSTRACT: In this article, we bring together academic literature tracing contemporary social movements centered on food, unpacking the discourses of local food, community food security, food justice, and food sovereignty. This body of literature transcends national borders and draws on a rich genealogy of studies on environmental justice, the intersections of race, class, and gender, and sustainable agro-food systems. Scholars have emphasized two key issues that persist within these movements: inequalities related to race and class that shape the production, distribution, and consumption of food, and the neoliberal constraints of market-based solutions to problems in the food system. This article claims that food movements in the United States would be strengthened through reframing their work within a paradigm of food sovereignty, an approach that would emphasize the production of local alternatives, but also enable a dismantling of the policies that ensure the dominance of the corporate food regime. The article concludes by offering a critical analysis of future research directions for scholars who are committed to understanding and strengthening more democratic and sustainable food systems.

KEYWORDS: food movements, food systems, inequality, neoliberalism, social justice

In regions across the United States, food movement activists are pursuing projects like community gardens, community-supported agriculture (CSA) projects, and farmers' markets as promising alternatives to the industrialized food system, or what Eric Holt-Giménez (2011) terms the "corporate food regime." Advocates comprising what Michael Pollan (2010) calls "the food movement" argue that increased production and consumption of local food is key to transforming a host of environmental problems (pesticide use, soil and water contamination, fossil fuel dependence) and social problems (increased consolidation of land ownership, corporate control of food system, high rates of diet-related health problems). In recent years, this movement has gained national attention as best-selling authors like Pollan, Marion Nestle, and Anna Lappè encourage us to connect our forks with our politics and eat close to home. Their impressive book sales signal that the idea of eating locally is no longer a fringe concept. Indeed, the word, "locavore," meaning one who eats local foods, was chosen as the Oxford Encyclopedia Dictionary's 2007 word of the year.

A wide range of scholarship has emerged alongside this movement, often penned by scholar-activists who are deeply engaged with on-the-ground actions, programs, and community orga-

Environment and Society: Advances in Research 2 (2011): 68–86 © Berghahn Books
doi:10.3167/ares.2011.020105

nizations. Some of this literature is quite celebratory, trumpeting the potential for the food movement to create environmental sustainability (Altieri 2000), vibrant local communities (Andreatta 2005; Goldschmidt 1978), a connection to place (DeLind 2002, 2006; Kloppenberg et al. 1996), and a renewed civic life (Lyson 2004). Others take a more critical approach, pushing the movement to engage with two key issues. First, critics have highlighted issues of inequality, examining the social, political, and cultural processes that determine who is drawn to and has the ability to produce and consume particular kinds of food (Alkon and McCullen 2010; Allen 1999, 2008; Allen et al. 2003; Guthman 2004, 2008; Patel 2007; Slocum 2007). Second, scholars argue that the food movement is characterized by neoliberal strategies that urge consumers to choose locally based economic alternatives rather than invite citizens to reform or even transform the food system itself (Alkon 2008a; Allen 2008; Allen and Guthman 2006; Allen and Wilson 2008; Brown and Getz 2008; Guthman 2008; Mares 2010; Pudup 2008).

Throughout this article, we refer to these challenges as those of inequality and neoliberalism, but maintain that these challenges and the spaces where they unfold are deeply entangled. The food movement's privileging of market-based strategies makes the alternatives they create and endorse less accessible to low-income communities and communities of color while conversely ignoring the ways that racial and economic privileges pervade both conventional and alternative food systems. Further, though alternative agro-food systems have been often conceptualized in oppositional terms, more contemporary investigations have problematized these binaries to reveal how the "local" or "sustainable" and the "global" or "industrial" are necessarily intertwined and not bound to concrete spatial locations. We follow this line of inquiry to argue that both localized food systems and the "corporate food regime" are complex and multidimensional social, political, and economic formations necessitating multidisciplinary and multisectoral research approaches.

As activists have reflected on these and other critiques, food movements have developed three concepts that build on the idea of local food in important ways. Community Food Security (CFS) broadens the unit of analysis from individuals to communities, inviting more structural perspectives on the need for and value of alternative food systems. In addition, it combines the local food discourse with an antihunger approach, arguing that all communities should have access to safe, culturally acceptable, nutritionally adequate, and sustainably produced diets (CFSC n.d.). Building on this framework, the concept of food justice speaks to the multiple ways that racial and economic inequalities are embedded within the production, distribution, and consumption of food. Activists working from these two perspectives often create local food system alternatives such as farmers' markets, CSA programs, urban farms, and cooperatively owned grocery stores in low-income communities of color, though food justice activists tend to strongly emphasize the need for these projects to be created not only for, but by members of these communities.

Although the CFS and food justice perspectives directly confront the issues of inequality described above, scholars assert that their strategies are still predominantly and problematically neoliberal in that they seek transformation through changes in individualized consumption practices rather than broader and more collective efforts. In contrast, the concept of food sovereignty has developed through the struggles of farmers in the Global South and embodies an explicit opposition to neoliberalism. Food sovereignty declares the rights of local peoples to define their own agro-food systems rather than remain subject to the consequences and demands of global trade policies. It includes calls for land reform, free access to and control over seeds, and the safeguarding of water as a public good. It also pays particular attention to the needs and desires of women and indigenous peoples, who do the majority of the food cultivation in the Global South. Local food activists in the United States and other industrialized nations have embraced food sovereignty's emphasis on the needs of farmers and opposition to genetically modified foods, and local food organizations have even joined transnational coalitions toward these goals. CFS and food justice activists often see their own aims reflected in the food sovereignty movement's focus on community self-sufficiency and self-determination. However, we argue that none of the US-based discourses engage with global food politics to the degree that food sovereignty activists have. Doing so may help these movements to scale up and build broad coalitions that can transform the food system in both the US and abroad.

In this article, we offer an overview of scholarship on the above-described four discourses, paying close attention to debates concerning inequality and neoliberalism. We begin by briefly reviewing studies of the local food and CFS discourses, outlining the critiques that scholars have detailed with respect to the limitations of these approaches. We then turn to the burgeoning literatures on food justice and food sovereignty to illuminate the ways that both concepts offer a deeper structural analysis of food system inequalities while highlighting food sovereignty's opposition to neoliberalism. Our review is necessarily uneven as the concept of local food is both the most prominent and most heavily critiqued. Indeed, the other concepts we review—community food security, food justice, and food sovereignty—are often described as addressing problematic aspects of local food discourses. These later discourses have not been critiqued to the degree that local food has, but we do review existing appraisals where appropriate. We conclude by offering a critical analysis of future research directions for scholar-activists who are committed to understanding and contributing to more just and sustainable food systems. Table 1 outlines the key points of these discourses to better situate their commonalities and differences and guide the reader through our review.

Local Food

The movement for local food is heavily guided by the philosophy and practice of "civic agriculture." In *Civic Agriculture: Reconnecting Farm, Food and Community,* Thomas Lyson defines civic agriculture as "a locally organized system of agriculture and food production characterized by networks of producers who are bound together by place [, which] … embodies a commit-

Table 1. Food Movement Discourses

Local Food	Community Food Security	Food Justice	Food Sovereignty
• Influenced by concept of civic agriculture and the notion of eating within a "foodshed" • Endorses market strategies including CSA projects, farmers' markets, market gardening, and farm-to-school programs • Frames solutions primarily within changing individual and familial consumption practices	• Endorses the same market strategies as local food, but also forges connections to emergency food programs, including food banks and meal programs • Emphasizes food systems approach and actions at the community scale, in addition to changing individual consumption practices • Frames solutions within improvements to food access and availability	• Emphasizes the connections among structural racism, classism, and inequalities in access to food and productive land • Endorses the same market strategies as local food, but emphasizes that these solutions should be designed and managed by those bearing the brunt of food inequalities, that is low-income communities of color	• Emphasizes the impacts of neoliberal trade and agricultural policies and the monopoly of the corporate agro-food sector • Frames solutions within agro-ecological production practices and redistribution of land and other natural resources • Draws attention to the rights and roles of women within agro-food systems

ment to developing and strengthening an economically, environmentally, and socially sustainable system of agriculture and food production that relies on local resources and serves local markets and consumers" (2004: 63). The promise of civic agriculture lies in its adherence to a democratic approach, which, according to Lyson, opens a space for people to transform from passive consumers into active "food citizens" (2004: 77).

Anthropologist Laura DeLind has claimed that the concept of civic agriculture gives "shape and legitimacy to a diverse and growing body of creative, socioeconomic relationships … [and] has the power to focus public attention on the contradictions within our industrially-modeled and corporately controlled agriculture, as well as on the potential of 'relocalized' food systems" (2002: 217; see also DeLind 2006). However, although DeLind finds value in civic agriculture's theoretical rejection of neoclassical economic thought and the commodification of food, she worries that, in practice, local food strategies like CSAs and farmers' markets do not escape the realm of private ownership and accumulation. In addition to its reliance on traditional market relationships, she is also troubled by the association of "good" consumption and production with "good" citizenship.

In addition to the framework of civic agriculture, the local food discourse draws on the notion of the "foodshed." In their defining article, "Coming in to the Foodshed," Kloppenburg and colleagues (1996: 34) argue that the term "becomes a unifying and organizing metaphor for conceptual development that starts from a premise of the unity of place and people, of nature and society." For these authors, relocalizing food systems offers a way to "reassemble our fragmented identities, reestablish community, and become native not only to a place but to each other" in the midst of an "unstable, post-modern world" (1996: 34). A critique of their use of the term "native" notwithstanding, these authors exalt the potential of localization and the ways that it could and should be embedded in a "moral economy." Although they note the need for participation of marginalized and disempowered groups, Kloppenburg et al. vacillate between the paths of autonomy and reform, disengagement and radical change, with no clear analysis of how these various strategies might facilitate or impede broader participation from marginalized groups.

In their attention to systems thinking and exploring the connections between people and place, local food advocates draw attention to the fact that agricultural production is not only an economic issue, but is also socially and politically relevant. Nevertheless, Allen (2004) argues that local food advocates' emphasis on the economic success of small farmers detracts from attention to issues of social justice such as farmworker rights and food access. For Allen, it is this attention to farm profits that orients activists toward solutions predicated on individual consumption rather than more collective change. Building on their earlier work on alternative agro-food systems, Allen (2004) and Guthman (2004) have offered extensive examinations of the connections between neoliberal economic formations and contemporary food politics. Together, they examine the growth of farm-to-school programs, showing how these initiatives are unique in that they unfold in the public realm of the educational system, yet "must work within the contemporary context of neoliberalization, in which solving social problems is seen as the domain of individuals and the market" (Allen and Guthman 2006: 402). The authors demonstrate how farm-to-school programs both "enable and reflect neoliberalization," specifically in their funding sources, labor practices, framings of academic performance and obesity, and emphasis on consumer choice. For an extended discussion of the interconnections between sustainable consumption and neoliberalism, see Cindy Isenhour's article in this volume.

In a related article, Guthman (2008) observes that in-depth analyses of neoliberalism have been largely absent from agro-food scholarship, and that when they do appear, "there is a marked tendency to treat neoliberalism (and globalization) as a source of 'impacts' on national and subregional food sectors, with a particular focus on the macro effects of neoliberal trade on

Third World agriculture" (2008: 1174). This treatment fails to address how First World areas are also sites of neoliberal practices and tends to generalize all actions that respond to neoliberalism as inherently good—resulting in a romanticization of the "local" and all forms of resistance. For Guthman, four themes in food activism reflect neoliberal rationalities that revolve around a highly individualized conceptualization of politics—consumer choice, localism, entrepreneurialism, and self-improvement. She states: "For activist projects, neoliberalization limits the conceivable because it limits the arguable, the fundable, the organizable, the scale of effective action, and compels activists to focus on putting out fires. Those who theorize neoliberalism as a mode of governmentality have recognized how material neoliberalizations themselves have shaped subjectivities" (2008: 1180).

Drawing on research on community gardening, Mary Beth Pudup (2008) supports Guthman's claims as she examines how organized community gardening projects operate as spaces of neoliberal governmentality, in that they are "spaces in which gardening puts individuals in charge of their own adjustment to economic restructuring and social dislocation through self-help technologies centered on personal contact with nature" (2008: 1228). In tracing how community gardening has been a recurring response to capital restructuring, Pudup distinguishes between community gardens that are sites of social resistance from those with an "ethos of personal responsibility." She reveals the tension between community gardens and new urban social movements beginning in the 1990s to argue that some gardens, those she terms "organized garden projects," are characterized by the work of "non-state and quasi-state actors who deliberately organize gardens to achieve a desired transformation of individuals in place of collective resistance and/or mobilization" (2008: 1230). More often than not, these projects are geared toward transforming the lives of poor and otherwise socially marginalized people through individualized experiences of direct contact with nature rather than explicitly challenging the systemic conditions that produce marginalization.

Another critique documenting how local food strategies articulate with neoliberalism is an analysis of how voluntary certification of sustainable foods supplants direct pressures on the state to restrict harmful production practices. Working in California and drawing on Peck and Tickell's work on "roll out neoliberalism," Brown and Getz (2008) argue that certification used as a means to achieve social justice is both shaped by a reaction to the material consequences of neoliberalism and reinforces neoliberal ideologies. Although they problematize the delineations that some scholars make between consumption politics and forms of organizing that seek structural change, the authors are concerned with the possibility of monitoring and labeling initiatives usurping the possibilities for collective action, especially because the voices of farmworkers have historically been absent from these initiatives. Brown and Getz conclude that "certification presents an incomplete and inadequate response to farm workers' grave situation, including eroding wages, exploitative conditions, and treacherous journeys across a militarized border zone" (2008: 1195).

Some scholars argue that local food's neoliberal strategies prevent it from challenging inequalities that pervade both the food system and social life more generally. Yet others argue that this discourse's romantic conceptualization of the local can unwittingly be a form of exclusion and xenophobia. (Allen 1999; Allen et al. 2003; DuPuis and Goodman 2005). Hinrichs notes the tension that persists even within "socially embedded" agricultural markets, arguing that economic concerns are inseparable from broader social concerns: "many direct agricultural markets involve social relations where the balance of power and privilege ultimately rests with well-to-do consumers. Struggling farmers and poor consumers, in contrast, must weigh concerns with income and price against the supposed benefits of direct, social ties" (Hinrichs 2000:

301). Winter traces the motivations behind buying organically and locally, arguing that "the more recent turn to quality is based, to some extent on a continuation and growth of demand for such luxury and positional goods and this continuity should not be forgotten" (2003: 25). Conversely, Born and Purcell (2006) emphatically caution those in the field of planning against falling into the "local trap," which they define as "the tendency of food activists and researchers to assume something inherent about the local scale. The local is assumed to be desirable, it is preferred a priori to larger scales" (2006: 195). Taken together, these critiques argue that local is not necessarily more sustainable, nor is it more just.

Through their work in rural Washington state, Joan Qazi and Theresa Selfa (2005) investigate how regional politics both impede and further the development of food system alternatives. They underscore that "the mapping of alternative agro-food provisioning in our central Washington case study is similarly challenged by spatial relations that do not coincide necessarily with local food networks, and by political discourse that is not necessarily progressive or transformative" (2005: 49). In a related study also based in Washington state, Lucy Jarosz (2008) demonstrates the importance of understanding both urban and rural contexts in analyzing the diversity and complexity of local, alternative food networks. She reveals how the increasing demand for seasonal, local, and organic produce in urban areas does not necessarily translate into better livelihoods for the rural farmers who attempt to meet these demands. This complicates the commonly held notion of food system localization as inherently transformational or progressive and demonstrates how local food systems "emerge from political, cultural and historical processes as they develop out of the interactions between rural restructuring and urbanization in metropolitan areas" (2008: 242).

In a similar vein, Melanie DuPuis and David Goodman (2005) call for a "reflexive politics of localism," arguing that "unreflexive localism" can "deny the politics of the local, with potentially problematic social justice consequences" and "can lead to proposed solutions, based on alternative standards of purity and perfection, that are vulnerable to corporate cooptation" (2005: 360). They do not seek to delegitimize localism, but rather illuminate the complexity of the spatial dynamics upon which much of the alternative food movement rests, a complexity that is necessarily bound up with dynamics of race, class, and gender. Though the local food movement has largely ignored how inequalities intersect with industrial and alternative food systems, the next two discourses discussed here ground local food system work in a desire to illuminate such relationships, but with an emphasis on race and class disparities. However, they are still constrained by their largely unrecognized reproduction of neoliberal subjectivities.

Community Food Security

The CFS discourse brings together local food's concern for environmentally sustainable production with the work of antihunger advocates responding to decreasing public sector support (Allen 1999; Gottlieb and Fisher 1996). The Community Food Security Coalition (CFSC; n.d.), the most prominent organization deploying this discourse, defines CFS as "a condition in which all community residents obtain a safe, culturally acceptable, nutritionally adequate diet through a sustainable food system that maximizes community self-reliance and social justice." Key to this perspective is an emphasis on the community scale. The inclusion of the term community demonstrates an understanding that hunger is not an individual phenomenon, but something created by wider, systemic social forces. In her analysis of the CFS movement, Allen describes its discursive merits:

> While the American food system has emphasized production with distribution and consumption considered separately, CFS seeks to re-link production and consumption with the goal of ensuring both an adequate and accessible food supply in both the present and the future. In its focus on consumption, CFS has prioritized the needs of low-income people; in its focus on production, it emphasizes local and regional food systems. (1999: 117)

Allen finds much promise in the CFS movement, but she also foreshadows recent critiques of local food. She expresses her concern about the effects of alternative economic strategies found in CSA and farmers' markets, raising a red flag that these types of "designer" food production schemes may create a two-tiered food system built on class differences. Given this analysis, Allen (2004) highlights contradictions between the CFS movement's twin goals of farm security and food security. She also critiques the movement's view that using food assistance programs is "dependence," pointing out that in antihunger perspectives food is viewed as a *right* to be fulfilled by the state if the market fails. In observing that so-called dependency programs may be part of the "normal" channels that women on assistance use, Allen asserts that "there will always be people who need food assistance as long as there is underemployment, unemployment, poverty-level wages, and inadequate pensions and access to food is based on ability to pay" (1999: 126). Although CFS makes issues of food access a central part of its discourse, Allen pushes those working from this perspective to take into better account class inequalities and the material realities of those who are unlikely to benefit from the economic arrangements that are most common in today's alternative food landscapes.

Another important critique comes from Molly Anderson and John Cook (1999) who argue that CFS lacks a uniting theoretical framework, which they believe is necessary for the movement to effectively guide policy and action. They assert that the CFS concept suffers from loose definitions and an absence of a clear theoretical basis, which could help to "improve understanding of the barriers to food security at several levels of analysis, and help policy-makers and practitioners improve food security in a given area" (1999: 142). They call for further study and evaluation of the CFS approach and a better understanding of its best practices. Anne Bellows and Michael Hamm (2003) extend these ideas, asserting that improvements in food security necessitate that the populace define and demand change rather than rely on institutional willingness to change. Bellows and Hamm see tremendous value in the community element of CFS, pushing for both national and international coalitions and mobilization. They also contend that the adoption of a human rights framework would be useful for the CFS movement as it would help to integrate an analysis of the systemic injustice that links food insecurity with other human rights concerns including housing, clean water, education, and employment with living wages.

Although issues of access are central to the CFS discourse, activists working from this perspective tend to be largely white and middle-class. Rachel Slocum (2007) investigates how whiteness is produced and embodied in practices that are framed within the CFS discourse. In her analysis, she rejects the notion that whiteness is inherently negative, but rather questions how the ethics and politics embedded within alternative food practices might move "the US, collectively, toward joy through food" (2007: 521). Slocum argues, "the desire for good and sufficient food and jobs and thriving economies is not white. It becomes white through what white bodies do in this effort" (2007: 521). Although Slocum argues that whiteness as it manifests in CFS activism is problematic, she simultaneously locates the knowledge of and commitment to "good food" primarily in white communities, and locates the possibility for transformation in how and when they might share this good food with "others." As a reaction to the whiteness embodied in the CFS discourse, people of color who were active in these projects (and some of whom remained active) began to embrace a food justice perspective, which is grounded in the lived experience of communities of color and foregrounds not only issues of access, but of racial and economic inequality.

Food Justice

The food justice discourse draws inspiration from both local food and CFS, but centralizes principles of racial and economic justice to argue that access to healthy, culturally appropriate, sustainably grown food is mediated by inequalities of race and class. More than any other strain of the food movement, US food justice activists mobilize at the grassroots level to dismantle the classist and racist structural inequalities that are manifest in the consumption, production, and distribution of food. Their dominant strategy is the creation of often-subsidized local alternative food provisioning strategies, such as farmers' markets, CSAs, and urban farms in low-income communities of color.

Alfonso Morales (2011) examines the work of the Growing Food and Justice for All Initiative (GFJI), which seeks to use local food system work as a tool to dismantle racism. GFJI developed in association with the work of Growing Power, perhaps the most well-known food movement organization led by people of color, in response to the inadequacies of the CSFC. Morales shows that, despite the coalition's emphasis on issues of cultural appropriateness and community responsibility, much of the early leadership was predominately white, even while the broader coalition had greater participation from people of color. He argues that the food justice movement has since been in "dynamic tension" with the CSFC, allowing for an overall productive and collaborative set of relationships.

People's Grocery is one of the most visible actors in the US food justice movement and their work delivering local chemical-free produce to poor and working class inner-city residents through their mobile market—CSA and other programs in Oakland, California—exemplifies the goals and objectives of food justice. According to People's Grocery (n.d.):

> Food justice asserts that no one should live without enough food because of economic constraints or social inequalities. Food justice reframes the lack of healthy food sources in poor communities as a human rights issue. Food justice also draws off of historical grassroots movements and organizing traditions such as those developed by the civil rights movement and the environmental justice movement. The food justice movement is a different approach to a community's needs that seeks to truly advance self -reliance and social justice by placing communities in leadership of their own solutions and providing them with the tools to address the disparities within our food systems and within society at large.

This definition illustrates the movement's emphasis on community self-reliance (particularly for the working class and communities of color), social justice, and fundamental human rights—all underdeveloped priorities in the movements for local food and CFS. Moreover, in its reference to the civil rights and environmental justice movements and its call for leadership by marginalized communities, People's Grocery seeks to ground its work in the racialized history of the predominantly African American neighborhoods in which it works.

Like all food movements, the food justice movement is increasingly garnering the attention of scholar-activists. *Food Justice,* co-authored by Robert Gottlieb and Anupama Joshi (2010), takes a wide-angled approach, describing injustices and activist responses in projects across the US. The authors define food justice "as ensuring that the benefits and risks of where, what, and how food is grown and produced, transported, and distributed, and accessed and eaten are shared fairly" (2010: 6). Gottlieb and Joshi argue that a number of injustices plague the industrial food system, including hazardous conditions faced by farm workers and the absence of healthy food in low-income neighborhoods. They also chronicle a series of community-based responses including community gardens and farmer training, youth empowerment, farm-to-school programs, and the campaign to persuade the Obama White House (and Michelle Obama in particular) to plant an organic garden.

Gerda Wekerle (2004: 378) argues that food justice should be of particular interest to urban planners because of its attention to local organizing, community development, and engaging citizens. Wekerle describes how the food justice movement grew out of an effort to reframe the food security movement into an "explicit critique of the global food system and a theoretical framing of local initiatives as both the practice of democracy and as means of de-linking from the corporate food system" (2004: 379). Following the work of Arjun Appadurai, Wekerle claims that food justice, in its simultaneous attention to local place and transnational networks, is best understood as a translocal movement. This translocality not only changes the terrain of relationships among community members, civil society institutions, and the state, but also creates new opportunities to form alliances and coalitions across geographical boundaries. Through Wekerle's assessment, we can clearly see the potential of food justice as an organizing tool to move beyond the neoliberal strategy of creating alternatives to argue for broader social change.

Charles Levkoe (2006) sees similar promise in the food justice movement for its potential to facilitate a shift toward more democratic food systems through transforming "consumers" into "citizens" and model a "healthy democracy" at the grassroots level. "Through food justice movements, a vision of food democracy has been adopted which directly challenges anti-democratic forces of control, exploitation, and oppression. Food democracy refers to the idea of public decision-making and increased access and collective benefit from the food system as a whole" (Levkoe 2006: 91). Levkoe believes that the benefits of developing civic virtues and critical perspectives through food justice organizing expand beyond the food system to build an overall stronger community.

Although food justice activists often foreground issues of racial inequality, these works celebrate it while eliding an explicit discussion of race. In doing so, they mirror many local food organizations' relabeling of themselves as food justice organizations, even without leadership and direction from communities of color. We certainly applaud any integration of issues of inequality and social justice into the local food and CFS perspectives but are also concerned that as these organizations rebrand themselves as advocates of food justice, they erase the discourse's original recognition of the need for a perspective grounded in the lived experiences of communities of color. Moreover, we fear that scholars seeking to celebrate a broad notion of food justice may be unwittingly complicit in this cooptation.

In contrast, we are both involved in the production of a new edited volume called *Cultivating Food Justice* (2011). This book explores the multiple ways that race and class are enmeshed in the food system, and seeks to highlight stories that might challenge or complicate the dominant narrative that has been told and retold within the food movement through examining environmental and social injustices. Citing the seminal work of Omi and Winant ([1986] 1994), the editors of the volume argue that contemporary food systems are implicated in *racial projects*, "political and economic undertakings through which racial hierarchies are established and racialized subjectivities created" (Alkon and Agyeman 2011: 7).

Cultivating Food Justice seeks to go beyond an examination of consumption politics to explore the structural inequalities that underlie contemporary food systems and food movements. Several chapters highlight the processes through which institutionally racist land use and agricultural policies have created widely unrecognized advantages with regard to whites' abilities to produce food. Norgaard et al. (2011) trace how racial projects of outright genocide, lack of recognition of land occupancy and title, and forced assimilation have resulted in a lack of access among the Karuk indigenous tribe to traditional food sources including salmon. In a parallel example, Green et al. (2011) argue that limited access to land and mistreatment on the part of federal institutions like the USDA have greatly reduced the number of African American farmers in the US. These authors argue that undermining these communities' land tenure stripped them of a means of sustainable food production, leading to present-day food insecurity.

Like Gottlieb and Joshi's book, *Cultivating Food Justice* also offers an array of responses, though the chapters highlight those from communities of color. For example, Mares and Peña (2011) provide an ethnographic examination of Latino/a community gardeners in Seattle and Los Angeles. The authors show how these urban farmers "weave their place-based identities into new landscapes" as they "negotiate their "social citizenship" in a "safe" and "self-made" space that can offer a buffer against elements of a nation that are increasingly hostile or ambivalent about their presence" (2011: 205). Similarly, Priscilla McCutcheon (2011) offers a fascinating analysis of the food politics of two black nationalist religious groups: the Nation of Islam and the Pan African Orthodox Christian Church, examining how both seek to address hunger and build empowerment among blacks by running organic farms in Georgia and South Carolina. These analyses open up an inquiry about the food practices of marginalized communities that operate not in opposition to, but autonomously from the mainstream alternative foods movement and demonstrate how issues of social and racial justice can be at the heart of community food work.

As is common within much of the scholarship engaging this perspective, *Cultivating Food Justice* is a largely celebratory volume aiming to give voice to and analyze this discourse. Its conclusion, however, pushes activists to acknowledge the limitations of the market-driven provision of alternatives that are at odds with this discourse's broad structural analysis of food system inequalities. This echoes Allen's assertion that

> effecting food justice is regularly constrained in actual practice regardless of the intentions of the actors ... because of the need to work within the constraints of the current political economic system along with a push towards neoliberal forms of governance. One result is that the alternatives being developed are much more accessible to relatively more privileged people, despite intentions to the contrary. (2008: 159)

Allen's critique signals that, despite food justice's radical rhetoric naming food a human right, the prioritization of market-based provisioning casts food as a commodity disproportionately accessible to those who can pay. Moreover, the belief that the creation of alternatives can address the broad structural critiques identified by activists signals a lack of critical understanding of neoliberalism that can undermine efforts to create greater access to local/sustainable food. A broad understanding of and response to neoliberalism is characteristic of the food sovereignty discourse.

The Food Sovereignty Movement

The food sovereignty discourse has developed largely through La Via Campesina, an international coalition that has devoted particular attention to the struggles and rights of Third-World peasant farmers (especially women), the impacts of neoliberal trade and agricultural policies, and the importance of working in solidarity across international borders. La Via Campesina defines food sovereignty as "the peoples', Countries' or State Unions' RIGHT to define their agricultural and food policy, without any dumping vis-à-vis third countries" (2003: 1). For an extended history of La Via Campesina and the concept of food sovereignty, see Hannah Wittman's article in this volume. As Wittman describes, food sovereignty is not necessarily "an established paradigm/concept but rather a potential new framework emerging from diverse sets of contemporary grassroots production practices and political approaches."

Food sovereignty's emphasis on rights and issues of control shifts the focus beyond the equitable provisioning of food to address more fundamental inequalities related to land distribution, resource management, and the commodification of food crops. In its engagement with global trade policy, there is a clear awareness of and direct opposition to many forms of neoliberalism, even as the buying and selling of food remains integral to the actions of food sovereignty activists. Eric Holt-Giménez argues that, "food sovereignty is a much deeper concept … because it proposes not just guaranteed access to food, but democratic control over the food system-from production to processing, to distribution, marketing, and consumption" (2009: 246). In a related article, Holt-Giménez and Annie Shattuck (2011) outline how food sovereignty reflects a "radical" trend in food movements, as it foregrounds entitlements to resources and calls for a fundamental redistribution of wealth and power. Hence, food sovereignty activists are not mere consumers "voting with their forks," to borrow a phrase common to the local food movement, but citizens demanding that their right to food trumps the demands of the corporate food regime.

Annette Desmarais (2007) outlines the ways that La Via Campesina has placed comprehensive agrarian reform back on the table through organizing for food sovereignty. In seeking to forefront the voices and experiences of peasant farmers, Desmarais respects their desires "to establish an international space in which they can articulate their needs, interests, and demands in their own voice" (2007: 18). Although La Via Campesina has worked in opposition to the WTO and other federations like the International Federation of Agricultural Producers (IFAP), it has also "demonstrated the importance of a delegitimization that stems from disengagement—a strategy first articulated by Gandhi almost a year ago" (2007: 21). In this way, the desire for autonomy from the corporate food regime that characterizes all of the discourses described in this review is paired with direct opposition to it.

Cuba is often cast as a food sovereignty success story (Alvarez et al. 2006; Funes et al. 2002). In the late twentieth century, the collapse of the Soviet Union and tightening of the US embargo prevented Cuba from importing the inputs required by its state-owned but export-oriented agribusiness system. Seeking to increase production while coping with limited inputs, the Cuban government reorganized its agricultural economy. Large portions of state-owned land were given to cooperatively organized farmers, thus increasing local control. Government research and extension focused on promoting biodiversity rather than specialized monocultures, and in developing low-cost inputs that could be produced on the farm such as compost and biological pest controls. *Campesinos* (farmers) who had farmed prior to the introduction of green revolution technologies played an important role in this research, as did *promotores* (promoters) from the Campesino a Campesino (farmer to farmer) movement. In addition, Cuba promoted urban agriculture and community gardens, and opened farmers' markets where producers could sell surplus (Alvarez et al. 2006). Yields fell initially, but recovered soon afterward and Cuba has maintained high standards of life expectancy and low rates of diet-related diseases (Amador and Peña 1991). For this reason, Rosset and Borque triumphantly describe recent Cuban history as "the overcoming of a food crisis through self-reliance, smaller farms, and agroecological technology" (2002: xv).

Given Cuba's success, it is not surprising that advocates for food sovereignty have played a leading role in analyzing and promoting solutions to the global food crisis. Analyses of the highly debated causes of this crisis include increased oil prices (Magdoff 2008), financial speculation (Lappè 2009), and the increased use of crops for biofuels (Elliot 2008). The global food crisis is largely a crisis of distribution rather than production. According to Josette Sheeran, the head of the UN's World Food Program, "There is food on shelves but people are priced out of the market" (quoted in Borger 2008: 18). Peter Rosset (2008) claims that food sovereignty is perhaps the only way out of the global food crisis. He argues that the current food crisis is the result of a systematic dismantling of the capacity of nations to produce food for themselves in the push to produce for export, a demand linked to the entry of speculative financial capital into food markets and the rise of agrofuels. According to Rosset (and other food sovereignty activists) food needs to be removed from the neoliberal trade regime imposed by the World Trade Organization and International Monetary Fund because it is not a mere commodity but necessary to human life. These trade regimes, Rosset argues, ensure that staples from wealthy countries such as the US, which are subsidized by the federal government, are exported to countries in the Global South where they bring down the prices available to farmers. Peasant farmers who can no longer feed themselves migrate to the cities, where they comprise much of the masses of urban and peri-urban poor (Davis 2006). Rosset finds promise in both the proposal of reregulating the food commodity markets that have been deregulated as a result of neoliberalism, which would allow peasants to remain in the countryside, and in applying agroecological principles so that their farming can be environmentally sustainable and independent from agribusiness inputs. Regulating these markets and promoting agroecology are two priorities of La Via Campesina.

In a similar vein, Michael Menser (2008) argues that the work of La Via Campesina and its members demonstrate the viability of transnational participatory democracy. For Menser, La Via Campesina pushes forward a framework that allows food activists to connect social and environmental issues through understanding democracy as intricately connected to self-determination and socioecological sustainable production. He shows how La Via Campesina's structure differs from the IFAP in their "horizontality" and "transnational regionalism" and their political program of food sovereignty where agroecology is as much a political program as a model of alternative agricultural production model (see also Holt-Giménez 2006).

Philip McMichael finds extreme importance in the framework of food sovereignty as he argues that the "current agrarian question and its resolution depend on the peasantry itself, in a politicized movement on a world scale to confront the international power, and socio-ecological impact, of capital" (2006: 408). For McMichael, the demand for food sovereignty is an attempt to "respatialize" social and economic relations against the constraints of a corporate food regime. He outlines the case of maize production in Mexico to show how changes connected to the North American Free Trade Agreement resulted in commercial foods displacing both local products and local producers, resulting in peasants becoming "semi-proletarianized." Thus, food sovereignty is part of a new *campesino* politics that are expressed "through reassertions of civil society through demanding territorial and/or ethnic autonomy, connecting "not simply with other agrarian and indigenous movements, but also those united with other agrarian and indigenous movements" (McMichael 2006: 414).

Although the food sovereignty movement has largely developed through the mobilization of rural peasant farmers, agro-food scholars and activists see great potential in furthering the movement in urban contexts. Raj Patel argues that the vision of food sovereignty is "important not only because it has been authored by those most directly hurt by the way contemporary agriculture is set up, but also because it offers a profound agenda for change for everyone ... [as it] aims to redress the abuse of the powerless by the powerful, wherever in the food system that abuse may happen" (2007: 302).

Christina Schiavoni (2009) agrees with Patel that the US does not have the same "peasant base" as many other countries. However, she finds potential in connecting the framework of food sovereignty to food justice activism that is already happening within urban centers across the country. Indeed, she describes several food justice activists in New York who feel the concept of food sovereignty resonates with them. For example, Yonnette Flemming of the Hattie Carthen Community Garden in Brooklyn describes food sovereignty as the overarching right of people to have "sovereign control over their own, culturally appropriate food ... Most people here in New York City don't have access to land, and lots of decisions over land use exclude the actual people most impacted by the decisions" (quoted in Schiavoni 2009: 687.). Similarly, Reverend DeVanie Jackson, whose Brooklyn Rescue Mission operates both a food pantry and an urban farm, regards the latter as "putting power in peoples' hands" (quoted in Schiavoni 2009: 687). We certainly agree with Schiavoni that the use of a food sovereignty discourse by New York food justice activists is evidence of the increased potential to build power through the creation of broad alliances. However, we also argue that food sovereignty differs from the above-described discourses, including food justice, in several ways.

Though the previous three discourses operate primarily at the local scale, food sovereignty transcends the boundaries of the local to demand consideration of the impacts of industrialization and centralization on local food economies everywhere, forging an interdependent connection between local food systems in the United States to local food systems around the world. In this way, food sovereignty follows a transnational, or even more so, a translocal approach. The local food discourse also pays significant attention to the global food regime as it tends to advocate resistance primarily through individual consumption while food sovereignty activists directly oppose the corporate food regime. In challenging the devastating impacts of unregulated trade and agricultural policies that privilege behemoth agro-food corporations, food sovereignty seeks to fundamentally rework the politics of food and agriculture and protect the rights of those who are actually working the land.

Further, within a framework of food sovereignty, it is of central importance that food sources are consistent with cultural identities and are embedded within community networks that promote self-reliance and mutual aid. Focusing on food sovereignty allows those who are working for more sustainable food systems to move beyond questions of access to a more comprehensive focus on entitlements to land, decision making, and control over natural assets. Therefore, food becomes a right, not a commodity to be traded on the open market or dumped on the emergency food system when it no longer holds value for food retailers. Adopting these values of food sovereignty would allow US activists to go beyond the framework of "cultural appropriateness," to really consider the cultural importance of food in sustaining social relationships and how food can be implicitly used to erode social relationships, cultural meanings, connections to place, and the exercising of rights. Further, it would allow US food activists to locate the sources of the injustices they rally against in the intersection of the corporate food regime, the transnational neoliberal policies that ensure its dominance, and the racist US policies that ensure that its harshest domestic effects are felt by communities of color. Such an analysis could guide an allied movement for food justice and food sovereignty that would not only emphasize the production of local alternatives, but also a dismantling of the policies that ensure the dominance of the corporate food regime.

Future Directions

The four reviewed bodies of literature cover a wealth of both celebratory and critical scholarship on various aspects of the food movement. Together, these pieces comprise an exciting field of research linking the production, distribution, and consumption of food to efforts for social change. Our review has focused on how four discourses within the food movement—local food, community food security, food justice, and food sovereignty—engage with problematics of inequality and neoliberalism. We conclude by offering several directions that we believe will further these lines of inquiry and strengthen this fertile field of activist scholarship.

First, much of the scholarship we have described above has been rightly critical of the US food movement for adopting market-driven strategies that envision individual, "ethical" consumers and producers as the locus of social change. Food movement actors, however, have also turned to food policy work as a route to transform the food system and the laws that govern it. Successful policy campaigns could potentially restrict the industry's most environmentally harmful and socially unjust practices, increase food access among marginalized communities, and mandate that all food meets higher food safety standards. The broadest of these campaigns attempts to make sweeping changes to the commodity title of the US farm bill, which subsidizes domestic staples such as corn and rice and is responsible for what food sovereignty activists call the "dumping" of cheap imports in the Global South. Changes to this bill could have sweeping effects on the corporate food regime, and are highly supported by food movement advocates around the world.

Although comprehensive changes may be far off, narrower US policy campaigns have seen recent victories. Examples include recent legislation increasing funding for school lunch and authorizing the Food and Drug Administration to demand, rather than merely request, the recall of contaminated products. Researchers following the food movement would do well to investigate campaigns such as these, offering insights on the relationship between individual, market-driven, and collective policy-oriented strategies. Questions of inequality and neoliberalism will remain paramount, as the food movement is forced to choose between competing priorities in their advocacy. For example, the increase in school lunch funding came at the expense of funds for the food stamp program, compromising the economic well-being and food decision-making power of those who qualify for these programs. Some in the food movement have also criticized the food safety legislation as placing undue burdens on mid-size businesses and small farmers, arguing that it fosters further industry consolidation. Researchers could examine the ways that various claims and counterclaims within food and agriculture policies are framed as the US food movement engages political work through the legislative process.

Attention to food policy is one way that various strains of the food movement can scale up and move beyond the local distribution of agroecologically grown food. A number of farmers who were early supporters of the local food movement have built businesses that include national, as well as local sales, and regional distribution networks are becoming increasingly common. In addition, national chains such as Whole Foods and even Walmart have signed contracts with growers to provide for particular local stores. Proponents of this trend argue that it increases food access, but we maintain that these developments call for critical assessments of the scaling up of alternative agriculture, investigating whether and how it maintains environmental sustainability and social justice. Is it possible for these networks to avoid evolving into the industrial organic sector documented by Guthman (2004)? Might this increased scale contain positive consequences for labor rights, as grocery chains are often unionized and Whole-Foods was the first national chain to sign an agreement with the Coalition of Immokalee Workers to increase farmworker pay? This is an exciting area for future work.

Another development, which exists in dynamic tension with the move toward regional and national distribution, is the embrace of a food sovereignty framework by US community activists. Researchers in the US can investigate how this framework articulates, and/or fails to articulate with US food discourses such as local food, CFS, and food justice. Local food activists often share food sovereignty's emphasis on local peoples' control over their food systems and the importance of agroecological principles. CFS shares an emphasis on issues of marginalization and since 2009, the CFSC has awarded a "Food Sovereignty Prize" at their annual conference. However, neither of these US-based discourses has developed the kinds of complex formal organizational practices deployed by food sovereignty activists to ensure democratic participation within the movement. As these activists increasingly build alliances, does this engagement radicalize US actors to look at how inequalities play out within the food system they seek to change as well as the processes through which they work to change it? Moreover, do collaborations with food sovereignty activists push US actors toward dismantling the US policies described above that uphold the corporate food regime? Or, will US activists merely celebrate the focus on autonomy and sovereignty while marginalizing food sovereignty's more collective and radical goals? These questions point to ripe areas of research in which food scholars can and should engage within and across national borders.

Related to the potential for deeper articulations between food sovereignty and US-based food movements, it is essential that future research integrates a critical analysis of gendered dynamics within food movement activism. With an emphasis on the rights and roles of women at all levels of the food system, the food sovereignty paradigm offers a beginning point for research in this area and has the potential to guide studies that explore building alliances between women within and across food movements. Although the topic of gender and food activism has recently made its way into popular treatments of contemporary food politics (e.g., Costa 2010) academic research on this connection is sparse. There is clearly a need for studies that are informed by an intersectional understanding of how race, class, and gender are mutually constitutive within food systems, particularly as feminist scholars (ourselves included) are increasingly drawn to studying and participating in food politics.

Another exciting area for investigation outside of the United States is exploring how global food sovereignty activists have fared in nations that have elected left-wing leaders. Several Latin American countries, including Venezuela, Brazil, Argentina, Uruguay, Bolivia, Peru, and Ecuador, have recently elected leaders explicitly concerned with reducing inequality and several have mentioned land reform and other issues relevant to food sovereignty activists. Of the four discourses discussed here, food sovereignty most explicitly recognizes the effects of national and global trade policy on local circumstances and is regularly invoked in these nations. These activists in the Global South seem well poised to advocate for measures that will increase their ability to feed themselves. However, even support from these national governments may not be enough to change the global political structures that favor the corporate food regime. Future research in this area could examine the specific processes by which these nations are pushing for change, the degree to which public discourses of transformation reflect on-the-ground realities, and the ways that these movements are represented in the United States and other nations central to the corporate food regime.

Finally, nearly all of the research summarized above takes the food movement itself, and its ability to create positive social change, as its object of study. Future studies may instead choose to examine the food movement as a way to speak to broader concerns about the workings of inequality and neoliberalism. In this recommendation we follow Brown and Getz who argue, "agro-food case studies can contribute to the growing body of literature exploring the unevenness and particularity of neoliberalization processes by considering how broader neoliberal technologies articulate with local circumstances" (2008: 1185). We also follow Minkoff-Zern and her coauthors (2011) who examine how policies restricting land ownership have helped to establish and reinforce racial categories and inequalities, a process they call "agricultural racial formation." Both of these avenues of research view food movements not as an end in themselves, but as elements of broader social processes and movements for justice. Studies in this area can ask not only whether and how food movements can be transformative, but what food movements can teach us about a host of subjects including race, neoliberalism, and the global political economy.

ACKNOWLEDGMENTS

The authors thank the editors of this volume and three anonymous reviewers for their constructive feedback on initial drafts of this article. An earlier version of this article was prepared while the first author was a Social and Behavioral Sciences postdoctoral fellow in the Department of Anthropology at Ohio State University.

TERESA MARIE MARES is assistant professor of anthropology at the University of Vermont and is affiliated with the new Transdisciplinary Research Initiative in Food Systems. She received her PhD in 2010 from the University of Washington in sociocultural anthropology. The title of her dissertation is *We Are Made of Our Food: Latino/a Immigration and the Practices and Politics of Eating.*

ALISON HOPE ALKON is assistant professor of sociology at the University of the Pacific. Her research examines food systems from an environmental justice perspective, focusing on issues of food access, racial identity formation, and inequality. She also uses efforts to create sustainable local food systems as a lens through which to understand the green economy more broadly. She is co-editor of *Cultivating Food Justice: Race, Class and Sustainability* (2011).

REFERENCES

Alkon, Alison. 2008a. "Black, White, and Green: A Study of Urban Farmers' Markets." PhD diss., University of California, Davis.

———. 2008b. "Paradise or Pavement: The Social Constructions of the Environment in Two Urban Farmers' Markets and Their Implications for Environmental Justice and Sustainability." *Local Environment: The Journal of Justice and Sustainability* 13 (3): 271–289.

Alkon, Alison, and Julian Agyeman. 2011. "Introduction: The Food Movement as Polyculture." Pp. 1–21 in *Cultivating Food Justice: Race, Class, and Sustainability*, ed. A. Alkon and J. Agyeman. Cambridge: MIT Press.

Alkon, Alison, and Christie McCullen. 2010. "Whiteness and Farmers' Markets: Performances, Perpetuations … Contestations?" *Antipode* 42 (5): 1–23.

Allen, Patricia. 1999. "Reweaving the Food Security Net: Mediating Entitlement and Entrepreneurship." *Agriculture and Human Values* 16 (2): 117–129.

———. 2004. *Together at the Table: Sustainability and Sustenance in the American Agrifood System.* University Park: Pennsylvania State University Press.

———. 2008. "Mining for Justice in the Food System: Perceptions, Practices, and Possibilities." *Agriculture and Human Values* 25 (2): 157–161.

Allen, Patricia, Margaret FitzSimmons, Michael Goodman, and Keith Warner. 2003. "Shifting Plates in the Agrifood Landscape: The Tectonics of Alternative Agrifood Initiatives in California." *Journal of Rural Studies* 19 (1): 61–75.

Allen, Patricia, and Julie Guthman. 2006. "From 'Old School' to 'Farm to School': Neoliberalization From the Ground Up." *Agriculture and Human Values* 23 (4): 401–415.

Allen, Patricia, and Alice Brooke Wilson. 2008. "Agrifood Inequalities: Globalization and Localization." *Development* 51 (4): 534–540.

Altieri, Miguel. 2000. "Ecological Impacts of Industrial Agriculture and the Possibilities for Truly Sustainable Farming." Pp. 77–92 in *Hungry for Profit,* ed. F. Magdoff, J. B. Foster, and F. H. Buttel. New York: Monthly Review Press.

Alvarez, Mavis, Martin Borque, Fernando Funes, Lucy Martin, Armando Nova, and Peter Rosset. 2006. "Surviving Crisis in Cuba: The Second Agrarian Reform and Sustainable Agriculture." Pp. 225–248 in *Promised Land: Competing Visions of Agrarian Reform,* ed. Peter Rosset, Raj Patel, and Michael Courville. Oakland, CA: Food First Books.

Amador, Miguel, and Manuel Peña. 1991. "Nutrition and Health Issues in Cuba: Strategies for a Developing Country." Food and Nutrition Bulletin 13 (4). http://www.unu.edu/Unupress/food/8F134e/8F134E08.html (accessed 1 July 2009).

Anderson, Molly D., and John T. Cook.1999. "Community Food Security: Practice in Need of Theory?" *Agriculture and Human Values* 16 (2): 141–150.

Andreatta, Susan. 2005. "Urban Connections to Locally Grown Produce." Pp. 117–140 in *Urban Place: Reconnecting With the Natural World,* ed. P. Barlett. Cambridge: MIT Press.

Bellows, Anne C., and Michael W. Hamm. 2003. "International Effects on and Inspiration for Community Food Security Policies and Practices in the USA." *Critical Public Health* 13 (2): 107–123.

Borger, Julian. 2008. "Feed the World? We are Fighting a Losing Battle, UN admits: Huge Budget Deficit Means Millions More Face Starvation." *The Guardian,* February 26, 2008, 18.

Born, Branden, and Mark Purcell. 2006. "Avoiding the Local Trap: Scale and Food Systems in Planning Research." *Journal of Planning Education and Research* 26 (2): 195–207.

Brown, Sandy, and Christy Getz. 2008. "Privatizing Farm Worker Justice: Regulating Labor Through Voluntary Certification and Labeling." *Geoforum* 39 (3): 1184–1196.

Community Food Security Coalition (CFSC). n.d. "What Is Community Food Security?" http://www.foodsecurity.org/views_cfs_faq.html (accessed 24 June 2010).

Costa, Temra. 2010. *Farmer Jane: Women Changing the Way We Eat.* Layton, UT: Gibbs Smith.

Davis, Mike. 2006. *Planet of Slums.* New York: Verso.

DeLind, Laura B. 2002. "Place, Work, and Civic Agriculture: Fields for Cultivation." *Agriculture and Human Values* 19 (3): 217–224.

———. 2006. "Of Bodies, Place, and Culture: Resituating Local Food." *Journal of Agricultural and Environmental Ethics* 19 (2): 121–146.

Desmarais, Annette A. 2007. *La Via Campesina: Globalization and the Power of Peasants.* London: Pluto.

DuPuis, E. Melanie, and David Goodman. 2005. "Should We Go 'Home' to Eat?: Toward a Reflexive Politics of Localism." *Journal of Rural Studies* 21 (3): 359–371.

Elliot, Kim. 2008. "Biofuels and the Food Price Crisis: A Survey of the Issues." Center for Global Development Working Paper No. 151, Washington, DC.

Funes, Fernando, Luis Garcia, Martin Borque, Nilda Perez, and Peter Rosset. 2002. *Sustainable Agriculture and Resistance: Transforming Food Production in Cuba.* New York: Food First Books.

Goldschmidt, Walter. 1978. *As You Sow: Three Studies in the Social Consequences of Agribusiness.* Montclair, NJ: Allanheld, Osmun.

Gottlieb, Robert, and Andrew Fisher.1996. "'First Feed the Face': Environmental Justice and Community Food Security." *Antipode.* 29 (2): 193–203.

Gottlieb, Robert, and Anupama Joshi. 2010. *Food Justice.* Cambridge: MIT Press.

Green, John J., Eleanor M. Green, and Anna M. Kleiner. 2011. "From the Past to the Present: Agricultural Development and Black Farmers in the American South." Pp. 47–64 in *Cultivating Food Justice: Race, Class, and Sustainability,* ed. A. Alkon and J. Agyeman. Cambridge: MIT Press.

Guthman, Julie. 2004. *Agrarian Dreams: The Paradox of Organic Farming in California.* Berkeley: University of California Press.

———. 2008. "Neoliberalism and the Making of Food Politics in California." *Geoforum* 39 (3): 1171–1183.

Hinrichs, C. Clare. 2000. "Embeddedness and Local Food Systems: Notes on two Types of Direct Agricultural Market." *Journal of Rural Studies* 16 (3): 295–303.

Holt-Giménez, Eric. 2006. *Campesino a Campesino: Voices from Latin America's Farmer to Farmer Movement for Sustainable Agriculture.* Oakland, CA: Food First Books.

———. 2009. "From Food Crisis to Food Sovereignty: The Challenge of Social Movements." *Monthly Review* 61 (3): 142–156.

———. 2011. "Food Security, Food Justice, or Food Sovereignty? Crises, Food Movements, and Regime Change." Pp. 309–330 in *Cultivating Food Justice: Race, Class, and Sustainability,* ed. A. Alkon and J. Agyeman. Cambridge: MIT Press.

Holt-Giménez , Eric, and Annie Shattuck. 2011. "Food Crises, Food Regimes and Food Movements: Rumblings of Reform or Tides of Transformation?" *Journal of Peasant Studies* 38 (1): 109–144.

Jarosz, Lucy. 2008. "The City in the County: Growing Alternative Food Networks in Metropolitan Areas." *Journal of Rural Studies* 24 (3): 231–244.

Kloppenburg, Jack Jr., John Hendrickson, and G.W. Stevenson. 1996. "Coming into the Foodshed." *Agriculture and Human Values* 13 (1): 33–42.

Kloppenburg, Jack Jr., Sharon Lezberg, Kathryn De Master, George W. Stevenson, and John Hendrickson. 2000. "Tasting Food, Tasting Sustainability: Defining the Attributes of an Alternative Food System with Competent, Ordinary People." *Human Organization* 59 (2): 177–187.

La Via Campesina. 2003. "What Is Food Sovereignty?" http://www.nyeleni.org/spip.php?article87 (accessed 24 June 2004).

Lappè, Frances Moore. 2009. "The City That Ended Hunger." *Yes Magazine,* February 13, 2009. http://www.yesmagazine.org/issues/food-for-everyone/the-city-that-ended-hunger (accessed 1 December 2010).

Levkoe, Charles Z. 2006. "Learning Democracy through Food Justice Movements." *Agriculture and Human Values* 23 (1): 89–98.

Lyson, Thomas A. 2004. *Civic Agriculture: Reconnecting Farm, Food and Community.* Lebanon: Tufts University Press.

Magdoff, Fred. 2008. "The World Food Crisis: Sources and Solutions." *Monthly Review* 60 (1). http://monthlyreview.org/2008/05/01/the-world-food-crisis-sources-and-solutions (accessed 5 June, 2009).

Mares, Teresa. 2010. "We Are Made of Our Food: Latino/a Immigration and the Practices and Politics of Eating." PhD diss., University of Washington, Seattle.

Mares, Teresa, and Devon Peña. 2011. "Environmental and Food Justice: Toward Local, Slow, and Deep Food Systems." Pp. 197–220 in *Cultivating Food Justice: Race, Class, and Sustainability,* ed. A. Alkon and J. Agyeman. Cambridge: MIT Press.

McCutcheon, Priscilla. 2011. "Community Food Security 'For Us, By Us': The Nation of Islam and the Pan African Orthodox Christian Church." Pp. 177–198 in *Cultivating Food Justice: Race, Class, and Sustainability,* ed. A. Alkon and J. Agyeman. Cambridge: MIT Press.

McMichael, Philip. 2006. "Peasant Prospects in the Neoliberal Age." *New Political Economy* 11 (3): 407–418.

Menser, Michael. 2008. "Transnational Participatory Democracy in Action: The Case of La Via Campesina." *Journal of Social Philosophy* 39 (1): 20–41.

Minkoff-Zern, Laura-Anne, Nancy Peluso, Jennifer Sowerwine, and Christy Getz. 2011. "Race and Regulation: Asian Immigrants in California Agriculture." Pp. 65–86 in *Cultivating Food Justice: Race, Class, and Sustainability,* ed. A. Alkon and J. Agyeman. Cambridge: MIT Press.

Morales, Alfonso. 2011. "Growing Food *and* Justice: Dismantling Racism through Sustainable Food Systems." Pp. 149–176 in *Cultivating Food Justice: Race, Class, and Sustainability,* ed. A. Alkon and J. Agyeman. Cambridge: MIT Press.

Norgaard, Kari Marie, Ron Reed, and Carolina Van Horn. 2011. "A Continuing Legacy: Institutional Racism, Hunger and Nutritional Justice on the Klamath." Pp. 23–46 in *Cultivating Food Justice: Race, Class, and Sustainability,* ed. A. Alkon and J. Agyeman. Cambridge: MIT Press.

Omi, Michael, and Howard Winant. [1986] 1994. *Racial Formation in the United States: From the 1960s to the 1990s.* London: Routledge.

Patel, Raj. 2007. *Stuffed and Starved: The Hidden Battle for the World Food System.* Brooklyn: Melville House.

People's Grocery. n.d. "People's Grocery." http://peoplesgrocery.org/brahm/peoples-grocery/why-we-call-it-food-justice (accessed 1 February 2010).

Pollan, Michael. 2010. "The Food Movement Rising." *New York Review of Books* June 10, 2010. http://www.nybooks.com/articles/archives/2010/jun/10/food-movement-rising/ (accessed 9 May 2011).

Pudup, Mary Beth. 2008. "It Takes a Garden: Cultivating Citizen-Subjects in Organized Garden Projects." *Geoforum* 39 (3): 1228–1240.

Qazi, Joan A., and Theresa L. Selfa. 2005. "The Politics of Building Alternative Agro-Food Networks in the Belly of Agro-Industry." *Food, Culture and Society* 8 (1): 46–72.

Rosset, Peter. 2008. "Food Sovereignty and the Contemporary Food Crisis." *Development* 51 (4): 460–463.

Rosset, Peter, and Martin Borque. 2002. "Lessons of Cuban Resistance." Pp. xiv–xx in *Sustainable Agriculture and Resistance: Transforming Food Production in Cuba,* ed. Funes, Fernando, Luis Garcia, Martin Borque, Nilda Perez, and Peter Rosset. Oakland, CA: Food First Books.

Schiavoni, Christina. 2009. "The Global Struggle for Food Sovereignty: From Nyéléni to New York." *Journal of Peasant Studies* 36 (3): 682–689.

Slocum, Rachel. 2007. "Whiteness, Space, and Alternative Food Practice." *Geoforum* 38 (3): 520–533.

Wekerle, Gerda R. 2004. "Food Justice Movements: Policy, Planning, and Networks." *Journal of Planning Education and Research* 23 (4): 378–386.

Winter, Michael. 2003. "Embeddedness, The New Food Economy, and Defensive Localism." *Journal of Rural Studies* 19 (1): 23–32.

Food Sovereignty

A New Rights Framework for Food and Nature?

Hannah Wittman

ABSTRACT: Food sovereignty, as a critical alternative to the concept of food security, is broadly defined as the right of local peoples to control their own food systems, including markets, ecological resources, food cultures, and production modes. This article reviews the origins of the concept of food sovereignty and its theoretical and methodological development as an alternative approach to food security, building on a growing interdisciplinary literature on food sovereignty in the social and agroecological sciences. Specific elements of food sovereignty examined include food regimes, rights-based and citizenship approaches to food and food sovereignty, and the substantive concerns of advocates for this alternative paradigm, including a new trade regime, agrarian reform, a shift to agroecological production practices, attention to gender relations and equity, and the protection of intellectual and indigenous property rights. The article concludes with an evaluation of community-based perspectives and suggestions for future research on food sovereignty.

KEYWORDS: agrarian citizenship, agroecology, food regimes, food security, food sovereignty, right to food

In 1996, an international coalition of peasant, farmer, rural women's, and indigenous people's movements met in Tlaxcala, Mexico to discuss their common concerns around the effects of an increasingly globalized and concentrated agri-food system on their livelihoods, communities, and ecologies. This coalition was consolidated in 1993 as La Via Campesina, or "peasant way" and now is one of the largest and most vibrant social movements in the world, encompassing more than 148 organizations in sixty-nine countries. At the Tlaxcala meeting, members of La Via Campesina proposed an alternative paradigm called "food sovereignty" as a concept and framework that both challenges the foundations of the current agri-food order and proposes a set of concrete alternatives for both theory and practice (La Via Campesina 1996).[1] Since the Tlaxcala conference, the concept of food sovereignty has garnered increasing attention, first from grassroots social movements and the non-governmental sector, then in policy arenas, notably the 2002 World Food Summit and counter-summit, the NGO/CSO Forum on Food Sovereignty, in Rome. In 2007, an international forum on food sovereignty held in Nyéléni, Mali attended by 500 representatives from eighty countries defined food sovereignty as:

Environment and Society: Advances in Research 2 (2011): 87–105 © Berghahn Books
doi:10.3167/ares.2011.020106

> The right of peoples to healthy and culturally appropriate food produced through ecologically sound and sustainable methods, and their right to define their own food and agriculture systems. It puts the aspirations and needs of those who produce, distribute and consume food at the heart of food systems and policies rather than the demands of markets and corporations. It defends the interests and inclusion of the next generation. It offers a strategy to resist and dismantle the current corporate trade and food regime, and directions for food, farming, pastoral and fisheries systems determined by local producers and users. Food sovereignty prioritizes local and national economies and markets and empowers peasant and family farmer-driven agriculture, artisanal fishing, pastoralist-led grazing, and food production, distribution and consumption based on environmental, social and economic sustainability. Food sovereignty promotes transparent trade that guarantees just incomes to all peoples as well as the rights of consumers to control their food and nutrition. It ensures that the rights to use and manage lands, territories, waters, seeds, livestock and biodiversity are in the hands of those of us who produce food. Food sovereignty implies new social relations free of oppression and inequality between men and women, peoples, racial groups, social and economic classes and generations. (Nyéléni Forum for Food Sovereignty 2007)[2]

Since first articulated in 1996, evaluations of the practice and potential of this emerging concept and its many sub-elements were quick in forthcoming, first emerging from civil society publications (e.g., Bagambanya 1999; Menezes 2001; Roberts 2003). Academic analyses began to appear shortly thereafter, mentioned in articles focused on the increasingly vocal, visible, and globally organized peasant movement acting in resistance to the corporate consolidation of agriculture in a neoliberal food regime (e.g., Borras 2004; Desmarais 2002, 2004; Patel 2005; Patel and McMichael 2004), and comprehensively examined in Desmarais (2007). Monographs focusing specifically on food sovereignty began to appear in the NGO sector in 2005 (Windfuhr and Jonsén 2005), including a multimedia presentation of the practice of food sovereignty (Pimbert 2008), with the first academic volumes appearing in 2009 and 2010 (Perfecto et al. 2009; Wittman et al. 2010).

Current literature on food sovereignty is rapidly expanding on a global scale, especially in grassroots publications from Latin America and Asia, but this article focuses on English-language publications coming from a wide range of academic approaches and disciplines. Authors reviewed here include anthropologists, sociologists, geographers, political scientists, agroecologists, nutritionists, and crop scientists, responsible for an increasing range of case study material from Africa (Bezner Kerr 2010; Msachi et al. 2009; Quaye et al. 2009), Latin America (Altieri 2010; Perfecto et al. 2009; Reardon and Perez 2010; Wittman 2009c), Canada (Wittman et al. 2011), and Asia (Kassam 2009, 2010; Ramdas 2009). Much of the most recent literature takes a world-historical and food-regime approach to understanding the theoretical potential of the food sovereignty concept for agri-food studies and its practical implications for addressing food and environmental crises. Current documentation of food sovereignty initiatives remains difficult to consolidate, however, because it is "based in the experience and knowledge of community groups, small farmers' organizations, and those working directly with them" (Ishii-Eiteman 2009: 691) and thus more likely to be published in local languages, consultant, NGO and donor reports, and MA and PhD theses. In addition, "scientific and academic representation of food sovereignty … remains fragmented among numerous journals that are not widely read beyond the source discipline" (Ishii-Eiteman 2009: 691). This review thus attempts to consolidate knowledge around food sovereignty as an "emergent science" (Kassam 2009), viewing it not as an established paradigm/concept but rather a potential new framework emerging from diverse set of contemporary grassroots production practices and political approaches. This consolidation of knowledge around the potential of food sovereignty is important because its proponents and practitioners—both in theory and practice—challenge conventional wisdom and policy on how to best "feed the world and cool the planet" and emphasize the importance of acknowledging communities of practitioners and indigenous knowledge in this agenda (CLOC-La Via Campesina 2010).[3]

In what follows, I examine the emergent framework of food sovereignty from a food regimes perspective, comparing its ontological and epistemological underpinnings with a "food security" agenda. This is followed by an exploration of rights-based approaches to food and food sovereignty, including the practice of rights and agrarian citizenship within an agrarian moral economy based on food sovereignty. I then review the demands and strategies of the food sovereignty movement: assessing the elements and substantive concerns of advocates for this alternative paradigm, including a new trade regime, agrarian reform, a shift to agroecological production practices, attention to gender relations and equity, and the protection of intellectual and indigenous property rights. I conclude with suggestions for future research on food sovereignty.

A Food Regimes Approach

In the late 1980s, Harriet Friedmann and Phil McMichael proposed the concept of a "food regime" as a constellation or cluster of class and interstate power relations, norms, and institutional rules, and socioecological/geographical specializations that link the global relations of food production and consumption to periods of capital accumulation (Friedmann 1987, 2009; Friedmann and McMichael 1989; McMichael 2009b). A food regimes approach offers an important lens to understand the historical development of agri-food systems and their transition/transformation within political ecologies and economies while also tracing "networks of actants—human, natural, discursive—from below" (Friedmann 2009: 342). Le Heron and Lewis suggest a food regimes approach as a "composite of political and intellectual projects" and a food regime as a category with "generative potential ... that places the multiple dimensions of food—as final product, intermediary, ingredient, nutrient, cultural performance, social relation, human necessity—into a wider recasting of what it means to be human, changing human-biophysical exchange and value creation" (2009: 345–346).

Friedmann and McMichael (1989) identified a first food regime as in force between 1870 and the 1930s, characterized by the transfer of basic grains and livestock from settler colonies to Europe and Great Britain, where they were used as "wage foods." This system of exchange included the consolidation of national agricultural sectors in settler states including the United States, Canada, and Australia (Friedmann and McMichael 1989). The second food regime—from 1950s to 1970s—sent subsidized, surplus food from the US via food aid to postcolonial "'development states"' to extend industrialization and attenuate the threat of communism, while an international division of specialized agricultural supply chains and commodity complexes was developed by agribusiness (McMichael 2009b). Part of this rationalization and shift toward a second food regime was the separation of agrarian components—a move from "agrarian-based policies" towards the emergence of politically differentiated environmental, food, and natural resource-based policies (Bonanno 1991).

Transition from one food regime to another stems from a series of contradictory relations resulting in crisis and transformation to a successor regime (Holt-Giménez and Shattuck 2011; Le Heron and Lewis 2009; McMichael 2009b). As norms and rules over the distribution of power and property are challenged, in many cases by social movements, stable relationships are disrupted and space opens for the emergence of alternative constellations of production and consumption practices, and mechanisms of control over food systems. For example, Pritchard (2009) notes the importance of multilateral negotiations around agricultural trade within the World Trade Organization (WTO) in facilitating the breakdown of the second food regime. When national governments joined the WTO in 1995, they "relinquished their powers to unilaterally set their own food and agricultural policies." WTO requirements caused a major restructuring of food security and rural livelihood programs in developing countries, but "the main effect of bringing agriculture into the WTO was not to reform global agriculture in line with market rationalities, but to aggravate already-existing uneven opportunities in the world food system" (Pritchard 2009: 300), as the effects of subsidy restructuring were felt to a much lesser extent in countries that were part of the Organization for Economic Cooperation and Development (OECD). Increasing resistance by peasants, farmers, and NGOs to the norms and institutional rules of the multilateral trading system were visibly marked by La Via Campesina's demand to "Get Agriculture out of the WTO" and the work of the international coalition Our World Is Not for Sale (OWINFS) during the Doha Round of WTO negotiations, which fell apart in 2008 (Desmarais 2007). Hegemonic patterns of financing, producing, and consuming food were thus challenged, leading to what several authors in the food regime literature express as a global agri-food system in a period of experimentation, tension, and contestation (e.g., Burch and Lawrence 2009; Friedmann 2009).

By the turn of the twenty-first century, this current but theoretically unconsolidated food regime, has been alternately described as "neoliberal" (Pechlaner and Otero 2008, 2010), "corporate" (McMichael 2005, 2009a, 2009b) and "corporate-environmental" (Friedmann 2005), as a product of the divergent tensions within, and resistance to, the ongoing financialization and corporatization of global food networks (Burch and Lawrence 2009). Resistance to a corporate food regime, characterized by what French activist-farmer José Bové has called "food from nowhere" (Bové and Dufour 2000) has most visibly been manifested in the food sovereignty movement, with its emphasis on rights, autonomy, and "food from somewhere" (Dixon and Campbell 2009; Fairbairn 2010; McMichael 2009b; Wittman 2009c). In this regard, although certainly not yet a consolidated food regime, food sovereignty can be regarded as a new, alternative paradigm and driver of change challenging the current food regime, in its efforts to re-embed economic, environmental, and equity-related concerns around agricultural production, consumption, and trade. Table 1 compares and contrasts areas of contested institutional and value relations within the current, corporate, or neoliberal food regime, contradictions that create openings for the potential emergence of a food sovereignty regime.

Food Security, the Right to Food, and Agrarian Citizenship

The food sovereignty approach can be distinguished as an "epistemic shift" in which value relations, approaches to rights, and a shift from an economic to an ecological calculus concurrently challenge the rules and relations of a corporate or neoliberal food regime (McMichael 2009b). One of the most salient shifts has been in the value relations, justification regimes, and frames around the concept of food security as it has been challenged by new conceptualizations of food sovereignty (Boyer 2010; Fairbairn 2010; Gonzalez 2010; Mooney and Hunt 2009; Windfuhr and Jonsén 2005). Food security—framed as a universal ideal to prevent world hunger—emerged as a post–World War II development principle enshrined in the 1948 Universal Declaration of Human Rights and the 1966 International Covenant of Economic, Social and Cultural Rights. As defined by the Food and Agriculture Organization (FAO 2010), food security exists when

Table 1: Food Sovereignty: An Emerging Alternative Food Regime?

	Corporate/Neoliberal Food Regime	**Food Sovereignty Regime**
Feeding the World	Food access/security through intensive production based on principle of comparative advantage and distributed through market mechanisms.	Food access/security through prioritizing local agricultural production and protecting local markets from dumping/subsidized food imports.
Role of Agriculture in Advancing National Development	Increase positive trade balances through increased exports of agricultural commodities. Economically successful communities will invest in infrastructure to improve community well-being (e.g., hospitals, schools).	Sustainable agriculture as part of a diversified economy will improve national well-being through improving food security and ensuring a healthy environment. Fair trade will spur economic growth.
Role of Technology in Advancing Agricultural Development	Increases in productivity come through scientific innovation, adoption of technology, and modern management. Problem solving based on a compartmentalized approach to problems of soil fertility, disease, pest infestation, etc.	Farmers must become efficient and competitive through diversifying production, using alternative technologies, and minimizing use of external inputs. Problem solving is based on a holistic approach to adoption of appropriate technology, including agroecology.
Environmental Stewardship	Protected areas, national parks, and environmental regulations are sufficient, as long as they do not harm the potential for the expansion of agricultural export crops.	Agriculture and environmental policy cannot be separated; sustainable agriculture protects biodiversity and leaves space for conservation areas.

Note: for other comparisons between the food sovereignty model and corporate/ neoliberal/ second food regime paradigms see Desmarais (forthcoming); Rosset (2003); Fairbairn (2010); Reardon and Perez (2010).

"all people, at all times, have physical, social and economic access to sufficient, safe and nutritious food that meets their dietary needs and food preferences for an active and healthy life."[4] As critiqued by several authors writing from a food sovereignty perspective, this definition of food security treats food as a problem of insufficient trade rather than hunger by privileging *access* to food rather than *control over* systems of production and consumption. In this conception, food is a tradable commodity rather than a right, and hunger simply a problem of distribution (McMichael 2004; Patel 2010b).

The way that food security is framed has significant implications for how agricultural and food policy is developed and challenged. For example, Mooney and Hunt (2009) suggest several distinct collective action frames around food security that are in concurrent use, including a *hunger frame*, which, corresponding to a corporate/neoliberal food regime, focuses on food aid and technological development to increase global food production and a *community frame* associated with food sovereignty, which addresses hunger by advocating more localized control over food and agricultural policy (see also Mares and Alkon in this volume). Food sovereignty also pushes an *ethical frame* based on control over and access to food as an element of the confluence of economic, social, cultural, political, and environmental rights (Anderson 2008; Gonzalez 2010). This frame connects food as a human right (a focus of consumer and aid policy) to the right to choose how and by whom that food is produced (a focus of agricultural/national food policies and of early proponents of the food sovereignty framework like La Via Campesina). Referring to a recent report by the United Nations Human Rights Council (UNHRC) on creating a human rights framework for world food and nutrition security (2008), which does not explicitly mention the concept of food sovereignty, Marcia Ishii-Eitemann (2009: 698) argues that "ultimately, the HRCR concludes that the right to food can only be realised where the conditions enabling food sovereignty are guaranteed". She further argues that the rights-based approach that is embedded in food sovereignty is "an explicitly moral enterprise that stands in contrast to the economic processes of market-driven globalization," noting that "this implies a radical shift from the existing hierarchical and increasingly corporate-controlled research system to an approach that devolves more responsibility and decision-making power to farmers, indigenous peoples, food workers, consumers and citizens for the production of social and ecological knowledge" (Dreyfus 2009: 114, cited in Ishii-Eiteman 2009: 691).

The UN-based right-to-food approach has been critiqued for focusing on the individual human right to food, rather than the structural problems of agricultural development, food production, and consumption within the world economic system. For example, Mazhar et al. suggest that

> The affirmation of individual rights to food, while a useful demand in the political environments of the North, has not been incorporated wholeheartedly into the food sovereignty discourse because it does not directly address the right of communities to produce food and to retain command and control over local food systems. From a food sovereignty perspective, a focus on egocentric rights diverts attention from concrete economic and political relations such as corporate control over agricultural inputs and knowledge and economic policies that structure the global food system. It also runs the risk of reducing the issue of hunger and malnutrition to a humanitarian problem for the rich countries to solve, a prospect unacceptable to societies with long and rich agrarian histories." (2007: 65)

Similarly, in her analysis of rights-based approaches to world hunger in international negotiations on the right to food at the intergovernmental level, Jacqueline Mowbray (2007: 561) critiques the 2004 *FAO Voluntary Guidelines to Support the Progressive Realization of the Right to Food in the Context of National Food Security* for focusing on the need for legal, political, and technical reforms at the level of the nation-state, while ignoring effects of the international economic system, and the need for redistributive change. In his analysis of a rights-based approach to food sovereignty, Raj Patel (2007, 2010b) argues that while rights-based challenges may not immediately produce progressive outcomes, they actively work to change social and political contexts and thus are a "means" rather than an end; thus, in the case of food sovereignty, they actually represent a call for a "right to a right." He suggests that the

> mass re-politicization of food politics, through a call for people to figure out for themselves what they want the right to food to mean in their communities, bearing in mind the community's needs, climate, geography, food preferences, social mix, and history ... is *transgressive*, insofar as it orients itself not toward the institutions that enshrine, enforce, and police rights, but toward the people who are meant to hold them. The approach operates not by pointing to extant rights and their violations but by using the language of rights to summon an active politics over a social domain that has, through progressive agricultural liberalization been technicized and rendered "anti-political." (Patel 2007: 91–92; emphasis added)

Although the "elaborate legal architecture" of international rights-based approaches to food may not yet be enforceable or effective in addressing problems of world hunger (Goulet 2009), the food sovereignty framework offers an alternative policy arena in which to discuss the rights and obligations around food production and consumption. In particular, Patel (2010a, 2010b) highlights the importance of understanding the overlapping geographies and jurisdictions in which demands for rights can be exercised at community, regional, national, and international levels, and incorporated into agrarian policy. Above all, food sovereignty proponents demand to participate in decisions and have a voice in establishing food system structures and particular, place-based conceptions of rights (Wittman 2009b, 2009c).

The roles of participation, citizenship, and democracy in a food sovereignty paradigm have been taken up by an increasing number of scholars. Desmarais (2007) provides a detailed critique of the attempts to garner the participation of some farmer organizations as the "fabrication of consent" in international fora such as the WTO, World Bank, and the Global Forum on Agricultural Research. Desmarais argues that at the international level, the spaces for farmer "consultation" can erase differences in political positions and dilute or silence opposition, thus reducing farmer demands to the "lowest common denominator." As such, participation is charged with political and economic consequences (2007: 118–121). Earlier attempts to erase peasant voices from international agricultural and food policy contexts can be contrasted with La Via Campesina's current presence in international policymaking fora and the case of the Citizen Space for Democratic Deliberation (or citizen's juries) in Mali, West Africa, which was organized to allow farmers to make policy recommendations on the adoption of genetically-modified organisms (GMOs) after hearing a range of expert opinions. Farmer jurors were selected to represent the diverse types of farmers in the region (with attention to farm size and gender). The jury voted in January 2006 against introducing GMOs to Mali; the deliberations were broadcast live throughout the region and subsequently delayed the introduction of genetically modified crops to Mali, spurring debates on the technology in the National Assembly (Pimbert et al. 2010).

The concept of agrarian citizenship creates explicit links between struggles for political and ecological rights and practices, bringing the rights of nature into the food sovereignty equation (Wittman 2009b, 2009c, 2010). The agrarian citizenship approach acknowledges a socioecological metabolism as a crucial law of motion in agroecological transformation, in which the advent of capitalism and relationships of unequal ecological exchange commodified nature, separated urban consumers from rural producers, disrupted traditional patterns of nutrient cycling, and contributed to both hunger and environmental degradation (Foster 1999; Marx [1939] 1973; Moore 2010). For example, agrarian citizenship acknowledges the diverse voices of human actors within the food system, but also considers how these voices and practices interact with nature's voice (such as changing weather patterns as a result of climate change), as a "lively" actor that shapes and constrains human activity (Wittman 2009c). Political and ecological voices are actively reshaping food policy and practice, especially in light of the implications of climate change for agricultural systems. By focusing on ecologically sustainable food production and reconnecting producers and consumers via the localization of "food from somewhere," food sovereignty as part of an "agrarian regeneration movement" is increasingly presented as having theoretical potential to rework (Wittman 2009c), repair (Schneider and McMichael 2010), or heal (Clausen 2007) the metabolic rift.

Moving Toward Food Sovereignty: Steps and Substantive Concerns

Trade Liberalization and Alternative Trade Regimes

The food sovereignty movement was born out of concerns about the effects of structural adjustment, trade liberalization, and a shift to an agricultural export orientation on local food economies, communities, and ecologies. Walden Bello and Mara Baviera (2010) summarize a large literature on the role of structural adjustment on weakening agricultural investment and support measures in developing countries, leading to supply constraints and an increase in hunger to over one billion people in 2009. WTO protocols, including the Agreement on Agriculture, prohibit price supports in the Global South while allowing developed countries to maintain key agricultural subsidies, leaving small farmers worldwide unable to "compete in markets where the prices for farm products fell substantially through the decade following implementation of WTO rules" (McMichael 2009a: 287). This period, along with its push to implement export-oriented agricultural systems in developing countries, was accompanied by a "race to the bottom" in terms of environmental and social policies (Rosset 2006). As demonstrated by Marta Rivera-Ferre (2009) in a review of the implementation of industrial, export-oriented shrimp farming in Asia, Latin America, and Africa, large companies benefitted from external financing and bilateral aid, while mangrove forest degradation and land conflicts reduced local food security and generated social exclusion. Similar reviews evaluating the social and environmental effects of expanded soybean production in Brazil (Steward 2007), biofuels expansion (Holt-Giménez and Shattuck 2010; McMichael 2010; Rosset 2009a), and the expanding livestock-grain complex (Jarosz 2009) highlight structural contradictions in the current food regime that set the stage for the emergence of a food sovereignty orientation aimed at localization and diversification of agricultural production and trade. Following the collapse of the Doha Round of the WTO, food sovereignty advocates argue that a subsequent proliferation of bilateral free trade agreements "will only serve to promote unfair trading practices, lock [countries] into even greater trade liberalisation than would be expected under the WTO, and negatively impact the majority of their fellow citizens and the local environment" (Smeltzer 2009: 14; see also Choudry (2007) for a comprehensive statement of grassroots perspectives on FTAs and food sovereignty).

While often perceived as "antitrade," the food sovereignty movement is actually engaged in deep, ongoing conversations about what kinds of trade relations will best serve the social, economic, political, and environmental principles of an alternative food paradigm. Themes and issues within this conversation include affirmative action, intellectual property rights, dumping, social, environmental and labor laws and regulations, labeling, denomination of origin, slow food, fair trade, regional networks, farmer's markets, and community-supported agriculture, many of which do not explicitly link to a food sovereignty framework but express many of the same social and environmental goals.[5] The vast alternative-food networks and local and regional food-supply chains literature emerging from Europe and North America covers these issues in great detail, but only rarely invokes food sovereignty. Conversely, little discussion of these issues is currently present in the extensive literature on trade and food sovereignty, most of which focuses on identifying trade as an obstacle to food sovereignty. A research challenge remains to connect these literatures and provide case-based examinations of alternatives to the dominant trade-liberalization system.

Rethinking Land and Nature: Food Production, Agrarian Reform and Indigenous Knowledge

A series of food crises since 2007 resulting in almost one billion people living in a state of hunger and even more facing malnourishment (FAO 2010) raises important questions for advocates of a food sovereignty framework: In the face of ongoing demographic shifts in food consumption practices, population increases, and the threat of climate change to agricultural productivity (Godfray et al. 2010; Jarosz 2009), what does a shift from a large-scale, export-oriented, and global trade-based system need to look like? How can a system based on small-scale, family-led farming, as food sovereignty's proposed alternative, "feed the world"?[6]

Globally since the 1960s, total area in agriculture has risen by about eleven percent (Pretty 2008), but land concentration (including land-grabbing [Zoomers 2010]), urbanization, and environmental degradation have reduced access to productive land for small-scale farmers in developing countries. The FAO estimates that more than thirty million peasants lost access to land in the decade after the establishment of the WTO (Madeley 2000, cited in McMichael

2009b: 154), following a more extensive period of depeasantization in the second half of the twentieth century (Araghi 1995). Just under half of the world population now lives and works in rural areas, but in most industrialized countries, agricultural producers comprise less than five percent of the population. For food sovereignty proponents, improving control over access to land for an increasing diversity of farmers worldwide is an essential step to implementing food sovereignty and challenging the consolidation of productive land evidenced in the corporate food regime. To this end, a number of recent articles and monographs summarize the case for redistributive land reform, highlighting its importance and implications for food sovereignty (Akram-Lodhi 2007; Borras 2001, 2007, 2008; Patel et al. 2006; Rosset 2009b; Teubal 2009; Wittman 2009b). But Borras and Franco (2010) suggest that the case for land reform is not as straightforward as it may seem—some land policies, including their dynamics, relations, and forms of implementation, are more conducive to food sovereignty than others. The authors argue that redistribution is a matter of degree, according to the extent to which "land-based wealth and power transfers from landed classes or state or community to landless or near-landless working poor" (Borras and Franco 2010: 109–110). This can be contrasted with land reform policies that serve to reconcentrate land, in which "land based wealth and power transfers from the state, community or small family farm holders to landed classes, corporate entities, states, or community groups" (110). This latter concern is increasingly an issue in emerging analyses of the global "land grab" in which corporations and wealthy states secure land in the global south to support their own domestic food supplies (Cotula and Vermeulen 2009; Lippman 2010; Zoomers 2010).

Sustainable agricultural intensification or "producing more food from the same area of land while reducing environmental impacts" (Godfray et al. 2010: 814) is a second area of concern for feeding the world under a food sovereignty framework. In its research on sustainable and ecologically sound production methods, the food sovereignty movement has sparked an increasing interest in agroecology, as a multifunctional approach to food production that incorporates livelihood provision, conservation of biodiversity, and ecosystem function and community well-being. Based on the theory of the inverse relationship between farm size and total agricultural output (Altieri 2009, 2010; Ishii-Eiteman 2009; Rosset 2008, 2009a), proponents of the smallholder, agroecological food production model argue that harnessing social and ecological diversity can lead to higher productivity and resilience to ecological vulnerability and climate change (Kassam 2010; Reardon and Perez 2010; Rosset et al. 2011). Multifunctional agriculture has been shown to produce certain ecological services more efficiently than monocropping systems (Jordan et al. 2007). Further, an important study by Badgley et al. (2007) has shown that organic agricultural production methods—while requiring higher labour inputs—can produce enough food to meet current food needs without expanding the agricultural land base.[7] A recent review of the literature on agroecology and the right to food (United Nations 2010) suggests that small-scale farmers can double food production within a decade in critical regions by using agroecological production methods, and research consistently indicates that agrobiodiversity based on indigenous farmer knowledge contributes to food security (e.g., Rerkasem et al. 2002, cited in Kassam 2009).

In addition to the potential of agroecological practices to reduce greenhouse gas emissions from agriculture and landscape contamination from agrochemicals and improve long-term soil fertility, food sovereignty has also been proposed as a new conservation paradigm involving "wildlife-friendly farming" (Isakson 2009; Mattison and Norris 2005; Zimmerer 2006). This model of agricultural production depends on the maintenance of an agricultural matrix (Perfecto et al. 2009; Vandermeer and Perfecto 2007) comprised of small, agrobiodiverse farms that preserve a variety of complementary agricultural niches (Chappell and LaValle 2011; Kassam 2010).

Advancing agroecology involves not just the promotion and maintenance of human-designed agroecosystems, but the recognition of the importance of preserving and utilizing local and indigenous seed and livestock varieties (Quaye et al. 2009; Ramdas 2009), indigenous food systems (Moncayo Marquez 2009; Morrison 2011), and uncultivated foods (Mazhar et al. 2007; Vazquez-Garcia 2008). These plants and the local knowledge necessary to cultivate them play important role in sustaining indigenous food sovereignty and are among the elements most threatened by encroachment of industrial agriculture. As indicated by the UN Special Rapporteur on the right to food, "agroecology is a knowledge-intensive approach. It requires public policies supporting agricultural research … private companies will not invest time and money in practices that cannot be rewarded by patents" (United Nations 2011). Food sovereignty practice and analysis has thus also focused on the practice of seed sovereignty and control over agricultural knowledge, technology and genetic resources (Borowiak 2004; Ishii-Eiteman 2009; Kloppenburg 2010a, 2010b), with strong theoretical linkages to the rights-based approach to food sovereignty and intellectual property regimes.

Unity in Diversity: Gender, Class, and Ideology

Since its inception, women within La Via Campesina have pushed the movement to address asymmetrical gender relations; this led to substantial structural changes within the movement in 2000 to ensure gender parity for regional and global representation (Desmarais 2004, 2005, 2007; Wiebe 2006). In 2008, La Via Campesina launched a world campaign "For an End to Violence Against Women" (La Via Campesina 2008), rearticulating its focus on gender as an integral component of strategic mobilization around food sovereignty (Desmarais and Hernández Navarro 2009; Martinez-Torres and Rosset 2010). Food sovereignty activists frequently cite that though over half of the world's agriculture work is performed by women (60 to 80 percent in developing countries), women own less than 2 percent of the land (FAO 2006, cited in Patel 2007). But reshaping gender relations is not just about shifting property relations, as Raj Patel (2010) points out, but rather in challenging base inequalities in power that include sexism, patriarchy, racism, and class. For example, significant class and cultural differences exist between "small-scale" farmers in different world regions; bridging these differences has posed a challenge to consolidating a way forward on achieving a food sovereignty regime. However, a continued focus on eradicating class and gender inequality, in addition to resource inequality, forms part of the food sovereignty's frames' particular attention to diversity and difference in the construction of an alternative food regime.

Some authors question, however, to what extent the food sovereignty paradigm offers a "coherent political economy of an alternative global agrarianism" as either a reconfiguration of capitalism or a non-capitalist alternative (Akram-Lodhi 2007). The question of differentiation of political, ideological, and class-based interests, and how these interests and positions affect both the experience of neoliberal globalization and the strategies and tactics to achieve food sovereignty, have thus continued as a topic of academic theorizing (Borras 2010; Borras et al. 2008; Wolford 2010). Many organizations associated with La Via Campesina and other alternative food movements experience internal tensions on how contradictions within the current global food system should be resolved. For example, although the food sovereignty movement as a whole expresses opposition to the use of genetically modified seeds, by 2007, 90 percent of the 13.3 million producers cultivating transgenic crops were small-scale farmers, mostly growing Bt cotton in China and India (ISAAA 2008). Debates about the relative merits of growing GM crops among smallholders reflect a complex set of issues having to do with economic survival, available markets, property rights, values, and politics (Scoones 2008). Finally, Borras (2010) notes that the shifting structures of global political economy have encouraged rural social movements to localize and internationalize at the same time, resulting in the emergence of "polycentric" rural social movements with different visions of development and demands for level of integration within and among local, regional, and international food systems.

As one former member of La Via Campesina's International Coordinating Commission argues,

> One of the principal characteristics of La Via Campesina is its social and cultural cohesion, within a comprehension of diversity. It is not a question of seeking out differences in order to synthesise or explain them. ... We have achieved a sensibility of these diverse cultures, in a common base. This common base is that we understand that the crisis of rural family agriculture is the same all over the world. The causes are the same, whether it be in Wisconsin or São Paulo. The reality is the same, and the same neoliberal, or more plainly, capitalist policies have caused this crisis (author interview with Paul Nicholson, cited in Wittman 2009a: 678).

In a similar vein, at the Nyeleni forum on food sovereignty, participants emphasized that "while it is critical to have a common framework, there is no single path or prescription for achieving food sovereignty. It is the task of individual regions, nations, and communities to determine what food sovereignty means to them based on their own unique set of circumstances" (Schiavoni 2009: 685). Thus, ongoing research into the framework of food sovereignty seeks to gain a richer and deeper understanding of that diversity as an element of niche complementarity leading to strong and resilient local food systems.

Community-Driven Research and Emerging Research Directions

A key challenge of theorizing food sovereignty lies in the problem of understanding the "diversity in unity." Food sovereignty actions and movements are vibrant, regionally, geographically, politically, and practically diverse, making generalizations difficult about what food sovereignty is or will be in a definitional sense. For example, semantic confusion related to the "stacked meanings" of multiple demands and scenarios has been noted by several scholars studying the ways that the terminology and usage of the food sovereignty frame, particularly in contrast to a globally more prevalent food security frame, plays out in political organizing and agrarian change.[8] In addition, the conflation of the human right to food with other goals for reforming the food system, particularly in the Global North, can confuse "means, ends and complementary goals" (Anderson 2008) related to food sovereignty as an element of, rather than framework for, overall food system reform. Thus it is important to note, as Annette Desmarais points out, that "the principles of food sovereignty are not a checklist of separate 'things to do' [but rather] integrative goals of a praxis that plays out differently from one organization, locale, region, country and transnational context to the next" (cited in Boyer 2010: 334).

The politics of scale pose a second theoretical conundrum to conceptualizing food sovereignty. Who 'gets to be sovereign' and who is responsible for conceptualizing and enforcing rights to food policy is an important emerging area of inquiry (Patel 2005). The food sovereignty movement focuses on local and regional autonomy in food system definition, but also depends on the enforcement of trade rules and supportive agricultural policy at the national level (see Scott et al. [2009] for a discussion of the challenges of institutionalizing agroecology in Cuba). These intersections of scale provide for some interesting potential contradictions, especially in areas where food sovereignty is now formally emerging in national constitutions (e.g., Ecuador, Bolivia, and Nepal) and national and municipal-level agricultural policies (e.g., Brazil, Maine).

Finally, the vast majority of research on food sovereignty has been focused on the relations of production, following the rural, agrarian, and productivist orientation of the social movements that gave rise to the concept. Many of these same movements are making or strengthening connections with urban consumers through alternative marketing relationships, including community-supported agriculture projects, farmers' markets, institutional procurement programs, and direct marketing. Yet, as mentioned above, the extant, primarily North American and UK-based literature, on the urban consumer in local food systems development and revitalization almost completely ignores the food sovereignty framework (see Mares and Alkon in this volume). In an important exception, in a case study of debates around sustainable of agriculture in British Columbia, Condon et al. (2010) advance the idea of "municipal enabled agriculture" as an opportunity to integrate issues of supply, the focus of food security, with control in municipal planning processes, as an element of food sovereignty in cities. Likewise, Moncayo (2009) examines the potential of school-based food sovereignty programs in urban Bolivia to address the "nutrition transition," where traditional diets comprised of healthy, indigenous foods have been replaced by the Western diet comprised of "junk food." The community garden movement in New York City has also been referenced as a potential driver of urban food sovereignty (Schiavoni 2009: 687), but much more research is needed in incorporating urban and consumer-driven elements into a food sovereignty framework.

In conclusion, the community-driven nature of food sovereignty conceptualization, practice, and more recently, research, has allowed the transformation of knowledge and ways of knowing in new and important ways. An excellent recent example of how peasant organizations and researchers have worked together to promote, analyze, and advance practices leading to food sovereignty is documented in Rosset et al. (2011), which documents a self-study using a "social process methodology" by La Via Campesina and the Campesino-a-Campesino Agroecology Movement in Cuba that involved identifying successful experiences and agroecological practices and disseminating lessons to other peasant organizations across the globe. Eric Holt-Giménez (2006, 2010) also documents the contemporary development of farmer-to-farmer movements as an important participatory method that acknowledges and prioritizes local and indigenous knowledge as well as local needs, culture, and conditions, rather than "replacing peasant knowledge with purchased chemical inputs, seeds and machinery, in a top-down process where education is more like domestication" (Rosset et al. 2011: 170). By documenting innovative practices and conceptualizations around the way that food, ecology, citizenship, and social organization are connected, communities of social practitioners have led a research agenda that has only recently been "noticed" by university-based communities of inquirers, but one that is sure to expand exponentially in the face of urgent demands for alternative agricultural and food policy models that can address the imminent effects of global climate change.

ACKNOWLEDGMENTS

I thank Annette Desmarais and the anonymous reviewers for helpful comments on an earlier version of this article.

HANNAH WITTMAN is assistant professor of environmental sociology and Latin American studies at Simon Fraser University in Vancouver, Canada. She conducts collaborative research on local food systems, agrarian social movements, and agrarian citizenship with the Landless Rural Workers Movement and La Via Campesina in Brazil and with local networks in British Columbia. She has published articles in the *Journal of Rural Studies, Journal of Peasant Studies,* the *Canadian Journal of Development Studies, Land Degradation and Development and Human Organization,* and is co-editor of *Food Sovereignty: Reconnecting Food, Nature and Community* (2010) and *Food Sovereignty in Canada: Creating Just and Sustainable Food Systems* (2011).

NOTES

1. For comprehensive examinations of La Via Campesina as an international peasant movement and progenitor of the concept of Food Sovereignty, see Desmarais (2002, 2004, 2005, 2007), Martinez-Torres and Rosset (2010), and Borras (2004).
2. There are many related conceptualizations of food sovereignty; the International Planning NGO/CSO Committee for Food Sovereignty definition is: "Food Sovereignty is the right of individuals, communities, peoples and countries to define their own agricultural, labour, fishing, food and land policies, which are ecologically, socially, economically and culturally appropriate to their unique circumstances. It includes the true right to food and to produce food, which means that all people have the right to safe, nutritious and culturally appropriate food and to food-producing resources and the ability to sustain themselves and their societies." See www.foodsovereignty.org.
3. Food sovereignty proponents are increasingly focusing on peasant agriculture as a way to prevent and mitigate further climate change; related strategies and statements proliferate in recent grassroots publications; e.g., La Via Campesina's Declaration in Cancún: The people hold thousands of solutions in their hands; Statement from the People's Movement Assembly on Food Sovereignty; international food summit in Madrid: Time to Change European Agricultural Policy. See www.viacampesina.org.
4. This definition has been in use by the FAO since 2001, as the latest of several modifications of the definition developed in 1974 at the World Food Summit: "availability at all times of adequate world food supplies of basic foodstuffs to sustain a steady expansion of food consumption and to offset fluctuations in production and prices."
5. For a discussion of the goals and objectives of the Slow Food movement, see Petrini (2010); for an extended discussion of the relation of a wide range of food movements with food sovereignty, see Holt-Giménez and Shattuck (2011).
6. It is important to note that current levels of hunger are closely linked to issues of poverty and distribution failure, rather than to a global failure to produce enough calories for human consumption (Sen 1981).
7. Organic agriculture is not a "proxy" for agroecology and food sovereignty; as important studies by Buck, Getz, and Guthman (1997), Hall and Mogyorody (2001), and Guthman (2004) have shown, organic production systems are also susceptible to potentially negative power disparities and environmental impacts in production and marketing.
8. For an excellent discussion of this issue in agrarian social movement organization in Honduras, see Boyer (2010).

REFERENCES

Akram-Lodhi, A. Haroon. 2007. "Land Reform, Rural Social Relations and the Peasantry." *Journal of Agrarian Change* 7 (4): 554–562.

Altieri, Miguel A. 2009. "Agroecology, Small Farms, and Food Sovereignty." *Monthly Review-an Independent Socialist Magazine* 61 (3): 102–113.

———. 2010. "Scaling Up Agroecological Approaches for Food Sovereignty in Latin America." Pp. 120–133 in *Food Sovereignty: Reconnecting Food, Nature and Community*, ed. Hannah Wittman, Annette Aurelie Desmarais, and Nettie Wiebe. Halifax: Fernwood.

Anderson, Molly D. 2008. "Rights-Based Food Systems and the Goals of Food Systems Reform." *Agriculture and Human Values* 25 (4): 593–608.

Araghi, Farshad. 1995. "Global Depeasantization: 1945–1990." *Sociological Quarterly* 36 (2): 337–368.

Badgley, Catherine, Jeremy Moghtader, Eileen Quintero, Emily Zakem, M. Jahi Chappell, Katia Avilés-Vázquez, Andrea Samulon, and Ivette Perfecto. 2007. "Organic Agriculture and the Global Food Supply." *Renewable Agriculture and Food Systems* 22 (2): 86–108.

Bagambanya, Jean Bakole. 1999. "The Food Security Initiative: An Interview with Jean Bakole Bagambanya." *Multinational Monitor* 20 (1/2): 27–29.

Bello, Walden and Mara Baviera. 2010. "Capitalist Agriculture, the Food Price Crisis and Peasant Resistance." Pp. 62–75 in *Food Sovereignty: Reconnecting Food, Nature and Community*, ed. Hannah Wittman, Annette Aurelie Desmarais, and Nettie Wiebe. Halifax: Fernwood.

Bezner Kerr, Rachel. 2010. "Unearthing the Cultural and Material Struggles over Seed in Malawi." Pp. 134–151 in *Food Sovereignty: Reconnecting Food, Nature and Community*, ed. Hannah Wittman, Annette Aurelie Desmarais, and Nettie Wiebe. Halifax: Fernwood.

Bonanno, Alessandro. 1991. "From an Agrarian to an Environmental, Food, and Natural Resource Base for Agricultural Policy: Some Reflections on the Case of the EC." *Rural Sociology* 56 (4): 549–564.

Borowiak, Craig. 2004. "Farmers' Rights: Intellectual Property Regimes and the Struggle over Seeds." *Politics & Society* 32 (4): 511–543.

Borras, Saturnino M. 2001. "State-Society Relations in Land Reform Implementation in the Philippines." *Development and Change* 32 (3): 545–575.

———. 2004. "La Via Campesina: An Evolving Transnational Social Movement." Amsterdam: Transnational Institute.

———. 2007. *Pro-Poor Land Reform: A Critique*. Ottawa: University of Ottawa Press.

———. 2008. "La Via Campesina and its Global Campaign for Agrarian Reform." *Journal of Agrarian Change* 8 (2–3): 258–289.

———. 2010. "The Politics of Transnational Agrarian Movements." *Development and Change* 41 (5): 771–803.

Borras, Saturnino M., Marc Edelman, and Cristóbal Kay. 2008. "Transnational Agrarian Movements: Origins and Politics, Campaigns and Impact." *Journal of Agrarian Change* 8 (2–3): 169–204.

Borras, Saturnino M., and Jennifer C. Franco. 2010. "Food Sovereignty and Redistributive Land Policies: Exploring Linkages, Identifying Challenges." Pp. 106–119 in *Food Sovereignty: Reconnecting Food, Nature and Community*, ed. Hannah Wittman, Annette Aurelie Desmarais, and Nettie Wiebe. Halifax: Fernwood.

Bové, José, and François Dufour. 2000. *The World Is Not for Sale: Farmers against Junk Food*. London: Verso.

Boyer, Jeff. 2010. "Food Security, Food Sovereignty, and Local Challenges for Transnational Agrarian Movements: The Honduras Case." *Journal of Peasant Studies* 37 (2): 319–351.

Buck, Daniel, Christina Getz, and Julie Guthman. 1997. "From Farm to Table: The Organic Vegetable Commodity Chain of Northern California." *Sociologia Ruralis* 37 (1): 3–20.

Burch, David, and Geoffrey Lawrence. 2009. "Towards a Third Food Regime: Behind the Transformation." *Agriculture and Human Values* 26 (4): 267–279.

Chappell, Michael Jahi, and Liliana A. LaValle. 2011. "Food Security and Biodiversity: Can We Have Both? An Agroecological Analysis." *Agriculture and Human Values* 28 (1): 3–26.

Choudry, Aziz. 2007. "Not Under the Same Sky: Bilateral Free Trade Agreements (FTAs), Agriculture and Food Sovereignty." Penang, Malaysia Pesticide Action Network, Asia and the Pacific. http://www.panap.net/en/fs/post/food-sovereignty/136. Access November 15, 2011.

Clausen, Rebecca. 2007. "Healing the Rift: Metabolic Restoration in Cuban Agriculture." *Monthly Review* 51 (1): 40–52.

CLOC-La Via Campesina. 2010. "Statement by the CLOC-Via Campesina on the Climate Summit in Cancun." November 29, 2010. http://www.viacampesina.org/en/index.php?option=com_content&view=article&id=983:statement-by-the-cloc-via-campesina-on-the-climate-summit-in-cancun-&catid=48:-climate-change-and-agrofuels&Itemid=75. Accessed November 15, 2011.

Condon, P. M., K. Mullinix, A. Fallick, and M. Harcourt. 2010. "Agriculture on the Edge: Strategies to Abate Urban Encroachment onto Agricultural Lands by Promoting Viable Human-Scale Agriculture as an integral Element of Urbanization." *International Journal of Agricultural Sustainability* 8 (1–2): 104–115.

Cotula, L., and S. Vermeulen. 2009. "Deal or No Deal: The Outlook for Agricultural Land Investment in Africa." *International Affairs* 85 (6): 1233–1247.

Desmarais, Annette Aurélie. 2002. "The Via Campesina: Consolidating an International Peasant and Farm Movement." *Journal of Peasant Studies* 29 (2): 91–124.

———. 2004. "The Via Campesina: Peasant Women on the Frontiers of Food Sovereignty." *Canadian Woman Studies* 23 (1): 140–145.

———. 2005. "'You Are Mostly Promised You Will Not Be Alone.' Women Farm Leaders Speak About Resistance and Agrarian Activism." *Canadian Woman Studies* 24 (4): 6–11.

———. 2007. *La Vía Campesina: Globalization and the power of peasants.* Halifax: Fernwood.

———. Forthcoming. "Food Sovereignty: A Radical Alternative." in *Critical Perspectives in Food Studies,* ed. Mustafa Koc, Jennifer Sumner, and Tony Winson. Toronto: Oxford University Press.

Desmarais, Annette Aurélie and Luis Hernández Navarro. 2009. "Feeding the world and cooling the planet." Briar Patch 38 (1): 10–13.

Dixon, Jane, and Hugh Campbell. 2009. "Introduction to the special Symposium: Reflecting on Twenty Years of the Food Regimes Approach in Agri-Food Studies." *Agriculture and Human Values* 26 (4): 261–265.

Dreyfus, F. et al. 2009. "Historical Analysis of the Effectiveness of AKST Systems in Promoting Innovation." Pp. 57–144 in *International Assessment of Agricultural Knowledge, Science and Technology for Development: Global Report,* ed. B.D. McIntyre et al. Washington, DC: Island Press.

Fairbairn, Madeleine. 2010. "Framing Resistance: International Food Regimes and the Roots of Food Sovereignty." Pp. 15–31 in *Food Sovereignty: Reconnecting Food, Nature and Community,* ed. Hannah Wittman, Annette Desmarais, and Nettie Wiebe. Halifax: Fernwood.

Food and Agriculture Organization (FAO). 2010. *The State of Food Security in the World 2010: Addressing Food Insecurity in Protracted Crises.* Rome: Food and Agriculture Organization of the United Nations.

Foster, John Bellamy. 1999. "Marx's Theory of Metabolic rift: Classical Foundations for Environmental Sociology." *American Journal of Sociology* 105 (2): 366–405.

Friedmann, Harriet. 1987. "International Regimes of Food and Agriculture since 1870." Pp. 258–276 in *Peasants and Peasant Societies,* ed. Theodor Shanin. Oxford: Basil Blackwell.

———. 2005. "From Colonialism to Green Capitalism: Social Movements and Emergence of Food Regimes." Pp. 227–264 in *Research in Rural Sociology and Development: New Directions in the Sociology of Global Development,* edited by Fred Buttel and Philip McMichael. Amsterdam: Elsevier.

———. 2009. "Discussion: Moving Food Regimes Forward: Reflections on Symposium Essays." *Agriculture and Human Values* 26 (4): 335–344.

Friedmann, Harriet, and Philip McMichael. 1989. "Agriculture and the State System: The Rise and Decline of National Agricultures, 1870 to the Present." *Sociologia Ruralis* 29 (2): 93–117.

Godfray, H. Charles J., John R. Beddington, Ian R. Crute, Lawrence Haddad, David Lawrence, James F. Muir, Jules Pretty, Sherman Robinson, Sandy M. Thomas, and Camilla Toulmin. 2010. "Food Security: The Challenge of Feeding 9 Billion People." *Science* 327 (5967): 812–818.

Gonzalez, Humberto. 2010. "Debates on Food Security and Agrofood World Governance." *International Journal of Food Science and Technology* 45 (7): 1345–1352.

Goulet, Richard. 2009. "Food Sovereignty: A Step Forward in the Realisation of the Right to Food." *Law, Social Justice and Global Development Journal* 1 (13): 1–21.

Guthman, Julie. 2004. *Agrarian Dreams: The Paradox of Organic Farming in California.* Berkeley: University of California Press.

Hall, Alan, and Veronika Mogyorody. 2001. "Organic Farmers in Ontario: An Examination of the Conventionalization Argument." *Sociologia Ruralis* 41 (4): 399–422.

Holt-Giménez, Eric. 2006. *Campesino a Campesino: Voices from Latin America's Farmer to Farmer Movement for Sustainable Agriculture.* Oakland, CA: Food First Books.

———. 2010. "Grassroots Voices: Linking Farmers' Movements for Advocacy and Practice." *Journal of Peasant Studies* 37 (1): 203–236.

Holt-Giménez, Eric, and Annie Shattuck. 2010. "Agrofuels and Food Sovereignty: Another Agrarian Transition." Pp 76–90 in *Food Sovereignty: Reconnecting Food, Nature and Community,* ed. Hannah Wittman, Annette Aurelie Desmarais, and Nettie Wiebe. Halifax: Fernwood.

———. 2011. "Food Crises, food Regimes and Food Movements: Rumblings of Reform or Tides of Transformation?" *Journal of Peasant Studies* 38 (1): 109–144.

International Service for the Acquisition of Agri-Biotech Applications (ISAAA). 2008. "Global Status of Commercialized Biotech/GM Crops: 2008, the First Thirteen Years, 1996 to 2008." Brief 39-2008.

Isakson, S. R. 2009. "No hay ganancia en la milpa: The Agrarian Question, Food Sovereignty, and the On-Farm Conservation of Agrobiodiversity in the Guatemalan Highlands." *Journal of Peasant Studies* 36 (4): 725–759.

Ishii-Eiteman, Marcia. 2009. "Food Sovereignty and the International Assessment of Agricultural Knowledge, Science and Technology for Development." *Journal of Peasant Studies* 36 (3): 689–700.

Jarosz, Lucy. 2009. "Energy, Climate Change, Meat, and Markets: Mapping the Coordinates of the Current World Food Crisis." *Geography Compass* 3 (6): 2065–2083.

Jordan, N., G. Boody, W. Broussard, J. D. Glover, D. Keeney, B. H. McCown, G. McIsaac, M. Muller, H. Murray, J. Neal, C. Pansing, R. E. Turner, K. Warner, and D. Wyse. 2007. "Sustainable Development of the Agricultural Bio-Economy." *Science* 316 (5831): 1570–1571.

Kassam, KA. 2009. "Viewing Change through the Prism of Indigenous Human Ecology: Findings from the Afghan and Tajik Pamirs." *Human Ecology* 37 (6): 677–690.

———. 2010. "Pluralism, Resilience, and the Ecology of Survival: Case Studies from the Pamir Mountains of Afghanistan." *Ecology And Society* 15 (2): 1–20.

Kloppenburg, Jack. 2010a. "Impeding Dispossession, Enabling Repossession: Biological Open Source and the Recovery of Seed Sovereignty." *Journal of Agrarian Change* 10 (3): 367–388.

———. 2010b. "Seed Sovereignty: The Promise of Open Source Biology." Pp. 152–167 in *Food Sovereignty: Reconnecting Food, Nature and Community,* ed. Hannah Wittman, Annette Aurelie Desmarais, and Nettie Wiebe. Halifax: Fernwood.

La Via Campesina. 1996. "Tlaxcala Declaration of the Via Campesina II International Conference of the Via Campesina." Tlaxcala, Mexico, 18–21 April.

———. 2008. "Declaration of Maputo: V International Conference of La Via Campesina: Food Sovereignty Now! Unity and Struggle of the People!" Maputo, Mozambique. http://viacampesina.org/en/index.php?option=com_content&view=article&id=623:declaration-of-maputo-v-international-conference-of-la-via-campesina&catid=47:declarations&Itemid=70. Accessed November 15, 2011.

Le Heron, R., and N. Lewis. 2009. "Discussion. Theorising food regimes: intervention as politics." *Agriculture and Human Values* 26 (4): 345–349.

Lippman, Thomas W. 2010. "Saudi Arabia's Quest for 'Food Security.'" *Middle East Policy* 17 (1): 90–98.

Madeley, John. 2000. *Hungry for Trade: How the Poor Pay for Free Trade.* New York: Zed Books.

Martinez-Torres, Maria Elena, and Peter M. Rosset. 2010. "La Via Campesina: The Birth and Evolution of a Transnational Social Movement." *Journal of Peasant Studies* 37 (1): 149–175.

Marx, Karl. [1939] 1973. *Grundrisse: Foundations of the Critique of Political Economy.* New York: Vintage Books.

Mattison, E. H. A., and K. Norris. 2005. "Bridging the Gaps between Agricultural Policy, Land-Use and Biodiversity." *Trends In Ecology & Evolution* 20 (11): 610–616.

Mazhar, Farhad, Daniel Buckles, P.V Satheesh, and Farida Akhter. 2007. *Food Sovereignty and Uncultivated Biodiversity in South Asia: Essays on the Poverty of Food Policy and the Wealth of the Social Landscape.*" Ottawa, ON: International Development Research Centre.

McMichael, Philip. 2004. *Development and Social Change: A Global Perspective.* Thousand Oaks, CA: Pine Forge Press.

———. 2005. "Global Development and the Corporate Food Regime." Pp. 265–300 in *New Directions in the Sociology of Global Development,* ed. F.H. Buttel and P. McMichael. Amsterdam: Elsevier.

———. 2009a. "A Food Regime Analysis of the 'World Food Crisis.'" *Agriculture and Human Values* 26 (4): 281–295.

———. 2009b. "A Food Regime Genealogy." *Journal of Peasant Studies* 36 (1): 139–169.

———. 2010. "Agrofuels in the Food Regime." *Journal of Peasant Studies* 37 (4): 609–629.

Menezes, Francisco. 2001. "Food Sovereignty: A Vital Requirement for Food Security in the Context of Globalization." *Development* 44 (4): 29–33.

Moncayo Marquez, Luis Carlos 2009. "Advancing towards Food Sovereignty in El Alto, Bolivia: Revitalizing the Consumption of Native, Nutritious and Agroecological food in Urban Centers." MA thesis, Simon Fraser University, Vancouver, BC.

Mooney, P. H., and S. A. Hunt. 2009. "Food Security: The Elaboration of Contested Claims to a Consensus Frame." *Rural Sociology* 74 (4): 469–497.

Moore, Jason W. 2010. "The End of the Road? Agricultural Revolutions in the Capitalist World-Ecology, 1450–2010." *Journal of Agrarian Change* 10 (3): 389–414.

Morrison, Dawn. 2011. "Indigenous Food Sovereignty: A Model for Social Learning." Pp. 97–113 in *Food Sovereignty in Canada: Creating Just and Sustainable Food Systems,* ed. Hannah Wittman, Annette Aurelie Desmarais, and Nettie Wiebe. Halifax: Fernwood.

Mowbray, Jacqueline. 2007. "The Right to Food and the International Economic System: An Assessment of the Rights-Based Approach to the Problem of World Hunger." *Leiden Journal of International Law* 20 (3): 545–569.

Msachi, R., L. Dakishoni, and R. B. Kerr. 2009. "Soils, Food and Healthy Communities: Working towards Food Sovereignty in Malawi." *Journal of Peasant Studies* 36 (3): 700–706.

Nyéléni Forum for Food Sovereignty. 2007. Declaration of the Forum for Food Sovereignty. Selingué, Mali, 23–27 February.

Patel, Raj. 2005. "Global Fascism Revolutionary Humanism and the Ethics of Food Sovereignty." *Development* 48 (2): 79–83.

———. 2007. "Transgressing Rights: La Via Campesina's Call for Food Sovereignty." *Feminist Economics* 13 (1): 87–93.

———. 2010a. *The Value of Nothing: How to reshape Market Society and Reclaim Democracy.* New York: Picador.

———. 2010b. "What Does Food Sovereignty Look Like?" Pp. 186–196 in *Food Sovereignty: Reconnecting Food, Nature and Community,* ed. Hannah Wittman, Annette Aurelie Desmarais, and Nettie Wiebe. Halifax: Fernwood.

Patel, Raj, Michael Courville, and Peter Rosset. 2006. *Promised Land: Competing Visions of Agrarian Reform.* Oakland, CA: Food First Books/Land Action Research Network.

Patel, Raj, and Philip McMichael. 2004. "Third Worldism and the Lineages of Global Fascism: The Regrouping of the Global South in the Neoliberal Era." *Third World Quarterly* 25 (1): 231–254.

Pechlaner, Gabriela, and Gerardo Otero. 2008. "The Third Food Regime: Neoliberal Globalism and Agricultural Biotechnology in North America." *Sociologia Ruralis* 48 (4): 351–370.

———. 2010. "The Neoliberal Food Regime: Neoregulation and the New Division of Labor in North America." *Rural Sociology* 75 (2): 179–208.

Perfecto, Ivette, John Vandermeer, and Angus Wright. 2009. *Nature's Matrix: Linking Agriculture, Conservation and Food Sovereignty.* London: Earthscan.

Petrini, Carlo. 2010. "Slow Food: An Interview with Carlo Petrini." *Development* 53 (3): 447–448

Pimbert, Michael. 2008. *Towards Food Sovereignty: Reclaiming Autonomous Food Systems.* London: IIED.

Pimbert, Michael, Boukary Barry, Anne Berson, and Khanh Tran-Thanh. 2010. *Democratising Agricultural Research for Food Sovereignty in West Africa.*" Bamako and London: IIED, CNOP, Centre Djoliba, IRPAD, Kene Conseils, URTEL.

Pretty, Jules. 2008. "Agricultural Sustainability: Concepts, Principles and Evidence." *Philosophical Transactions of the Royal Society B: Biological Sciences* 363 (1491): 447–465.

Pritchard, Bill. 2009. "The Long Hangover from the Second Food Regime: A World-Historical Interpretation of the Collapse of the WTO Doha Round." *Agriculture and Human Values* 26 (4): 297–307.

Quaye, W., K. Adofo, Y. E. Madode, and A. R. Abizari. 2009. "Exploratory and Multidisciplinary Survey of the Cowpea Network in Tolon-Kumbungu District of Ghana: A Food Sovereignty Perspective." *African Journal of Agricultural Research* 4 (4): 311–320.

Ramdas, S. R. 2009. "Reclaiming Endangered Livelihoods: Untold Stories of Indigenous Women and Backyard Poultry." *World's Poultry Science Journal* 65 (2): 241–249.

Reardon, J. A. S., and R. A. Perez. 2010. "Agroecology and the Development of Indicators of Food Sovereignty in Cuban Food Systems." *Journal of Sustainable Agriculture* 34 (8): 907–922.

Rivera-Ferre, Marta. 2009. "Can Export-Oriented Aquaculture in Developing Countries be Sustainable and Promote Sustainable Development? The Shrimp Case." *Journal of Agricultural & Environmental Ethics* 22 (4): 301–321.

Roberts, Wayne. 2003. "The Whole Enchilada." *Alternatives Journal* 29 (4): 3.

Rosset, Peter. 2003. "Food Sovereignty: Global Rallying Cry of Farmer Movements." *Institute for Food and Development Policy Backgrounder* 9 (4).

———. 2006. *Food Is Different: Why We Must Get The WTO Out of Agriculture.* London: Zed.

———. 2008. "Food Sovereignty and the Contemporary Food Crisis." *Development* 51 (4): 460–463.

———. 2009a. "Agrofuels, Food Sovereignty, and the Contemporary Food Crisis." *Bulletin of Science, Technology & Society* 29 (3): 189–193.

———. 2009b. "Fixing our Global Food System Food Sovereignty and Redistributive Land Reform." *Monthly Review* 61 (3): 114–128.

Rosset, Peter Michael, Braulio Machín Sosa, Adilén María Roque Jaime, and Dana Rocío Avila Lozano. 2011. "The Campesino-to-Campesino Agroecology Movement of ANAP in Cuba: Social Process Methodology in the Construction of Sustainable Peasant Agriculture and Food Sovereignty." *Journal of Peasant Studies* 38 (1): 161–191.

Schiavoni, C. 2009. "The Global Struggle for Food Sovereignty: From Nyeleni to New York." *Journal of Peasant Studies* 36 (3): 682–689.

Schneider, Mindi, and Philip McMichael. 2010. "Deepening, and Repairing, the Metabolic Rift." *Journal of Peasant Studies* 37 (3): 461–484.

Scoones, I. 2008. "Mobilizing against GM Crops in India, South Africa and Brazil." *Journal of Agrarian Change* 8 (2–3): 315–344.

Scott, Steffanie, Erin Nelson, Ángel Leyva Galán, and Judie Cukier. 2009. "Institutionalizing Agroecology: Successes and Challenges in Cuba." *Agriculture and Human Values* 26 (3): 233–243.

Sen, Amartya Kumar. 1981. *Poverty and Famines: An Essay on Entitlement and Deprivation.* Oxford: Oxford University Press.

Smeltzer, S. 2009. "A Malaysia-United States Free Trade Agreement: Malaysian Media and Domestic Resistance." *Asia Pacific Viewpoint* 50 (1): 13–23.

Steward, Corrina. 2007. "From Colonization to 'environmental soy': A Case Study of Environmental and Socio-Economic Valuation in the Amazon Soy Frontier." *Agriculture and Human Values* 24 (1): 107–122.

Teubal, Miguel. 2009. "Agrarian Reform and Social Movements in the Age of Globalization Latin America at the Dawn of the Twenty-first Century." *Latin American Perspectives* 36 (4): 9–20.

United Nations. 2010. "Agroecology and the Right to Food." Report submitted by the Special Rapporteur on the right to food, Olivier De Schutter A/HRC/16/49, Geneva.

United Nations Human Rights Council (UNHRC). 2008. "Building Resilience : A Human Rights Framework for World Food and Nutrition Security: Report of the Special Rapporteur on the Right to Food, Olivier De Schutter." A/HRC/9/23, 8 September 2008, Geneva.

———. 2011. "Eco-Farming Can Double Food Production in 10 Years, Says New UN Report." United Nations Office of the High Commissioner for Human Rights, GenevaPress Release, 8 March.

Vandermeer, John, and Ivette Perfecto. 2007. "The Agricultural Matrix and a Future Paradigm for Conservation." *Conservation Biology* 21 (1): 274–277.

Vazquez-Garcia, Veronica. 2008. "Gender, Ethnicity, and Economic Status in Plant Management: Uncultivated Edible Plants among the Nahuas and Popolucas of Veracruz, Mexico." *Agriculture and Human Values* 25 (1): 65–77.

Wiebe, Nettie. 2006. "Women Reversing Desertification: Via Campesina Takes on Aracruz Corporation in Brazil." Canadian Woman Studies 25 (3/4): 167–172.

Windfuhr, Michael, and Jennie Jonsén. 2005. *Food Sovereignty: Towards Democracy in Localized Food Systems.* Bourton-on-Dunsmore, UK: FIAN-International; ITDG Publishing.

Wittman, Hannah. 2009a. "Interview: Paul Nicholson, La Vía Campesina." *Journal of Peasant Studies* 36 (3): 676–682.

———. 2009b. "Reframing Agrarian Citizenship: Land, Life and Power in Brazil." *Journal of Rural Studies* 25 (1): 120–130.

———. 2009c. "Reworking the Metabolic Rift: La Vía Campesina, Agrarian Citizenship, and Food Sovereignty." *Journal of Peasant Studies* 36 (4): 805–826.

———. 2010. "Agrarian Reform and the Environment: Fostering Ecological Citizenship in Mato Grosso, Brazil." *Canadian Journal of Development Studies* 29 (3): 281–298.

Wittman, Hannah, Annette Aurélie Desmarais, and Nettie Wiebe, eds. 2010. *Food Sovereignty: Reconnecting Food, Nature and Community.* Halifax: Fernwood.

———. 2011. *Food Sovereignty in Canada: Creating Just and Sustainable Food Systems.* Halifax: Fernwood.

Wolford, Wendy. 2010. *This Land Is Ours Now: Social Mobilization and the Meanings of Land in Brazil.* Durham: Duke University Press.

Zimmerer, Karl. 2006. "An Expanding Interface with Agriculture will change Global Conservation." Pp. 26–33 in *Agroecology and the Struggle for Food Sovereignty in the Americas*, ed. Avery Cohn, Jonathan Cook, Margarita Fernández, Kathleen McAfee, Rebecca Reider, and Corrina Steward. London; Tehran; New Haven: International Institute for Environment and Development (IIED), the Yale School of Forestry and Environmental Studies (Yale F&ES), and the IUCN Commission on Environmental, Economic and Social Policy (CEESP).

Zoomers, Annelies. 2010. "Globalisation and the Foreignisation of space: Seven Processes Driving the Current Global Land Grab." *Journal of Peasant Studies* 37 (2): 429–447.

Shared Meals and Food Fights

Geographical Indications, Rural Development, and the Environment

Fabio Parasecoli and Aya Tasaki

ABSTRACT: The article highlights relevant issues within the global debate on geographical indications, as they relate to food products. Geographical indications, a form of intellectual property designated by considering principally the place of origin of products, have become a hot topic among producers, activists, economists, and politicians worldwide. Commercial and legal issues related to them have generated complex negotiations in international organizations and national institutions, while their cultural aspects have stimulated theoretical debates about the impact of global trade on local identities. Geographical indications could become a valid tool to implement community-based, sustainable, and quality-oriented agriculture, depending on the sociopolitical environment and whether they are relevant for the producers involved, affordable in terms of administrative and management costs, and applicable on different scales of production. The article also explores the environmental impact of geographical indications and their potential in ensuring the livelihood of rural communities in emerging economies and promoting sustainable agricultural models.

KEYWORDS: biodiversity, cultural heritage, geographical indications, global trade, intellectual property, indigenous knowledge, rural development

In the past decades, markets worldwide have witnessed the development of categories, referred to as value-based food, that define products not according to flavor or biochemical characteristics, but rather their marketing, cultural, and political connotations. Among these, we can mention "local," "fair trade" "sustainable" and, from a certain point of view, "organic" (Barham 2002, 2003). This article focuses on one of such value-based categories—geographical indication. This is a form of intellectual property that, when referring to food products, designates them principally by their places of origin, where origin is considered to be the main factor behind the products' intrinsic quality and reputation. We examine this specific intellectual property category because so far it is the only one to be legally defined at the international level, namely in the 1994 World Trade Organization's Agreement on Trade-Related Aspects of Intellectual Property Rights (TRIPS), which mainly covers issues related to copyright, trademark, and patent. However, despite the legal recognition, the cultural elements that are often deemed intrinsic to geographical indications—local customs, traditions, and artisanal know-how—are

Environment and Society: Advances in Research 2 (2011): 106–123 © Berghahn Books
doi:10.3167/ares.2011.020107

quite difficult to define in terms of individual property rights, as they are generally considered to be common goods. As we demonstrate, the tension between the existing legal categories of intellectual property and geographical indications is generating juridical controversies, political negotiations, and theoretical debates at the global trade level.

Currently, the literature on geographical indications is quite fragmented among disciplines and practices that seldom dialogue among each other. Covered mostly in discussions about economics, development, trade (especially international trade), and the law (in particular about intellectual property), geographical indications have become objects of interest in rural sociology, political science, and anthropology, due to their multifaceted connections to food, identities, and place. Although the potential impact of geographical indications on environment and sustainability, and their relation to indigenous knowledge, have been somewhat explored, scholarship specifically addressing these issues remains scarce.

Due to the extent and complexity of the debates surrounding geographical indications, this article does not aim to be comprehensive or to provide a definitive evaluation about the topic. While reviewing the current state of the research and highlighting a number of relevant legal, political, and economic issues, this article also identifies areas for future investigation, especially about the possible impact of geographical indications on the environment. Because of the connection of geographical indications to contemporary economic, political, and trade-related issues, this article does not exclusively focus on scholarly literature, but also takes into consideration material produced by administrative entities, international institutions, private business, and marketing organizations.

Geographical Indications: Past and Present

The concept of geographical indications and related trade and juridical controversies have received attention in scholarly literature from the legal point of view, generating a great number of articles, documents, and a few full-length books (Echols 2008; Giovannucci et al. 2009; Heath and Sanders 2005; Kongolo 2008). As we show, their vast majority consists of either examinations of the category from the legal point of view and its possible impact on global trade, or case studies analyzing particular occurrences. Due to the relative novelty of the topic, it might be useful to offer a brief overview of what geographical indications are, their history, and the circumstances of their growing relevance.

The concept of geographical indication has a long history in Europe. It originates from the classification that France established in 1855, ranking sixty wine makers (*château*) from the Bordeaux area in five growths (*crus*) on both their quality and the price they commanded (Colman 2008; Stanziani 2004). The Appellation d'Origine Contrôlée (AOC; Controlled Denomination of Origin) as a legal category was officially established in 1919 by the Law for the Protection of the Place of Origin, which led in 1935 to the foundation of the National Institute for Controlled Denominations (Institut National des Appellations d'Origine) with the goal of managing the system. The first AOC was Côtes du Rhône, approved in 1937, for the Rhône wine region. Each wine-producing area was entitled to create rules to discipline its viticulture (grape varieties that could be used, their proportion in the allowed blends, the aging methods, and so on), within general guidelines imposed by the central authority. Wine makers had to meet specific requirements in order to receive the coveted AOC denomination, which was perceived as a sign of higher quality. The system paid off, with consumers ready to pay more for wines that had received some sort of recognition from the state. The concept of AOC was particularly important at a time when urban consumers were increasingly removed from food production, a situation that made them victims of frequent frauds, provoking anxieties about quality and safety. It provided reassurance against the modern anonymity of science and technology (Atkins et al. 2007; Stanziani 2006).

AOCs, conceptual ancestors of what are now geographical indications, were protected within the framework of national laws and limited in effect to the state territory. However, with the development of international commerce in the second half of the nineteenth century, it became clear that control was needed outside the national borders. Efforts were made to create forms of international protection, starting with the 1883 Paris Convention on Intellectual Property and the 1891 Madrid Agreement for the Repression of False or Deceptive Indications of Source of Goods. The next step was the 1958 Treaty of Lisbon, which further defined the protection of the appellation of origin to include cases in which "the appellation is used in translated form or accompanied by terms such as 'kind,' 'type,' 'make,' 'imitation' or the like" (art. 6). In 1989 a Protocol Concerning the International Registration of Marks was signed to allow protection of certain denominations and appellations as trademarks, followed by a formal agreement in 1991. According to the protocol, an international registration made with the International Bureau of the World Intellectual Property Organization (WIPO), one of the specialized agencies of the United Nations, produces the same effects as an application for registration of the mark made in each of the countries designated by the applicant. The protocol also made it possible to record necessary changes and to renew registration through the same office.

By the late 1980s many European states had already set up national systems for food registration and classification. Therefore when the European Union (EU) started issuing regulations on this matter, the existing quality denominations had to be acknowledged and coordinated by the EU authorities, and rules had to be set on how to establish new ones. Every country rushed to have as many protected products as possible, making negotiations difficult and long. Finally in 1992 the EU issued regulation 2081, which allowed the registration of products under three categories: the PDO (Protected Designation of Origin), the PGI (Protected Geographical Indication), and the TSG (Traditional Specialty Guaranteed). Each category allowed different connections with their place of origin, traditions, and methods of production, with PDO being the most binding category and TSG the loosest.

The new European system accelerated the international process of recognition and systematization of the legal concept of geographical indication. In 1994, the TRIPS agreement, one of the founding treaties of the newly founded World Trade Organization (WTO), codified the term geographical indications (besides regulating copyrights, trademarks, and patents). Article 22 of the TRIPS agreement states that "Geographical indications are, for the purposes of this Agreement, indications which identify a good as originating in the territory of a Member, or a region or locality in that territory, where a given quality, reputation or other characteristic of the good is essentially attributable to its geographical origin" (WTO 1994).

In this legal definition, reference to tradition, history, culture, or any other human factor is conspicuously absent, and the text does not specify who is entitled to enjoy the intellectual property rights of the good. Conversely, the insistence on the uniqueness of products connected to specific places runs against the free, smooth, and predictable flow of goods that is one of the explicit goals of the WTO. Due to this intrinsic tension, WIPO acknowledged the unusual characteristics of geographical indications within global trade negotiations, examining their functioning as specific cases of intellectual property (Broude 2005b).

Multilateral and regional agreements are important to the international protection of geographical indications, which vary according to each country's legal framework and safeguard system. Though it is relatively easy to obtain protection in countries that have some form of legislation on this specific category of intellectual property, it is more complicated when a country only recognizes general trademarks or collective marks, as is the case with the United States. The status of geographical indication is one of the hot issues of the Doha Development Agenda (Doha Round) of the WTO multilateral trade negotiations, which has brought agriculture to the forefront around issues of global market accessibility, technical barriers to trade, export subsidies, domestic support to agriculture, and implementation of the WTO agreements by developing economies. The EU, which has linked geographical indications to the reform of its common agricultural policy and to the protection of its farmers from price-based competition, is pushing to establish an internationally recognized geographical indication system similar to its own. The adoption of a multilateral register would extend the protection granted to wines and spirits by article 23 of TRIPS to all products, prohibiting the use of translated geographical indication names and even expressions such as "style" or "type" after the geographical indication denomination (Lang 2006). However, a group of countries (including the United States, Australia, New Zealand, and several Latin American and Asian countries from the Cairns Group of agricultural exporting countries) proposes instead a voluntary system based on the creation of an international database (WIPO 2007; WTO 2010). These countries seem determined to uphold the primacy of trademarks in global trade. In January 2011, for the first time in thirteen years of negotiations, negotiators started working on a draft for a multilateral geographical indications register for wines and spirits, which could potentially be used in the future as the model for an international register for all kinds of goods (WTO 2011a). Shortly after, in April 2011, a report on the issues related to the extension of the provisions in article 23 of TRIPS to other products was issued, expressing all of the opposing views. Many developing countries have continued to voice concerns with regard to the special level of protection for wines and spirits, that they disproportionately benefit the more developed countries rather than those whose main products are agriculture and textile. There also exist strong opinions that an expanded geographical indication protection beyond wines and spirits would negatively affect some export goods of developing countries (WTO 2011b).

The debate is both economic and cultural since geographical indications, as a form of collective intellectual property, challenge the law and spirit of US companies oriented toward individual ownership (Daviron and Ponte 2005: 37–43). In the European system geographical indications do not belong to individual producers but to producers' associations, regulated by public regulations at the regional, national, and EU levels. This approach goes against both the practice and the legal framework of American business, which considers brands as intellectual property protected under a trade name. Trademarks belong to individuals or private companies, and can be bought and sold as a business asset. If violated, it is up to the natural or legal persons to defend their rights to the name before a court of law. The legal contrast between trademarks, WTO geographical indications and the stricter EU system, together with the very disparate ways WTO members have implemented the commitment to protect national and foreign geographical indications, has elicited growing interest among legal experts in international regulations on trade and commerce (Babcock and Clemens 2004; Creditt 2009; Doster 2006; Echols 2003; Gervais 2010; Hughes 2006; Ilbert and Petit 2009; Kireeva and O'Connor 2010: Wattanapruttipaisan 2009).

Because the principle "first in time, first in right" can be applied to trademarks, European producers of geographical indication goods often have to deal with trademarks registered in other countries that contain their products' names. Another problem is that most trademark laws prohibit the registration of geographical names because they consider them merely as an indication of the place of origin, a description, or even a generic name (e.g., Chablis in the US). In those cases geographical indication names have to be protected as a collective or a certification mark, when such legal concepts exist, or the producers have to obtain limited safeguard simply for their logo through the registration of the visual trademark (O'Connor 2004). The trademark registration often does not cover translations, the use of the geographical indication with the so-called delocalisers (i.e., "Wisconsin Parmesan"), and tags such as "like," "style," and so on.

Geographical Indications and Development

The establishment of geographical indications as a possible tool for the development of rural communities is stimulating reaction not only in academia, but also among political and economic decision makers in local, national, and international institutions, as reflected in a wide variety of literature. The commercial and cultural relevance of the perceived connections between specific foods and their place of origin has led to the debate on whether geographical indications can help in implementing just and innovative forms of community-based, quality-oriented agriculture, or whether they are an expression of agrarian utopias based on romanticized ideas of the past that in fact contribute to reproduce local and global inequalities.

By banking on differentiation and the exclusive rights of benefiting from the products' reputation of origin, food manufacturers and growers in Europe are fully aware that geographical indications have the potential to increase the value of their goods, avoid oversupply, protect them from competitors selling similar products at lower prices under the same name, and create entry barriers for producers who do not have the means to comply with the often complex regulations. Geographical indication products have been discussed as "club goods"; a subtype of public goods, such as cable TV or social clubs, which are excludable but nonrivalrous in that they might be expensive to access but can equally be enjoyed by all users at least until they reach a point where congestion occurs and excessive demand may turn them into competitive goods (Langinier and Babcock 2006; Torre 2002).

Geographical indications are supposed to partly make up for the information asymmetries, the misinformation, and the high search costs that are often referred to as possible failures in the neoclassic theoretical model of perfect competition markets. Their clear regulations are meant to make both consumers and administrative authorities responsible for food systems and safety feel more protected from fraud. However, it can also be argued that the respect for production protocols does not inherently guarantee sensory excellence, thus keeping crucial information from the consumer (Josling 2006). Furthermore, once a geographical indication acquires recognition there is always the possibility that producers from the same or other regions may try to come up with similar goods of lesser quality that could profit from the fame of the geographical indication and obtain premium prices from consumers, jeopardizing the reputation of the whole group (Anania and Nisticò 2004). Yet other researchers have pointed out that all the producers involved may enjoy advantages, albeit of different kinds. These dynamics have been examined through the concept of the "avatar," referring to the copies and variants of a traditional product that can coexist in a structurally complementary relationship with the original. In such environments geographical indication producers can take advantage of their reputation, of consolidated channels of distribution, and of consumers ready to pay premium prices, while the producers of the avatars can benefit from the spillover effect of the geographical indication product notoriety (Ceccarelli et al. 2010). The debate has turned into a transatlantic controversy, with all parties involved trying to get as many countries as possible on their side in order to influence decisions within the WTO (Raustiala and Munzer 2007). The EU maintains that geographical indications should be protected from these occurrences, but the US trademark system gives preeminence to entrepreneurship and creativity. New World countries have also pointed out that some of their geographical names, although deriving from places originally located in the colonial powers that controlled them, have actually been used for autonomous products with deeply rooted traditions.

The limitations imposed on the availability of a product by its recognition as geographical indication can move its supply curve toward increases in price, even with production costs remaining stable, which a neoliberal framework of analysis could interpret as a monopolistic position within. The profit resulting from it could be criticized as a form of nonproductive rent granted to those who own the geographical indication rights, a distortion of competitiveness, and a noneconomic barrier to free trade and the emergence of new industries (Linnemer and Perrot 2000). Conversely, high prices could be justified as a compensation for the considerable costs involved in developing high-quality products and in obtaining and managing a geographical indication, which require either a great number of producers sharing the costs or forms of external subsidies (Moschini et al. 2008).

This set of debates seems quite remote from the realities of many farmers and small producers outside Europe and other industrialized countries where the states' role has been steadily declining, increasingly leaving the establishment of new geographical indications to producers' associations. In emerging economies it is often development agencies, nongovernmental organizations, and national institutions—both private and public—that are examining geographical indications as a possible tool for rural development in disenfranchised areas, working with producers to create forms of cooperation and to ensure protection and promotion of their goods. Geographical indications have been criticized for being a "development adverse" instrument since, unlike trademarks and patents, they are a form of intellectual property that is not based on innovation but rather on the reputation of traditions and artisanal skills that are transmitted over generations in specific places (Vittori 2010). It is also undeniable that this very aspect offers growth opportunities for emerging economies where traditional foods are often produced in marginal or unfavorable areas by small farmers who are unable to reach economies of scale that would allow them to free themselves from a commodity-based export model (CIRAD 2006; Tregear et al. 2004). However, farmers need to be able to ensure the minimum production volume and the surplus necessary to access any market activity, requiring at times the intervention of external aid and development investment (Larson 2010).

Under this set of conditions, the economic potential of geographical indications has been examined within the research on global value chains that focuses on the vertical relationships among buyers, distributors, and producers by following the movement of goods or services from the origin to the final consumer, as well as through the analysis of their governance dynamics. According to this approach, the final consumers' perception of the unique qualities attributed to local specialties produced in limited amounts can differentiate them from homogeneous and mass-produced commodities, pushing them up along the global value chain and ensuring greater portions of the final sale value for the producers. Value-chain analysis suggests that by using certain characteristics of global trade to their advantage, producers and administrative bodies in emerging economies can actually take advantage of the renewed interest in immaterial, value-based qualities in the final markets, usually located in postindustrial societies (Kaplisnky 2004; Nadvi 2004; Ponte and Gibbon 2005). However, this raises the problem of establishing products in developing economies whose commercial viability depends on consumers whose preferences can easily change and who are culturally and geographically far from the very traditions that geographical indications are supposed to enhance.

The few aspects discussed so far suggest how the debate on geographical indications is extremely complex and varies greatly in different situations. The economic impact of geographical indications has been widely studied in Europe, but the literature regarding their implementation in other areas is still relatively limited, although quickly growing (Kireeva and Vergano 2006; Marie-Vivien 2010; Roussel and Verdeaux 2007; Suh and MacPherson 2007; Teuber 2010; Wang and Kireeva 2007). Overall, research indicates that geographical indications may present several advantages for rural development depending on the sociopolitical environment and whether they are relevant for the producers involved, affordable in terms of administrative and management costs, and applicable on different scales (Bowen 2010; Bramley et al. 2009; Giovannucci 2008; Grote 2009; Rangnekar 2004; Reviron 2009; van de Kop et al. 2006). Geographical indications tend to discourage possible registrations of traditional local products as commercial trademarks by third parties. For example, in 1997 the Texas-based company RiceTec was granted a patent for basmati rice by the US Patent and Trademarks Office. The patent was so broad it would have granted RiceTec the exclusive use of the term basmati and a monopoly on farm-bred basmati varieties from South Asia, including proprietary rights on seeds from hybrids (Lightboume 2003; Watal 2001). The international patent was revoked while the national one is still valid, but the question of the registration of basmati as a geographical indication is still pending for a name that refers to several varieties grown in the northern part of Western Punjab, on both sides of the Indo–Pakistani border. Although India has implemented geographical indication legislation in 2003 and a Trademarks Ordinance came into force in Pakistan in 2004 (with the application for the registration of basmati as geographical indication filed in 2005), political tensions and pressures from local traders keep national authorities on both sides of the border from resolving the situation (Chandola 2006; Giraud 2008). Similarly, an attempt to trademark the name "Darjeeling" was thwarted by the Tea Board of India, which succeeded in having the trademark canceled (Le Goffic 2008).

Another aspect is that by employing local knowledge and farmers' know-how, geographical indications can stimulate entrepreneurial attitudes within rural communities that have the potential to assure higher revenues and to limit the migration toward urban centers and to the Global North. However, this is not a guaranteed outcome; depending on the social and political structures surrounding the geographical indication production, the new income can potentially increase inequality. Some research has evaluated the impact of geographical indications on the real estate market of the area of origin, to assess whether increased commercial values can exclude the less affluent producers (Blackwell 2007). Alternatively, if geographical indication property rights were considered as collectively owned and a common good that cannot be delocalized or sold, the resources derived from their commercial value could be reinvested in the development of local communities and in the conservation of the environment through sustainable agriculture (Zografos 2008).

In any case, the actual incidence of these potential advantages depends on the social and political relations among the producers of geographical indication specialties. A study on tequila has indicated how the profits deriving from the international success of the beverage and its protection as geographical indications have been absorbed mostly by the bottlers, while very little income has actually trickled down to agave farmers (Bowen and Gerritsen 2007; Bowen and Zapata 2009). The evaluation of the impact of geographical indications on specific communities also needs to consider gender relations issues, including who grows and produces what in very diverse family and social structures (Parasecoli 2010).

Not all countries are following the geographical indication road. For example, in 2007 Ethiopia's pro bono lawyers at Arnold & Porter became involved in a very public debate with Starbucks about the trademark registration of three of Ethiopia's more famous coffee varieties (Sidamo, Harar, and Yirgacheffe) in more than thirty countries (Barraclough 2007). Starbucks polemically pointed out that Ethiopia should have protected its coffees with geographical indications rather than under intellectual property legislation. Rather than charging royalty fees, the

Ethiopian government is issuing royalty-free licenses to international companies for the use of the three trademarks in exchange for their promotion and marketing, hoping that the increase in demand will also lead to a rise in price.

Geographical Indications, Cultural Heritage, Indigenous Knowledge, and the Environment

As mentioned in the introduction, very little attention has been paid specifically to the impact of geographical indications on the environment. This section suggests areas for future research, based on the literature available on cultural heritage, indigenous knowledge, agrobiodiversity, and sustainability. It needs to be emphasized that most of the suggestions raised in this section are mostly speculative and only partly research-based and that diverse approaches and competing hypotheses on the topic exist.

Although not directly connected with environmental issues, the legal concept of "cultural heritage," which also covers aspects of material culture of specific communities, could be extended to include local agricultural varieties, food production, as well as food-related traditions and techniques. Consequently, it might be applied to promote agrobiodiversity conservation and stewardship of potentially endangered territories. The protection of cultural artifacts has been taken up by the United Nations and in particular UNESCO (United Nations Educational, Scientific, and Cultural Organization). At first the safeguard was focused on architectural treasures, sites, and landscapes, but the 1972 UNESCO Convention concerning the Protection of the World Cultural and Natural Heritage expanded international protection to specific natural habitats as well, reflecting growing concerns about the environment. This recognition acknowledges the relevance of landscapes and natural characters for certain communities and cultures. Article 47 of the convention defines cultural landscapes as "cultural properties and combined works of nature and of man. … They are illustrative of the evolution of human society and settlement over time, under the influence of the physical constraints and/or opportunities presented by their natural environment and of successive social, economic and cultural forces, both external and internal." In 1989 UNESCO issued a Recommendation on the Safeguarding of Traditional Culture and Folklore, which underlines the human aspects of cultural heritage. In 1995 the European Conference of Ministers responsible for the Cultural Heritage in Helsinki embraced the concept of cultural landscapes to include peoples' tradition, their cultural identity, and also their interaction with the environment. The 2003 UNESCO Convention for the Safeguarding of the Intangible Cultural Heritage defined the latter as "the practices, representations, expressions, knowledge, skills—as well as the instruments, objects, artifacts and cultural spaces associated therewith—that communities, groups and, in some cases, individuals recognize as part of their cultural heritage." The preamble of the 2005 UNESCO Convention on the Protection and Promotion of the Diversity of Cultural Expressions underlines that cultural diversity can be considered per se as part of the common heritage of humanity and should be preserved as such. The protection under the heading of cultural heritage has been extended to whole cuisines. Based on the provisions of the 2003 UNESCO convention, in 2010 "The gastronomic meal of the French," "The Mediterranean diet," and "Traditional Mexican cuisine—ancestral, ongoing community culture, the Michoacán paradigm" were added to the Representative List of the Intangible Cultural Heritage of Humanity, confirming the inclusion of food and food traditions under the convention's provisions.

These legal texts could potentially play an important role in helping to define and safeguard food productions as cultural expressions of specific communities and of their interaction with the environment. In developing economies, the UNESCO conventions would protect producers who might find difficulties in establishing geographical indications due to financial costs and administrative problems, and from unwarranted use of their cultural resources. However, the interpretation of the conventions on cultural heritage and the very principles constituting their framework of reference are object of debate. The most fundamental problems derive from the concept of culture itself, too vast to become the basis for a legal definition of cultural heritage. Is folklore "part of the heritage of humanity" or part of cultural identity of a specific community or people? Because applying the concept of intellectual property to cultural resources would imply a separation of ownership from control and access, and as a consequence the possibility of commodifying cultural elements, the legal harmonization of these elements has solicited scholarly interest especially in emerging economies (Arewa 2006; Long 2006; Srinivas 2008; Sunder 2007). It is particularly difficult to define the extent of natural heritage to be protected. Although article 3 of the 2003 UNESCO convention clearly affirms that "nothing in this convention can be interpreted as … affecting the rights and obligations of States Parties deriving from any international instrument relating to Intellectual Property rights or to the use of biological and ecological resources to which they are parties," some authors have suggested that the extension of article 23 of TRIPS to all geographical indications could be based on the relevance of cultural identity as public domain (Broude 2005a, 2005b).

The same tensions are evident also when it comes to indigenous knowledge, also referred to as traditional knowledge. Potentially relevant in the geographical indication debate, the category is broadly identified as "the knowledge that an indigenous (local) community accumulates over generations of living in a particular environment. This definition encompasses all forms of knowledge—technologies, know-how, skills, practices and beliefs—that enable the community to achieve stable livelihoods in their environment" (UNEP 2010). Although there is still no international convention that provides a precise legal framework and specific modalities for the protection of traditional products that could fall under this category, the 1992 Convention for Biological Diversity makes frequent reference to indigenous and traditional knowledge, explored also by scholars in its legal and political potential (Antons 2009; Bratspies 2006; Langton and Ma Rhea 2005; Munzer and Raustiala 2009; Oguamanam 2006). In 2000 WIPO established the Intergovernmental Committee on Intellectual Property and Genetic Resources, Traditional Knowledge and Folklore (IGC-GRTKF) precisely to examine the complex relationship between traditional knowledge, intellectual property rights, and agrobiodiversity.

Indigenous knowledge as a possible juridical category raises issues with regard to the control of national states over natural resources, to the point that the preamble to the 1992 UN Convention on Biological Diversity used the expression "common concern of humankind" for the conservation of biological diversity, all while reaffirming the states' "sovereign rights over their own biological resources." The convention further states that each contracting party, "subject to its national legislation, respect, preserve and maintain knowledge, innovations and practices of indigenous and local communities embodying traditional lifestyles relevant for the conservation and sustainable use of biological diversity and promote their wider application with the approval and involvement of the holders of such knowledge, innovations and practices and encourage the equitable sharing of the benefits arising from the utilization of such knowledge, innovations and practices" (art. 8j). Indigenous knowledge is identified as relevant for national interest and as such subject to sovereign rights and national legislation, and states are supposed to "protect and encourage customary use of biological resources in accordance with traditional cultural practices that are compatible with conservation or sustainable use requirements" (art. 10c). The convention tries to strike a balance among national priorities, the interests of the private sector, and the objectives of intergovernmental and nongovernmental organizations in terms of access and benefit sharing, by affirming that "the authority to determine access to genetic resources rests with the national governments and is subject to national legislation," while "each Contracting Party shall endeavor to create conditions to facilitate access to genetic resources for environmentally sound uses by other Contracting Parties and not to impose restrictions that run counter to the objectives of this Convention" (art. 15). These goals, referred to as ABS (Access and Benefit Sharing), were further spelled out in 2010 with the Nagoya Protocol on Access to Genetic Resources and the Fair and Equitable Sharing of Benefits Arising from their Utilization, which at the time of writing is open to signature at the United Nations Headquarters in New York. In article 7, the protocol states that "in accordance with domestic law, each Party shall take measures, as appropriate, with the aim of ensuring that traditional knowledge associated with genetic resources that is held by indigenous and local communities is accessed with the prior and informed consent or approval and involvement of these indigenous and local communities, and that mutually agreed terms have been established." Furthermore, article 12 strengthens the ability of these communities to benefit from the use of their knowledge, innovations, and practices. In fact, developing countries have been lobbying to obtain disclosure of the origin of genetic materials or associated indigenous knowledge claimed in patent applications, also connecting it to the negotiations about the geographical indication registry (Correa 2010).

The concept of indigenous knowledge has mostly been employed within the framework of social and economic development projects, but it has potential to be used as a legal weapon in the battle to maintain collective intellectual property. In India, indigenous knowledge-related techniques and products have been granted patents and the protection connected to them even though they did not fulfill the traditional legal requirements of novelty and invention. For instance, a pharmaceutical patent in the US for the use of turmeric in the healing of wounds and rashes was canceled by the United States Patent and Trademark Office due to the previous existence in India of similar traditional medicine procedures and products, while the European Patent Office (EPO) revoked the patent for neem (*Azadirachta indica*), whose extracts are traditionally employed as insect repellent. In 1995 a trust fund was built to commercialize agent compounds from the medicinal plant arogyapaacha (*Trichopus zeylanicus*) and a treaty was signed between the Tropical Botanic Garden and Research Institute and the Kani tribal community that traditionally used it to transfer 2 percent of the revenue to the tribe (Subba Rao 2006). The 2001 International Treaty on Plant Genetic Resources for Food and Agriculture seems to be moving the international community in a similar direction, guaranteeing food security through conservation, exchange, and sustainable use of world's plant genetic resources, as well as their fair use and equitable benefit sharing in harmony with the Convention of Biological Diversity. The 2003 Cartagena Protocol on Biosafety also underlines the relevance of agrobiodiversity by requiring "safe transfer, handling and use of living modified organisms resulting from modern biotechnology that may have adverse effects on the conservation and sustainable use of biological diversity" (art. 1), "especially with regard to the value of biological diversity to indigenous and local communities" (art. 26).

Currently however, indigenous knowledge needs to be backed by documentation and proof showing its prior existence in order to acquire relevance in the international legal system of intellectual property and to reap and distribute the commercial benefits connected to its utilization. India is well positioned in this sense due to its millennial of written culture, but other civilizations that did not adopt writing find themselves in a much more difficult situation. The Third World Network, an independent nonprofit international network of organizations and individuals involved in issues related to development, proposed a Community Intellectual Rights Act, according to which local communities would be the "custodians" (or "stewards") of their traditions, prohibiting the concession of any rights of exclusive use of such innovations (Nijar 1994).

Articles 215 and 216 of the federal constitution of Brazil states that the government shall protect "the expressions of popular, Indian and Afro-Brazilian cultures" and "promote and protect the Brazilian cultural heritage, by means of inventories, registers, vigilance, monument protection decrees, expropriation and other forms of precaution and preservation" (Santilli 2006). The harmonization between the highly diverse social and political organizations of local communities and the legal system of the nation-states within which intellectual property laws operate, presents juridical and practical challenges.

Depending on legal interpretations and international negotiations on the issue, and in the case that certain foods and products are proved to be a fundamental part of a population's identity, the legal extension of cultural heritage and cultural property might have the potential to be used to maintain food and products of populations that lack the bureaucratic structures necessary to implement their protection as geographical indication. Of course, specific multilateral instruments should be elaborated that would require reduced legal and administrative interventions from local authorities, also with the goal of limiting costs. In some cases geographical indication status for specific products might enhance their protection as indigenous knowledge and vice versa under the conditions that negotiations on TRIPS, ABS, and development achieve a higher degree of coordination, which is an important factor for countries that cannot afford to send delegations to multiple diplomatic negotiations (Correa 2010; Downes 2002; Panizzon 2006). At any rate, all the parties involved need to consider the inevitable impact of the marketing of geographical indication and indigenous knowledge products on the international market. The innovation that this exposure requires would affect the social and political structures of the local communities involved, even when ensuring viable livelihoods. For this reason, the dynamics of governance of the value chain built around the products are crucial in ensuring cultural survival.

Geographical indications were not explicitly designed to ensure the conservation of agrobiodiversity, and the discussion regarding the assessment of the effects of geographical indication products on sustainability and the environment is still limited. However, authors that address the issue tend to state that these legal protections might contribute to maintaining and developing genetic resources connected with cultivation, shepherding, and also foraging of specific species (Berard and Marchenay 2006). Many of the traditions related to these resources are in fact quite specific to limited areas and usually not produced on a large scale. The diffusion of these legal categories not only in the Western world but also in emerging economies would contribute to maintaining and defending biodiversity, which could turn into an economic asset with a potentially noticeable impact on rural development. At times, research and collaboration among local communities, scientists, academics, and institutional actors might be necessary to identify, study, and to recover local resources with geographical indication potential that might have remained invisible—both to consumers and policymakers—due to cultural and social hurdles. These initiatives can directly reinforce agrobiodiversity through the protection and valorization of crops and animal species, and through a sustainable use aimed to achieve long-term economic viability (Thrupp 2000). Furthermore, the promotion of rural livelihoods that depend on the sustainable use of geographical indication resources can indirectly affect biodiversity through the diffusion of agricultural techniques with a positive impact on the conservation of local ecosystems and landscapes. At the same time, the homogenization of the varieties or breeds specifically required by quality standards in geographical indication regulations and reinforced by market and productivity demands can cause the marginalization of other varieties and, as a consequence, a reduction in diversity (Larson 2007, 2010). Moreover, climate change could possibly affect the productivity or even the survival of traditional or local crops identified as potential geographical indications, making it necessary for the communities involved to maintain a variety of productions.

Various factors need to be taken into consideration when evaluating the long-term viability of geographical indications. Especially in emerging economies they tend to be economically profitable because they can count on an international audience of gourmets, which often implies export to far destinations. The environmental impact of the fuel necessary to transport the products to their final markets is the object of debates that focus especially on the concept of "food miles" (MacGregor and Vorley 2006; Weber and Matthews 2009; Wynen and Vanzetti 2008). Very often geographical indications become a travel motivation increasing the inflow of visitors interested in culinary specialties and tradition. What is the environmental impact of this kind of culinary tourism on the area of production? Do geographical indications inherently tend to promote forms of sustainable tourism or could they become also appealing for mass tourism? (Hall et al. 2003; Long 2004; Smith and Costello 2009; Spurlock 2009). To this date, few studies are available on the environmental effects of geographical indications and their impact on the sustainability of local food systems. Theoretically, the higher commercial value of their crops could push farmers to grow them more intensively and in larger areas, increasing the risks of soil degradations and water scarcity. At the same time, the growing market value of traditional crops could make them too expensive for the very communities where they originate, excluding locals from their consumption. The cases of Andean crop quinoa and the Brazilian berry açai have attracted the attention of the international press (Colapinto 2011; Romero and Shahriari 2011). Depending on the social dynamics within the different communities, variable amounts of the geographical indication products should be kept for local use at accessible prices in order to maintain their cultural significance.

Conclusions

This brief overview of the role and relevance of geographical indications for sustainable rural development and the safeguard of cultural heritage as a bastion against the loss of agrobiodiversity leaves many questions unanswered, while suggesting directions for future research. The contemporary debates on these issues and the interest they are eliciting in a variety of fields indicate their relevance not only from the theoretical point of view, but also in terms of governance, trade, and economics, with real-life impact on many communities all around the world.

Geographical indications could have the potential to become a valid tool in implementing community-based, sustainable, and quality-oriented agriculture, when given an accurate analysis of depending on the socio-political environment and whether they are relevant for the producers involved, affordable in terms of administrative and management costs, and applicable on different scales of production. However, in reality, these benefits cannot be considered automatically inherent to the establishment of a geographical indication for a local traditional crop, because its regulation may only refer to its place of origin and its characteristics without necessarily mentioning the environmental aspects of the agricultural practices connected with its production. On the one hand, the implementation of geographical indications could increase the commercial value of traditional crops and avoid their disappearance due to low yields, high costs associated with labor-intensive methods of production, and lack of transmission of the necessary know-how. On the other hand, it is necessary to assess the impact of the popularity of a geographical indication, which often entails increases in price and in quantities reserved for sales in distant but more profitable markets, thus reducing its availability and affordability for the producing communities.

Future research should also focus on the consequences for local ecosystems in case geographical indications are extended to unsuitable areas or for the adoption of intensive techniques aiming to increase yields to reap larger revenues. Furthermore, because the final consumers of sought-after geographical indications might be far from the places of production, is it necessary to consider the environmental externalities connected to long distance transportation. Last, the risk of dependency is also something to be considered when exporting a large percentage of sales to foreign markets, which are often influenced by trends determined by media and local sociopolitical dynamics outside the control of the producing communities.

The legal status of geographical indication products and crops that could fall under the definition of indigenous knowledge as either collective or private rights continues to constitute a main source of contention in the ongoing international negotiations. Many countries argue that this is not the only available model to ensure protection of the intellectual property rights connected with agricultural products, pushing instead to uphold the trademark system as a valid alternative. As the 2010 Nagoya Protocol of Access and Benefit Sharing and the other conventions discussed in the article indicate, this controversy is far from being theoretical. Its outcome will have a relevant impact on how the stream of revenue potentially connected to the production, distribution, promotion, and sales of the geographical indications or of products falling under the category of indigenous knowledge are distributed along the value chain.

The review of the literature reveals that the research on geographical indications falls under different disciplines, and the topic is also discussed in political and economic venues that at times do not communicate enough with each other. There is need for more case studies that monitor implementation processes and evaluate the actual impact of geographical indications on specific communities, not only in terms of trade opportunities and value chains, but development, human rights, and social dynamics, especially in developing countries. Their analysis should also consider the sustainability of the products involved, by applying life cycle assessment methods from production to distribution to final consumption. Qualitative and quantitative approaches are equally necessary to provide all the actors involved in the establishment of new geographical indications with reliable data and effective policy tools, contributing to the elaboration of regulations that are more sensitive to cultural, social, and environmental aspects. Building on the cooperation of scholars, practitioners, policymakers, and activists, together with the direct input of the communities involved, a truly multidisciplinary approach would be better suited to explore the connections among the various features of this new form of protection of intellectual property in all its ramifications, including its relation with indigenous knowledge and agrobiodiversity.

FABIO PARASECOLI is associate professor and coordinator of food studies at The New School in New York City. His research focuses on the intersections among food, media, and politics. He is program adviser at Gustolab, a center for food and culture in Rome, and collaborates with other institutions such as the Universitat Oberta de Catalunya in Barcelona, Spain and the University of Gastronomic Sciences in Pollenzo, Italy. His recent publications include *Food Culture in Italy* (2004), the introduction to *Culinary Cultures in Europe* (2005) and *Bite me! Food in Popular Culture* (2008). He is general editor with Peter Scholliers of a six-volume *Cultural History of Food* (2012).

AYA TASAKI is a queer immigrant activist and native of Japan with bicultural roots. She is completing her MA in international affairs from The New School. Her thesis focuses on US immigration policies, inspired by the work she does with the Audre Lorde Project surrounding the queer people of color community. She has contributed to the *Kathmandu Post* and *Y! Magazine* while doing research on migrants in Nepal.

REFERENCES

Anania, Giovanni, and Rosanna Nisticò. 2004. "Public Regulation as a Substitute for Trust in Quality Food Markets: What if the Trust Substitute Cannot Be Fully Trusted?" *Journal of Institutional and Theoretical Economics* 160 (4): 681–701.

Antons, Christoph. 2009. *Traditional Knowledge, Traditional Cultural Expressions, and Intellectual Property Law in the Asia-Pacific Region.* Alphen aan der Rijn: Kluwer Law International.

Arewa, Olufunmilayo. 2006. "TRIPS and Traditional Knowledge: Local Communities, Local Knowledge, and Global Intellectual Property Frameworks." *Marquette Intellectual Property Law Review* 10 (2): 155–180.

Babcock, Bruce A., and Roxanne Clemens. 2004. "Geographical Indications and Property Rights: Protecting Value-Added Agricultural Products." MATRIC Briefing Paper 04-MBP 7, Iowa State University.

Barham, Elizabeth. 2002. "Towards a Theory of Values-Based Labeling." *Agriculture and Human Values* 19 (4): 349–360.

———. 2003. "Translating Terroir: The Global Challenge of French AOC Labeling." *Journal of Rural Studies* 19 (1): 127–138.

Barraclough, Emma. 2007. "From Commodity to IP asset." *Managing Intellectual Property* 8 (173): 18–19.

Berard, Laurence, and Philippe Marchenay. 2006. "Local Products and Geographical Indications: Taking Account of Local Knowledge and Biodiversity." *International Social Science Journal* 58 (187): 109–116.

Blackwell, Martin. 2007. "The Relationship of Geographical Indications with Real Property Valuation and Management." *Property Management* 25 (2): 193–203.

Bowen, Sarah. 2010. "Development from Within? The Potential for Geographical Indications in the Global South." *Journal of World Intellectual Property* 13 (2): 231–252.

Bowen, Sarah, and Peter R. W. Gerritsen. 2007. "Reverse Leasing and Power Dynamics among Blue Agave Farmers in Western Mexico." *Agriculture and Human Values* 24 (4): 473–488.

Bowen, Sarah, and Ana Velenzuela Zapata. 2009. "Geographical Indications, *Terroir,* and Socioeconomic and Ecological Sustainability: The Case of Tequila." *Journal of Rural Studies* 25 (1): 108–119.

Bramley, Cerkia, Estelle Bienabe, and Johann Kirsten. 2009. "The Economics of Geographical Indications: Toward a Conceptual Framework for GI Research in Developing Countries." Pp. 109–141 in *The Economics of Intellectual Property,* ed. WIPO. Geneva: WIPO.

Bratspies Rebecca. 2006. "The New Discovery Doctrine: Some Thoughts on Property Rights and Traditional Knowledge." *American Indian Law Review* 31 (2): 315–340.

Broude, Tomer. 2005a. "Culture, Trade and Additional Protection for Geographical Indications." *Bridges* 9 (9): 20–22.

Broude, Tomer. 2005b. "Taking Trade and Culture Seriously: Geographical Indications and Cultural Protection in WTO Law." *University of Pennsylvania Journal of International Economic Law* 26 (4): 623–692.

Ceccarelli. Giovanni, Alberto Grandi, and Stefano Magagnoli. 2010. Avatar: La Costruzxione di un modello storico-economico per i prodottu tipici." Paper presented at the La Tipicita' nella storia conference Parma-Langhirano, 9–11 September.

Chandola, H. V. 2006. "Basmati Rice: Geographical Indication or Misindication." *Journal of World Intellectual Property* 9 (2): 166–188.

CIRAD. 2006. CIRAD Annual Report. http://www.cirad.fr/en/publications-resources/science-for-all/rapports-annuels/cirad-2006/innovating (accessed 4 January 2011).

Colapinto, John. 2011. "Strage Fruit: The Raise and Fall of Açai." *New Yorker,* 30 May. http://www.newyorker.com/reporting/2011/05/30/110530fa_fact_colapinto (accessed 5 July 2011).

Colman, Tyler. 2008. *Wine Politics: How Governments, Environmentalists, Mobsters, and Critics Influence the Wines We Drink.* Berkeley and Los Angeles: University of California Press.

Correa, Carlos M. 2010. *Geographical Indications and the Obligation to Disclose the Origin of Biological Materials: Is a Compromise Possible under TrIPs? ICTSD Policy Brief 8.* Geneva: ICTSD.

Creditt, Emily C. 2009. "Terroir vs. Trademarks: The Debate over Geographical Indications and Expansions to the TRIPS Agreement." *Vanderbilt Journal of Entertainment and Technology Law* 11 (2): 425–459.

Daviron, Benoit, and Stefano Ponte. 2005. *The Coffee Paradox.* London: Zed Books.

Doster, Ivy. 2006. "A Cheese by Any Other Name: A Palatable Compromise to the Conflict over Geographical Indications." *Vanderbilt Law Review* 59 (3): 873–903.

Downes, David. 2002. "How Intellectual Property Could Be a Tool to Protect Traditional Knowledge." *Columbia Journal of Environmental Law* 25: 253–282.

Echols, Marsha A. 2003. "Geographical Indications for Foods, Trips and the Doha Development Agenda." *Journal of African Law* 47 (2): 199–220.

———. 2008. *Geographical Indications for Food Products: International Legal and Regulatory Perspectives.* Alphen aan den Rijn: Kluwer Law International.

Gervais, Daniel J. 2010. "Reinventing Lisbon: The Case for a Protocol to the Lisbon Agreement (Geographical Indications)." *Chicago Journal of International Law* 11 (1): 67–126.

Giovannucci, Daniele, ed. 2008. *Guide to Geographical Indications: Approaches to Value.* Geneva: International Trade Centre.

Giovannucci, Daniele, Tim Josling, William Kerr, Bernard O'Connor, and May T. Yeung. 2009. *Guide to Geographical Indications: Linking Products and Their Origins.* Geneva, Switzerland: International Trade Center.

Giraud, Georges. 2008. "Range and Limit of Geographical Indication Scheme: The Case of Basmati Rice from Punjab, Pakistan." *International Food and Agribusiness Management Review* 11 (1): 51–76.

Grote, Ulrike. 2009. "Environmental Labeling, Protected Geographical Indications and the Interests of Developing Countries." *Estey Centre Journal of International Law and Trade Policy* 10 (1): 94–110.

Hall, Michael C., Liz Sharples, Richard Mitchell, Niki Macionis, and Brock Cambourne. 2003. *Food Tourism around the World: Development, Management and Markets.* Burlington, MA: Butterworth-Heinemann.

Heath, Christopher, and Anselm Kamperman Sanders. 2005. *New Frontiers of Intellectual Property Law: IP and Cultural Heritage, Geographical Indications, Enforcement and Overprotection (IIc Studies).* Oxford and Portland, OR: Hart.

Hughes, Justin. 2006. "Champagne, Feta, and Bourbon: The Spirited Debate about Geographical Indications." *Hastings Law Journal* 58: 299–386.

Ilbert, Hélène, and Michel Petit. 2009. "Are Geographical Indications a Valid Property Right? Global Trends and Challenges." *Development Policy Review* 27 (5): 503–528.

Josling, Tim. 2006. "What's in a Name? The Economics, Law and Politics of Geographical Indications for Foods and Beverages." Institute for International Integration Studies Discussion Paper No. 109, Dublin.

Kaplisnky, Raphael. 2004. "Spreading the Gains from Globalization : What Can Be Learned from Value-Chain Analysis?" *Problems of Economic Transition* 47 (2): 74–115.

Kireeva, Irina, and Bernard O'Connor. 2010. "Geographical Indications and the TRIPS Agreement: What Protection Is Provided to Geographical Indications in WTO Members?" *Journal of World Intellectual Property* 13 (2): 275–303.

Kireeva, Irina, and Paolo Vergano. 2006. "Geographical Indications and the Interface between Trade Mark Protection and Sui Generis Protection: The Example of China, Thailand and Vietnam." *International Trade Law and Regulation* 12 (4): 97–108.

Kongolo, Thsimanga. 2008. *Challenges in Respect to Protection of Geographical Indications.* Kluwer Law International, Alphen aan den Rijn: Kluwer Law International.

Lang, Aaron C. 2006. "On the Need of Expanding art. 23 of the TRIPS Agreement." *Duke Journal of Comparative and International Law* 16 (2): 487–510.

Langinier, Corinne, and Bruce Babcock. 2006. "Agricultural Production Clubs: Viability and Welfare Implications." CARD Working Paper 06-WP 431, Iowa State University.

Langton, Marcia, and Zane Ma Rhea. 2005. "Traditional Indigenous Biodiversity-Related Knowledge." *Australian Academic & Research Libraries* 36 (2): 47–72.

Larson, Jorge. 2007. *Relevance of Geographical Indications and Designations of Origin for the Sustainable Use of Genetic Resources.* Maccarese: Global Facilitation Unit for Underutilized Species.

———. 2010. *Geographical Indications, in Situ Conservation and Traditional Knowledge.* ICTSD Policy Brief 3. Geneva: ICTSD.

Le Goffic, Caroline. 2008. "Cancellation of a Trade Mark Based on a Prior Foreign Geographical Indication Related to Different Products." *Journal of Intellectual Property Law & Practice* 3 (3): 152–154.

Lightboume, Muriel. 2003. "Of Rice and Men—An Attempt to Assess the Basmati Affair." *Journal of World Intellectual Property* 6 (6): 875–894.

Linnemer, Laurent, and Anne Perrot. 2000. "Une Analyse économique des signes de qualité." *Revue économique* 51 (6): 1397–1418.

Long, Doris Estelle. 2006. "Traditional Knowledge and the Fight for the Public Domaine." *John Marshall Review of intellectual Property Law* 5: 317–329.

Long, Lucy, ed. 2004. *Culinary Tourism.* Lexington: University Press of Kentucky.

MacGregor, James, and Bill Vorley. 2006. *Fair Miles? The Concept of "food miles" through A Sustainable Development Lens.* London: IIED.

Marie-Vivien, Delphine. 2010. "The Role of the State in the Protection of Geographical Indications: From Disengagement in France/Europe to Significant Involvement in India." *Journal of World Intellectual Property* 13 (2): 121–147.

Moschini, Giancarlo, Luisa Menapace, and Daniel Pick. 2008. "Geographical Indications and the Competitive Provision of Quality in Agricultural Markets." *American Journal of Agricultural Economics* 90 (3): 794–812.

Munzer, Stephen, and Kal Raustiala. 2009. "The Uneasy Case for Intellectual Property Right in Traditional Knowledge." *Cardozo Arts & Entertainment Law Journal* 27: 37–97.

Nadvi, Khalid. 2004. "Globalisation and Poverty: How can Global Value Chain Research Inform the Policy Debate?" *IDS Bulletin* 35 (1): 20–30.

Nijar, G.S. 1994. "Third World Network Discussion Paper: Towards a Legal Framework for Protecting Biological Diversity and Community Intellectual Rights: A Third World Perspective." Second Session of the ICCBD, Nairobi, 20 June–1 July.

O'Connor, Bernard. 2004. "The Legal Protection of Geographical Indications." *Intellectual Property Quarterly* 1: 35–57.

Oguamanam, Chidi. 2006. *International Law and Indigenous Knowledge: Intellectual Property, Plant Biodiversity, and Traditional Medicine.* Toronto: University of Toronto Press.

Panizzon, Marion. 2006. "Traditional Knowledge and Geographical Indications: Foundations, Interests and Negotiating Positions." NCCR Working Paper No. 2005/01, Berne.

Parasecoli, Fabio. 2010. "The Gender of Geographical Indications: Women, Place, and the Marketing of Identities." *Cultural Studies/Critical Methodologies* 10 (6): 467–478.

Ponte, Stefano, and Peter Gibbon. 2005. "Quality Standards, Conventions and the Governance of Global Value Chains." *Economy and Society* 34 (1): 1–31.

Rangnekar, Dwijen. 2004. "The Socio-Economics of Geographical Indications, UNCTAD-ICTSD Project on IPRs and Sustainable Development." ICTSD Issue Paper No. 8, Geneva.

Raustiala, Kal, and Stephen R. Munzer. 2007. "The Global Struggle over Geographical Indications." *European Journal of International Law* 18 (2): 337–365.

Reviron, Sophie. 2009. *Geographical Indications: Creation and Distribution of Economic Value in Developing Countries.* Zurich: ETH.

Romero, Simon, and Shahriari, Sara. 2011. "Quinoa's Global Success Creates Quandary at Home." *New York Times,* March 19. http://www.nytimes.com/2011/03/20/world/americas/20bolivia.html (accessed 5 July 2011).

Roussel, Bernard, and François Verdeaux. 2007. "Natural Patrimony and Local Communities in Ethiopia: Advantages and Limitations of a System of Geographical Indications." *Africa: The Journal of the International African Institute* 77 (1): 30–150.

Santilli, Juliana. 2006. "Cultural Heritage and Collective Intellectual Property Rights." *IK Notes* 95. http://siteresources.worldbank.org/INTINDKNOWLEDGE/Resources/iknt95.pdf (accessed 13 January 2011).

Smith, Sylvia, and Carol Costello. 2009. "Segmenting Visitors to a Culinary Event: Motivations, Travel Behavior, and Expenditures." *Journal of Hospitality Marketing & Management* 18 (1): 44–67.

Spurlock, Cindy. 2009. "Performing and Sustaining (Agri)Culture and Place: The Cultivation of Environmental Subjectivity on the Piedmont Farm Tour." *Text and Performance Quarterly* 29 (1): 5–21.

Srinivas, Krishna Ravi. 2008. "Traditional Knowledge and Intellectual Property Rights." *AJWH Asian Journal of WTO & International Health Law and Policy* 3 (1): 81–120.

Stanziani, Alessandro. 2004. "Wine Reputation and Quality Controls: The Origin of the AOCs in 19th Century France." *European Journal of Law and Economics* 18 (2): 149–167.

———. 2006. "Histoire de la qualité alimentaire." *Food and History* 4 (1): 307–344.

Subba Rao, Siriginidi. 2006. "Indigenous Knowledge Organization: An Indian Scenario." *International Journal of Information Management* 26 (3): 224–233.

Suh, Jeongwook, and Alan MacPherson. 2007. "The Impact of Geographical Indication on the Revitalisation of a Regional Economy: A Case Study of 'Boseong' Green Tea." *Area* 39 (4): 518–527.

Sunder, Madhavi. 2007. "The Invention of Traditional Knowledge." *Law and Contemporary Problems* 70 (2): 97–124.

Teuber, Ramona. 2010. "Geographical Indications of Origin as a Tool of Product Differentiation: The Case of Coffee." *Journal of International Food & Agribusiness Marketing* 22 (3/4): 277–298.

Thrupp, Lori Ann. 2000. "Linking Agricultural Biodiversity and Food Security: The Valuable Role of Sustainable Agriculture." *International Affairs* 76 (2): 265–281.

Torre André. 2002. "Les AOC Sont-elles des Clubs? Réflexions sur les conditions de l'action collective localisée, entre coopération et règles formelles." *Revue d'économie industrielle* 100 (30): 39–62.

Tregear, Angela. Filippo Arfini, Giovanni Belletti, and Andrea Marescotti. 2004. "The Impact of Territorial Product Qualification Processes on the Rural Development Potential of Small-Scale Food Productions." Paper presented at the Eleventh World Congress of Rural Sociology Trondheim, Norway, 25–30 July.

UNEP. 2010. *About IK*. http://www.unep.org/ik/Pages.asp?id=About%20IK (accessed 13 January 2011).

van de Kop, Petra, Denis Sautier, and Astrid Gerz, eds. 2006. *Origin-Based Products: Lessons for Pro-Poor Market Development*. Amsterdam: KIT and CIRAD.

Vittori, Massimo. 2010. "The International Debate on Geographical Indications (GIs): The Point of View of the Global Coalition of GI Producers—oriGIn." *Journal of World Intellectual Property* 13 (2): 304–314.

Wang, Xiaobing, and Irina Kireeva. 2007. "Protection of Geographical Indications in China: Conflicts, Causes and Solutions." *Journal of World Intellectual Property* 10 (2): 79–96.

Watal, Jayashree. 2001. *Intellectual Property Rights in the WTO and Developing Countries*. The Hague, Netherlands: Kluwer Law International.

Wattanapruttipaisan, Thitapha. 2009. "Trademarks and Geographical Indications: Policy Issues and Options in Trade Negotiations and Implementation." *Asian Development Review* 26 (1): 166–205.

Weber, Christopher L., and H. Scott Matthews. 2008. "Food-Miles and the Relative Climate Impacts of Food Choices in the United States." *Environmental Science & Technology* 42 (10): 3508–3513.

WIPO. 2007. "Geographical Indications: From Darjeeling to Doha." *Wipo Magazine* 4 (3). http://www.wipo.int/wipo_magazine/en/2007/04/article_0003.html (accessed 20 December 2010).

WTO. 1994. "Agreement on Trade-Related Aspects of Intellectual Property Rights, Part II, Section 3: Geographical Indications. "http://www.wto.org/english/tratop_e/trips_e/t_agm3b_e.htm#3 (accessed 20 December 2010).

WTO. 2010. "Geographical Indications Talks Gear Up for 2011 Endgame." http://www.wto.org/english/news_e/news10_e/trip_10dec10_e.htm (accessed 14 January 2011).

———. 2011a. "Geographical Indications Talks Produce First Single Draft." http://www.wto.org/english/news_e/news11_e/trip_ss_13jan11_e.htm (accessed 18 March 2011).

———. 2011b. "Report on Issues Related to the Extension of the Protection of Geographical Indications Provided for in the Article 23 of the TRIPS Agreement to Products Other Than Wines and Spirit and Those Related to the Relationship between the TRIPS Agreement and the Convention on Biological Diversity." WT/GC/W/633, TN/C/W/61, 21 April. WTO General Council, Trade Negotiations Committee.

Wynen, Els, and David Vanzetti. 2008. "No Through Road: The Limitations of Food Miles." Asian Development Bank Institute Working Paper No. 118, Tokyo.

Zografos, Daphne. 2008. "Geographical Indications & Socio-Economic Development." IQsensato Working Paper 3, Geneva.

Rethinking the Food-versus-Fuel Debate

An Appraisal of International Perspectives and Implications for the South African Industrial Biofuels Strategy

Shaun Ruysenaar

ABSTRACT: The global rush toward a biofueled future (and subsequent apprehension concerning unintended consequences) has met with powerful and wide-ranging critique. Bolstered by globally increasing food prices peaking in 2008, food insecurity has become a central concern when considering pursuing biofuels. Arguments in the wider literature propose a number of perspectives with which to evaluate the biofuels-food security nexus. In South Africa, however, the debate is largely configured around maize-for-ethanol and polarized between two antagonistic camps. A host of agricultural lobbies and industrial interests argue in support of biofuels while some politicians, civil society, and NGOs argue against it. Both groups draw their arguments from various domains of the food security discourse in support of their cause. This article considers the merits of these opposing arguments in relation to wider perspectives in the literature, in many cases highlighting non-holistic assumptions made by the opposing claimants. This article seeks to rekindle a waning dialogue and provide a more robust outline of the major concerns that need to be addressed when considering biofuels production from a food security perspective. Only then can South Africa expect to weigh up accurately the value of pursuing biofuels production.

KEYWORDS: biofuels, food security, food-versus-fuel debate, rural development, South Africa

Biofuels, or agrofuels as La Via Campesina movement have labeled them, are fuels specifically derived from agricultural produce. As an alternative energy source, these fuels present particularly complex problems. Promises of sustainable and rural development, renewable energy-security, mitigating climate change and a host of other benefits (Davison et al. 2010; Hazell and Pachauri 2006; Peters and Thielmann 2008; World Bank 2008) have been eclipsed by controversy and failure (Oxfam International 2008). After gaining substantial momentum, the cure soon threatens to become worse than the disease (Doornbosch and Steenblik 2007), a crime against humanity according to a United Nations expert on food security, Jean Ziegler. Nowhere has this trajectory been more dramatic than in the contests between food and fuel (Ziegler 2007), which have become the subject of a prolific international debate after spiraling food prices and the onset of an international food crisis (Eide 2008). At its most extreme, the food-versus-fuel

Environment and Society: Advances in Research 2 (2011): 124–148 © Berghahn Books
doi:10.3167/ares.2011.020108

debate is characterized as food crops being snatched from the mouths of the poor (South) to fuel the cars of the rich (North); although as I discuss it is increasingly recognized as a far more nuanced but nonetheless vexing issue (see also Rosillo-Callé and Johnson 2010).

Prior to the 2008 global food crisis biofuels' production surged, tripling between 2000 and 2007 (Coyle, 2007). From origins supposedly beset by uncertainty, biofuels were suddenly "booming" (Mol 2007: 297; see also Cadenas and Cabezudo 1998: 83), resulting in the emergence of a globally integrated biofuel network (Mol 2007). As Smith (2010: 65) points out, we are witnessing the "unfolding of a global, socio-technical assemblage,[1] which gives momentum, meaning and legitimacy to a biofueled future," even in the face of seemingly intractable challenges in their widespread adoption. Biofuels notably engage with multiple increasingly globalized systems in ways not yet fully understood (Smith 2010). A variety of policy objectives that characterize such an assemblage have led to a scramble for biofuels, based more on mobilizing metaphors (or propaganda as Moore [2008] calls it) than on achievable demands within the market. Sharman and Holmes (2010: 309) highlight, for example, that target-setting around biofuels in the European Union was more a result of political imperatives and a process of "policy-based evidence gathering" than anything else. Meeting these targets, particularly in the United States and Brazil, has resulted in a significant international trade regime; one in which restrictions are in place designating who benefits and gains access (Dauvergne and Neville 2010; Oosterveer and Mol 2010), and where institutional arrangements to ensure that food remains accessible are lacking (Koning and Mol 2009).

Within the biofuels assemblage, multiple perspectives have been used to interpret and understand the implications that biofuels—in the form of first generation agrofuels—present to food security. The debates are complex and contested with food security topically constructed, deconstructed, and reconstructed by constituent interests. Seemingly, the promises around biofuels are suggestive of development narratives (Roe 1991). These are storylines used by academics, development practitioners, bureaucrats, and policymakers to articulate and make sense of uncertainty in the real world and simplify ambiguity. Through their enticing simplicity these storylines can become blueprints for development. Yet as Roe (1991) suggests, it is beneficial to examine ways in which the employment of these narratives can be improved or superseded before calling for their outright dismissal. This is especially important given that biofuels appear to be steadfast and their production increasing. Who employs what kind of narrative is also important.

This article presents a review of the thematic elements or basic story lines of the food-versus-fuel debate, while also providing a broad theoretical underpinning of food security with which to interrogate such a debate. In so doing I trace the origins and outline current understandings of food security, highlighting the importance of systems and critical perspectives in addition to more traditional perspectives that isolate access and availability. These offer valuable insight into the complex nature of the agro-food system and the possible implications of biofuels. To illustrate the nuances, if not difficulty, of subscribing to more holistic approaches, I present food security arguments within South Africa, contextualizing the South African maize-to-ethanol proposals within this broader intellectual framework.

Since the promulgation of the Industrial Biofuels Strategy in 2007 the South African debate has been hinged around the exclusion of maize-to-ethanol production. Although maize is a staple food crop for South Africa and surrounding countries, there is a potential to grow more than what is required for national self-sufficiency. Yet the South African debate has entered something of a theoretical quagmire. Proponents of maize-to-ethanol—agribusinesses, commercial farmers, and agricultural lobbyists—are eying and rely on new markets for their surplus production to turn greater profits. They argue from neoclassical agronomist perspectives that producing biofuels will create employment, rejuvenate rural economies, and increase food security through increased income. Those arguing against biofuels—politicians, civil leaders, and the media—argue along the lines of food price increases reflecting the global furor after the 2008 food crisis. Many of the arguments within the current debate draw on narrow domains within the food security discourse, eliding more holistic perspectives.

The South African case highlights the way the debate, specifically the economic perspective, removes the importance and interaction of local contexts and sociopolitical and economic circumstances, as well as potential environmental externalities. What remains necessary is a comprehensive and critical deconstruction of the situation and systematic identification of the major challenges the rise of a South African biofuels regime presents to food security. As I argue for the South African case, making agro-food systems work for the poor, or transforming rural economies using biofuels as a basis, are unlikely to have the professed or desirable outcomes for food security until a thorough understanding of the workings of the (global) agro-food system in such contexts is acquired. Scholarly attention to the issues surrounding biofuels from a South African perspective has also been relatively modest, despite the ramifications for development in general and food (in)security in particular. I contend that recent proposals within the international literature have important lessons for the South African context (and other countries facing similar dilemmas), especially in moving past the current impasse. Similarly, the debate that has taken place in South Africa so far sheds a contextual light on the more abstract elements of the global biofuels assemblage, the nature of the prevailing discourses, and their supposed rationalizing effect in local contexts.

Food Security and Biofuels: An Uncomfortable Couple?

First-generation technologies used to produce biofuels rely on sugar- and starch-rich crops (e.g., sugarcane, maize) to produce ethanol, or oil-rich crops (e.g., canola/rapeseed) to produce biodiesel (Naik et al. 2010). Their production competes with that of food as many of these are food crops as well. Second-generation technologies, using crop residues and other plant material may circumvent many of the challenges faced by the first generation (e.g., CO_2 emissions, competition with food) but maintain some distance from becoming mainstream (Vermeulen et al. 2009)[2]. First-generation biofuel production therefore remains resilient and its momentum largely unchecked (Smith 2010; Walker et al 2009). For this reason, the food-versus-fuel debate and the conceptual frameworks underpinning opposing arguments remains important, especially as we move into an increasingly institutionalized and growing biofuels market.

Food Security and the Changing Global Food System: From Traditional Perspectives to Global Food Regimes

South Africa's definition of food security, which resonates with declarations of the 1996 World Food Summit, recognizes food security as "physical, social and economic access to sufficient, safe, and nutritious food by all South African [*sic*] at all times to meet their dietary and food preferences for an active and healthy life" (NDA 2002: 15; see also FAO 2002; World Bank 1986).

Food security has been a central issue within development thinking since the 1970s. Concerns that declining global food supplies were facing a neo-Malthusian crisis dominated the earlier stages. Once (global) production stabilized, persistent hunger highlighted the linkage between poverty and malnutrition as more important than economic growth and increased production (Reutlinger and Selowski 1976). Drawing on these insights, Sen (1981) highlighted that entitlements at the household and individual level were actually key to ensuring food security. Through the entitlements framework, food security became easily overlapped with employment and was then drawn into wide-ranging issues from rural development to industrial policy (Maxwell 1990) and to some extent has now been overtaken by more robust frameworks around, for example, sustainable livelihoods (Chambers 1992). Through these various shifts it is understood that food security depends on *availability* of food, *accessibility* to food, *utilization* of food, and the *stability* thereof (Devereux and Maxwell 2001; Webb and Rogers 2003).

The increasing interconnectedness of a global agro-food system, including increasing awareness of both social and ecological dimensions, and its impacts on food security at all levels necessitated a further paradigm shift. By emphasizing system-based approaches, food security then becomes the primary but not the only outcome of a food system, dominated by increasingly sophisticated and complex market chains, which are sutured together through processes of urbanization, globalization, and industrialization (Maxwell and Slater 2003). Such a food system encompasses a range of economic, environmental, and social features undergoing rapid change (Lang and Heasman 2004, cited in Ericksen 2007; see also Maxwell and Slater 2003). These require new adaptive policy frameworks, robust enough to deal with new complexities and shifting the focus more broadly around food systems activities *and* their food security outcomes (Ericksen et al. 2009; Liverman et al. 2009).

Evaluating the food system requires an understanding of multicausality resulting from interactions among interdependent components and comprises four sets of activities, namely producing, processing and packaging, distributing and retailing, and consuming food. The interaction of biofuels with these components of the food system opens up a hitherto neglected opportunity to evaluate the implications of biofuels production for food security, especially at national and subnational levels. A full examination of the agro-food system is beyond the scope of this article and only the production element is later considered for the South African case.

Biofuels force a rethinking of priorities within the system; or rather crystallize the importance of systematic failures and potential contradictions across all levels. First, fears that drastic increases in biofuels production may outstrip global food supplies have prompted the re-emergence of neo-Malthusian anxiety that preoccupied thinking in the 1970s. These fears now have a global resonance in the form of market speculation and increasing volatility. Second, local aspects of biofuels production have important ramifications not only for local populations, but due to the interconnectedness of the global system changes in production in one region can cause massive repercussions globally. Such is the nature of what Evans (2011) calls an increasingly fragile agro-food system, in which vulnerabilities include the nature of the system itself but more so the marginalized and poor who are dependent on it. Food security therefore needs to be, but is not necessarily easily considered within the context of glocal agro-food systems (Robertson 2001).

Leading food security theorists such as Simon Maxwell, Polly Ericksen, and John Ingram emphasize the importance of understanding food systems, yet others provide more critical perspectives of such systems. Scholars looking at the political economy of agriculture have analyzed the changing context of the food system and globalization in terms of food regimes.[3] A food regime constitutes the system of food production and consumption and its relations to forms of accumulation and regulation during specific historical periods (Friedmann 1990, 1993; Friedmann and McMichael 1989).

The third regime is distinctly global and corporate, focusing on a neoliberalizing[4] capitalist project (e.g., Mares and Alkon, this issue). Young (2004: 13) succinctly highlights how regime analyses of food draw attention to fundamental contradictions at the heart of the globalization project; "food security is fundamentally incompatible with shifts towards liberalization and globalization." In other words, even if the system appears legitimate, it may—through its very nature—renege on its most basic and important outcome, namely food security. Young (2004) argues that "liberalization policies exported across the globe with such vigor since the 1980s have dangerously undermined the ability of national governments to protect their population's health and nutritional status." In addition, whereas focus on the household and individual (accessibility) shaped food security discourses over much of the past thirty years at local levels, the dominant approach to development followed a rising tide of antistate sentiment and a focus that essentially locked agrarian reform and food systems into the vagaries of the globalized free market (Kent 1999), reinforcing an increasingly criticized industrial agriculture paradigm[5] (McMichael 2000). This system benefits some but also marginalizes many. Wealth creation and its equitable distribution are neither automatic nor sometimes even existent, paralleled by differential effects on food security.

These leftist critiques complement a rich intellectual commitment to understanding agro-food systems in general. Considerable debate persists with regard to the costs and benefits of deregulating and liberalizing agro-food markets (Bardhan 2005; Gibbon 2003; Hertel and Winters 2006; Raynolds 2004, cited in Jacobs 2009; von Braun and Diaz-Bonilla 2008) and the benefits of a global food systems in general (Kent 1999). Whereas it is hard to deny that innovations in the current system have created unprecedented efficiencies and surpluses, the benefits have not been ubiquitous and there are always externalities. It is then within the local contexts of agro-food systems where other analytical foci—for example, rural development or livelihoods frameworks—remain important to understanding local dynamics that could affect food security in the face of such global imperatives and theorizing. These frameworks illustrate that it is not only the nature of global agro-food systems (availability) but also their counterpart and localized contexts (in which access occurs) that allow us to interpret the widespread effects biofuels may have on food security.

Food versus Fuel: International Debates

Two salient narratives have dominated the discussion of biofuels and food security. The most obvious and important are concerns of the impact biofuels production have on global food prices and thereby poverty (Eide 2008; Royal Society 2008). This so-called battle between food and fuel is outlined within the food-versus-fuel debate (Doornbosch and Steenblik 2007), directed largely by notions consistent with availability and accessibility.

The reduction of available food by using it for fuel seems readily apparent. To qualify such an assumption it is first necessary to ask how much food (or land or resources for food production) has been and will be directed into energy crops, and whether this is likely to deplete food surpluses to levels below aggregate demands? The overall thinking here is that globally there is enough available food at aggregate levels. This presumption is becoming increasingly contentious. In a 2008 speech, director-general of the Food and Agriculture Organization (FAO) Jacques Diouf has projected that food supplies will need to double by 2050 to maintain aggregate supplies. This figure might, however, be hugely inflated as it is based on spurious projections that have acquired a subsequent discursive currency, shaping the discourse without being verified by rigorous analysis (Tomlinson 2010). Also neglected is the significant but preventable waste of available food within the food system (Stuart 2009). Nevertheless, increasing populations and a potential rise in wage earners/incomes, especially in developing countries, signal the need for increasing not decreasing food production in order to meet future demands. How to ensure this is achieved sustainably is also of growing importance (Godfray et al. 2010).

Two main arguments refute claims that biofuels diminish available food. First, biofuel production has only consumed modest amounts of the global food supply. Pfuder et al. (2009), for example, illustrate that the proportion of maize used for bioethanol increased from 4 percent in 2001/2 to only 12 percent in 2007/8 globally. Food and animal feed still dominate agro-food commodity markets and therefore biofuels cause only a limited impact on aggregate levels compared to other factors, such as drought, trade policies, declining reserves, increasing incomes, and changing food preferences in developing countries). Second, most grain grown is used as animal feed and not for human consumption. The production of biofuels has even less of a negative effect on this market as much of the nutrition and monetary value is recuperated (if not dictated) through demand for nutritious by-products, such as distiller's dried grains and solubles (DDGS) in the case of maize and oilcake from Soya(Currie et al. 2007). Maize used for ethanol, therefore, incurs a proportional but not absolute loss to the overall agro-food commodity market. Any increases in biofuels production are considered likely to add to the global feedstock supplies through increased supply of these by-products and thus improve availability.

These arguments do not necessarily take into account future demands, which could outstrip food supplies, especially in the face of already declining storage and available surpluses (Peskett et al. 2007). Existing surpluses are important (although controversial) as food aid to many dependent, developing countries and as a protection against global shortfalls and price shocks.[6] Moreover, there is no institutional architecture able to ensure that aggregate levels of food are maintained and the market has failed to do so (Koning and Mol 2009). Finally, by increasing available feedstock and thereby decreasing input costs for livestock, meat and dairy prices might be reduced but the majority of food insecure people are not in a position to take advantage of such increased availability.

As has been highlighted elsewhere (e.g., Young 2004) the case for availability working through a market system has only limited effects on improving overall food security. Although having more food available is obviously beneficial, this ostensible benefit from biofuels production suffers from all the shortcomings of previous global oversupplies in food, with the added risk of destabilizing that supply. It is also important to acknowledge that at local levels especially, land used for food and/or livelihoods services might be displaced[7] by energy crops with potentially dire impacts on both availability and accessibility (Molony and Smith 2010). Others have suggested that such impacts are better determined within broader ecosystem services assessments to ensure multiple benefits are assured and all negative ones accounted for (Gasparatos et al. 2011), at the same time recognizing that such impacts may not be as clear cut as these authors suggest.[8] Such contention can be seen in the most profound impact of even a modest use of biofuels globally, the spike in food prices since 2007/8.

In 2008, massive increases in food prices were accompanied by widespread food riots in the developing world (Brown 2008). According to the FAO (2008), export prices of wheat increased by 130 percent and corn by 38 percent between 2007 and 2008. Because grain prices increased the most, developing countries reliant on these staples suffered greatly (Clements 2008) as did the poor who spend 50–60 percent of their income on food (Von Braun 2008). Oxfam, for example, estimates that "the livelihoods of at least 290 million people are immediately threatened by the food crisis, and the [World] Bank estimates that 100 million people have already fallen into poverty as a result" (Oxfam International 2008: 3).

Interpreting the crisis remains a matter of debate. The conclusion that food security has been compromised presents less of an issue than how it has occurred and how much biofuels are to blame. Three broad types of review, which include historical reviews, projections (modeling) and commentaries from experts, have offered some explanation (Gerber et al 2008). Most assessments are ad hoc and consensus among the three depends on the assumptions made, the types of calculations or models used, and the experts or institutions doing them. Early findings differed considerably and, given the multitude of causal factors and their relative influences, may be indeterminate (Abbott et al. 2008; Eide 2008; Gerber et al. 2008; Lipsky 2008; Mitchell 2008; Molony and Smith 2010).

Although most reports agree that biofuels played some part in the price spikes, predicting the future implications of biofuels production is challenging. Several studies forecast significant future increases in world food prices due to the demand for feedstock for biofuel production (Rosegrant 2008; Von Braun 2008; Wiggins et al. 2008). Other experts use longer-term modeling to predict that markets will correct themselves; the current shock should be considered only a short-term fluctuation representative of the price inelasticity commonly associated with food commodities (Pfuderer et al. 2009). The only certainty is that the low food prices we have become accustomed to in the twentieth century are a thing of the past with biofuels exacerbating the situation in complex ways (FAO 2008; Koning and Mol 2009).

Linking Availability and Accessibility: Food versus Fuel or Food and Fuel in the Agro-Food–Fuel System

In a simplified picture, the new food-fuel nexus has been characterized by two things: a depletion of what food is available (not necessarily in a neo-Malthusian sense but more in a speculative market sense), and a subsequent rise in food prices, which has eroded accessibility. However, the picture is more complicated than simply assuming the accessibility and availability arguments. The distinction between them is also arbitrary; biofuels do not simply create a trade-off between one and the other. Essentially the price fluctuations are symptomatic of the interconnected and market basis of the global agro-food system. As discussed earlier, such a system has many benefits but also has many faults, which through increasing interconnectedness and concentration has only thin margins for error and is growing increasingly vulnerable to economic, environmental, and other shocks (Evans 2011). At a global level, biofuels production extends the risk in the system by creating new dependencies while potentially reducing supply.

Phillip McMichael, one of the main contributors to the food regime discourse has articulated a useful understanding of biofuels within the framework of the third food regime. Biofuels signify a crisis in the existing (corporate) food regime, by displacing food with fuel and reneging on corporate promises to provide the world with food. More so, it exemplifies the nature of capitalist industrial agriculture, now also a global assemblage, continuing the externalization of capitalism's costs but through "the distraction of green fuel" (McMichael 2010: 609; see also Smith 2010). Ultimately we should be critical of a system in which the currency of food security rhetoric is in its ability to disguise or propagate contradictory objectives. These objectives are manifest as part of a new agro-food–fuel system.[9]

The restructuring and pressures currently facing the global agro-food–fuel commodity market has met with an urgent call for institutions capable of stabilizing prices (which existing hedge fund mechanisms and markets structures have failed to achieve), while monitoring the commodities flowing into the energy and food markets (Koning and Mol 2009). If biofuels are not to deplete available food, institutions and policies are needed to limit any misuse of existing supplies. The existing provisions for such frameworks are severely limited given the lackluster, if not nonexistent governance of food security in a globalized context (Evans 2011; Mohamed-Salih 2009). Paarlberg (2002) points out that food security is generally better governed at local and national levels, which may suggest a similar response for biofuels (see also, Young, 2004). At the very least, local contexts are as important as the global perspective.

Diaz-Chavez et al. (2010) have, in some ways, renewed optimism for biofuels by stating that they can coexist with food security, and do so in many African countries. Drawing on the experiences of Tanzania, Kenya, Mozambique, Zambia, Mali, and Senegal they conclude that biofuels and food can be grown without major impacts on food security; however, this requires direction from well thought-out policies. Questions of the real benefits of biofuels and implications for food security are therefore intrinsic to local contexts, as articulated in the narratives underlying the rural development benefits of biofuels. At the same time, such optimism is tempered by similar work that comes to different conclusions, questioning such win-win outcomes (Matondi et al. 2010). Matondi and colleagues point out complex local circumstances in which opposing narratives, be they empirically founded or not, can dismantle biofuel projects and their potential benefits, even when food security benefits were beginning to accrue (see specifically the Ghanaian example by Prosper Boamah [2010). Although global aggregations and generalizations hide important nuances within local levels, the reality is mixed.

Biofuels, Rural Development, and Agrarian Reform

As noted earlier, the food system comprises four main activities, with production being one of them. It is within the realms of production that food security as an outcome is often misconstrued with the production of food and more so rural development in general.

Ambitions of rural development, drawing on the mantra of an unlimited market for biofuels, have been prevalent within the biofuels lexicon for some time. Prophetically, by increasing market opportunities and therefore increasing employment or income, food security will be enhanced. Some consider this a paradigm shift associated with biofuels that would transform hitherto neglected rural economies in the developing (and developed) world, stabilizing commodity—and thus food—prices in the long run (see Peskett et al. 2007 for a useful overview). Similar to many blueprint development narratives in the past, the beauty of biofuels is the ease with which they confront multiple challenges including rural underdevelopment. The popular assumption insists that such underdevelopment is well understood, suffering only the absence of suitable agrarian reforms, or what Scoones et al. (2005) critique as quick fixes approaches. Scoones and colleagues (2005: 11) aptly remind us that "there is of course no magic bullet for the problems of African agriculture: no technical, market, institutional or policy fix."

The main problem with the biofuels equals rural-development narrative is that it is dominated by a laissez-faire approach to agrarian reform in which the so-called quick-fixes—commensurate with current hegemonic liberalization and modernization discourses—are relied upon. The efficacy, equity, and sustainability of market-based and industrial agrarian reform are questionable as markets also fail (Jacobs 2009).[10] Even recent proposals to transform the agro-food commodity chain, based on the World Banks' vision of making agro-food markets work for the poor, neglects that such transformation affects the poor in complex ways that differ from neoclassical textbook stories. Incorporating food security concerns more explicitly into agricultural marketing policies is vital to overcome the contradiction in the modern food system in which those working in the agro-food sector tend to experience higher levels of food insecurity and are more vulnerable when markets fail.

The type of agriculture being perpetuated for biofuels production serves as a useful example here, by falling largely into agrofuels or large-scale ergoculture (Mathews 2007). Agricultural reform in this instance is a modernizing campaign based largely on capital-intensive mechanization and industrial inputs, sometime framed within the context of green revolution. However, evidence suggests that such large-scale industrial enterprises and the pattern of biofuels production and consumption are unlikely to benefit developing countries and especially the rural poor within those countries (Ariza-Montobbio et al. 2010; Dauvergne and Neville 2010; Matondi et al. 2010; Richardson 2010; Wolde-Georgis and Glantz 2009; Woods 2006). These are especially problematic in the form of numerous so-called land grabs (mostly foreign acquisitions of land usually based on industrial-based monocrop farming for export), not just for biofuels production but for food as well. Others argue that, although the potential for harm does exist, foreign investment into agriculture has its benefits and need not be an exercise in exploitation (see Cotula et al. 2008 for a useful outline).

If anything, concern about biofuels production highlights "the highly questionable sustainability of most of modern agriculture" (Moore 2008: 6; see also the critical appraisal by the International Assessment of Agricultural Knowledge, Science and Technology for Development 2009a[11]). Any calls for a green revolution (Conway 1998; Eicher et al. 2006; Mosley 2002; Sanchez et al. 2009) should also consider the (infra)structural, agroecological, climatic, and geomorphologic constraints that are particularly common to agriculture in Africa, as well as dynamic social, economic, and political diversity at all levels. For one, political economy factors nearly always conspire to return weaker players to their marginalized positions regardless of any new technological advantage (Kent 1999). Similarly, benefits (essentially profits but food security as well) do not increase proportionately with increasing yields (Leisinger 1995, cited in Kent 1999; see also Richardson 2010).

Whereas green revolution technologies present technical fixes that need to be adapted to specific circumstances, regulating or fixing food commodity prices through increased demand in the case of biofuels have come to represent common practices as policy fixes. The FAO (2008) states that higher agricultural prices could revitalize the role of agriculture as an engine of economic growth over the medium to long term. However, they stipulate an important yet often neglected caveat—urban residents and the large number of net food buyers in rural areas are likely to be negatively affected. Such a conundrum is considered the food-price dilemma in which rising commodity prices benefit net producers of food but may harm net buyers (Ravallion 1989; Sah and Stiglitz 1992; Timmer et al. 1983). It is generally assumed that rural people are net growers of food so they are most likely to benefit from increased prices. This, however, bears little resemblance to what happens in reality (see Weber et al. 1988 for an African perspective; and Van Zyl and Kirsten 1992 for South African perspective).

Contextualizing the above, White and Dasgupta (2010) provide an informative perspective of biofuels and rural development by drawing on political economy analyses of nonfood agricultural commodity markets. They conclude that it is not so much biofuels as a commodity crop but the manner in which these crops are grown, under which forms of ownership and labor regimes, and in what kinds of commodity chains. Following Henry Bernstein's (1992: 24) neat framework that asks: "Who owns what?" "Who does what?" "Who gets what?" and "What do they do with it?" they ask more specific questions: "Where the land will come from?", "How will production be organized?" and "for whose benefit"? The simplicity of these questions is attractive but answering them, while giving specific and adequate attention to food security, can be a complex undertaking. The overall emphasis here overlaps with emphasizing and understanding rural heterogeneity, embodied by the post-Washington Consensus (Maxwell and Ashley 2001; Scoones et al. 2005). Underpinned by the continued importance of the agricultural sector as a driver of economic well-being and rural development, there are multiple livelihood strategies and circumstances of considerable flux, driven through forces such as globalization, deagrarianization (Bryceson 1996), or legislated agrarian labor reforms (Aliber 2003; Greenberg 2010).

Synopsis

Biofuels may present challenges and opportunities to food security. Initially, their interaction was defined by the impact on availability (more through speculation than actual quotas) and the concomitant rise in food prices, which eroded the food accessibility of the poor. This is precisely the situation encapsulated by the food-versus-fuel debate. The interaction between biofuels and food security needs to be considered within broader frameworks that acknowledge both their interconnected nature and characteristics common to their production. It is here that the conception of a biofuels assemblage is instructive, acknowledging the interaction of biofuels with multiple systems, at multiple levels where debates and discourse around these interactions are set substantially according to powerful entities within them, and used more to appraise praxis than guide practice. The reality is that in some cases the impact on food security becomes an abstraction around which rhetorical debates are fashioned rather than an outcome to be quantified, qualified, and any negative impacts eliminated. Yet, as we have seen, there are also theoretical approaches that exist with which to identify and ameliorate such concerns.

Within the production element of the agro-food system, biofuels at best present an opportunity to transform hitherto neglected rural economies and allow new entrants (i.e., emerging farmers) into the market; at worst they are set to reinforce existing disparities between the rich and the poor. Embodied within such proposals are assumptions of rural development, yet rural development itself is contested, involving a plurality of solutions rather than the simple quick fixes many biofuels proponents advocate. Understanding the context in which such rural development is to progress are necessary in order to understand the implications such development poses for food security, even if fueled by biofuels. The next section focuses on the debate around South Africa's (potential) uptake of biofuels and the implications for food security, especially within this production realm. Ideas drawn from the international arena may serve as a useful catalyst to transform the situation from a polemic debate, with hostile oppositions, to a more dialectic compromise.[12]

Food-versus-Fuel Debate in the South African Context

South Africa can generally be considered food secure at the national level, but local and individual food insecurity remains a persistent challenge (Altman et al. 2009; Van Zyl and Kirsten 1992). For example, at least 14.3 million people are defined as vulnerable to food insecurity and 43 percent of households as vulnerable to food poverty in South Africa (De Klerk et al. 2004). It should be noted that recent reviews of existing datasets show a considerable uncertainty when it comes to quantifying national and local food insecurity statistics (Labadarios et al. 2009). In 2007/8, South Africa experienced rapid food price inflation through knock-on effects in the international market, despite suffering no drastic changes in local supply, especially not due to biofuels production.

As is the case internationally, the South African biofuels regime comprises multiple stakeholders with their associated ideologies and interests. Historically, South Africa's relationship with biofuels began in the 1920s when ethanol derived from sugar cane was mixed with petrol (Blanchard et al. 2011). Following global trends, agricultural lobbies, agribusinesses, and commercial farmers in South Africa have recently latched onto biofuels as a potential new (and endless) market for a domestic oversupply of maize and new profit streams for existing commodity chains. During apartheid, maize farmers and agricultural lobbyists enjoyed considerable government support, which has slowly eroded due to restructuring and liberalization beginning in the 1980s. After full liberalization in 1996, concentration within the overall maize market at farmer, silo, and miller levels has created a powerful bloc within the market; with biofuels presenting them a lucrative opportunity. Commercial farmers can, and in some years have also produced approximately 11–12 million tons of maize, nearly 3 million tons more than what is required for local (food) consumption (Kupka and Lemmer 2007; Makenete et al. 2007). Drawing on the US experience, early lobbying for a maize-to-ethanol industry and feasibility was based on these potential surpluses, backed by powerful industry players (African Centre for Biosafety 2008).

A biofuels Feasibility Report (BTT 2006), which was dominated by a commercial agriculture focus, was finalized in 2006 and included maize-to-ethanol as a significant part, while only haphazardly addressing many issues, including food security (McDaid 2007). Thereafter, the South African government released the Draft Industrial Biofuels Strategy for public consultation. At the time, the international media were beginning to sensationalize rising food prices globally, stoking anxiety around the food-versus-fuel debate. Mounting domestic food security–related concerns and considerable objections from the public (including NGOs, labor unions, and environmental lobbies) dominated the consultative process. Even the political leadership began to question the biofuels proposals with the Reserve Bank governor Tito Mboweni warning against the use of maize[13] for biofuels due to potential food price increases (Swanepoel 2007).

Counterclaims soon followed, largely from agricultural economists with interests in maize-to-ethanol conversion, calling the exclusion of maize premature, ill-considered, and detrimental to food security in South Africa (Kupka and Lemmer 2007; Makenete et al. 2007). Nevertheless, the South African government decided formally to exclude maize as a feedstock,[14] inadvertently shackling in part the future development of a biofuels industry. Removing any mandatory blending in the final strategy, providing few incentives, and prohibiting existing commercial farming areas have also stifled the development of a biofuels industry (Funke et al. 2011). The decision was considered politically charged and not necessarily technically derived, given the original feasibility assessments. Notwithstanding potential distortions from vested interests within the feasibility study itself, the potential benefits and powerful agribusiness interest in pursuing biofuels has meant the exclusion of maize has since become the pivot around which the polemic food-versus-fuel debate in South Africa has been hinged. It is against this backdrop that one has to evaluate the benefits of a biofuel program in South Africa and its potential impacts on food security in the country and the region. In addition, it is important to uncover the nature of these debates in which food security concerns are now addressed largely as a matter of national supply, volatile markets, increased labor, and a response to the effects of not including maize. Furthermore, following the neoclassical approaches and a bias for modeling, the nature of food security seems to be abstracted from specific realities.

The Biofuels Strategy and Food Security in South Africa

Despite its potentially major impacts on both food security and the agricultural sector at large, the Industrial Biofuels Strategy has not encountered widespread academic attention thus far, especially with regard to widespread food security implications. With no large-scale agro-fuel projects in operation (Esterhuizen 2009), there has been little chance for empirical explorations into the subject but there is room for theoretical debates, and predictive modeling has at least provided some preliminary results. Much of the published academic material has focused on situational analyses, production potentials, and overall feasibility, with modeling undertaken to determine the macroeconomic effects, support measures, production potentials, impacts on the feed market, and so on (Funke et al. 2009; Haywood et al. 2009; Meyer et al. 2008; Ngepah 2010; Strydom 2009; Strydom et al. 2010; Von Maltitz and Brent 2008; Von Maltitz et al. 2009). Food security is largely considered through ancillary food price effects within the various modeling activities or labeled an issue requiring caution. Also notable is a succinct paper by Adeyemo and Wise (2009); however, they only consider sugarcane as a feedstock for bioethanol and sunflower for biodiesel. Should appropriate management be employed when using these crops there are not likely to be significant impacts on food security. There has been little debate over the use of other crops so focus here is on the various assumptions made around the food security concerns of maize. Following a similar pattern as the international debate, arguments in South Africa have coalesced around price issues and the narrative of improved rural economy meaning improved food security.

Within the initial Feasibility Report, potential food price rises were understood to be a potential concern, although were not considered in depth if not completely swept aside (McDaid 2007; Sugrue and Douthwaite 2007). Rather, a Social Accounting Matrix is used to calculate impacts on the national economy, showing considerable improvements to the gross domestic product, income, and employment (BTT 2006). Food prices are considered more explicitly through modeling undertaken by the Bureau for Food and Agriculture Policy (see Bureau for Food and Agricultural Policy (BFAP) 2005), which used baseline scenarios to establish the impact of biofuels production on the agricultural commodity markets in South Africa. According to the Feasibility Report, which draws on the BFAP work, "preliminary calculations show a marginal price increase as a result of the increased demand; namely 7.5 percent for milk, two percent for chicken, 9.6 percent for beef, and 2.5 percent for eggs per annum until 2015" (BTT 2006: 54). All of these products are high value and do not identify with potential impacts on the poorest of the poor, at least not in terms of their staple diet.

The BFAP models, such as the one mentioned in the Feasibility Report, are probably the most comprehensive modeling series predicting the macroeconomic impact of biofuel production in South Africa (see Bureau for Food and Agricultural Policy 2005, 2007; Funke et al. 2009; Meyer et al. 2008). The analyses also provide a summary of the support necessary to ensure a viable industry. In an optimistic 10 percent ethanol blend (in the national fuel supply) and 5 percent biodiesel blend scenario, under normal weather conditions, with government administering the requisite (but unlikely) tariffs to ensure feasibility, the prices of white and yellow maize increased by 12 percent and 18 percent, respectively (BFAP 2007: 23). Should shortages occur, during a drought for example,[15] the local maize industry can be expected to move quickly to import parity scenarios (higher prices for maize). Similarly, increased demand for yellow maize (used as animal feedstock and favored for bioethanol) will displace other field crops with a likely increase in the price of white maize (used for human consumption in the form of maize meal). In the second, less likely scenario (with no protective tariffs provided by government), the white and yellow maize prices are projected to increase by 7 percent and 11 percent, respectively (BFAP 2007: 30; see also Strydom 2009).

Models such as these should obviously be treated with caution, as even their developers hasten to suggest. The maize industry in South Africa is extremely vulnerable to price spikes influenced by international markets, drought, and oil prices, so similar changes in staples may actually occur regardless of biofuels. What seems apparent is that there has been little emphasis on redressing the exposure to international shocks and oil prices through the reliance on intensive inputs as well as better managing the risk of drought by switching to less vulnerable systems.

Other reviewers modeling the affect of biofuels show a similar trend of increasing food prices (Wiggins et al. 2008, cited in Gerber et al. 2008). While illustrating the importance of biofuels in reducing price volatilities, these models highlight that there are potentially dire food price implications. Again, one can anticipate significant variations depending on the scale of production, type of farming, and feedstock to be used.

Generally, these models have not been employed exclusively within the food security debate; such models are considered inadequate to base a rigorous argument for or against biofuels, although they already highlight potential challenges. Those objecting to biofuels have the proverbial upper hand in that global food price increases attest to the pervasive negative impacts biofuels can have. Those protesting against the use of maize on these grounds therefore welcome the so-called ban on maize. There is, for example, no reason to assume that a similar price inflation will not occur from domestic production, placing the poorest and most vulnerable at risk (Sugrue and Douthwaite 2007).

Proponents of biofuels emphasize the more positive aspects of these models to argue more broadly for the rural development potential of biofuels. The concomitant food security rhetoric is used to bolster their arguments. While the price of food *may* increase, they tend to move beyond price scenarios, in which the focus is purely on *potential* increases in prices, focusing rather on macroeconomic benefits (job creation) and stabilizing domestic agro-food commodity markets that are equally important to food security.

A central theme or familiar rationale underlying the prospects for rural development is the employment potential that a biofuels industry could create. The Feasibility Report suggests that "the agriculture sector is expected to create approximately 30,000 direct and indirect jobs due to the introduction of the biofuels enterprise, assuming that new crops are planted on new lands to make up the increased demand … the number of jobs created in the value chain is estimated at 34,300" (BTT 2006: 24). Including all direct, indirect, and induced jobs, this figure rises to 48,399 resulting in an increase of household income generation[16] of R1,700 million per annum (BTT 2006: 43). Similarly, agricultural economists dismissive of the exclusion of maize (e.g., Makenete et al. 2007) have subsequently made even stauncher claims about food security and the sustainable employment potential that a biofuels industry has to offer, as well as the huge benefits they present to the agroeconomy in general.

Taking an agronomist perspective, using maize makes sense for a few reasons. For one, a logistical analysis undertaken by ABSA bank (Makenete et al. 2007) indicates that existing agricultural areas are optimal and land and surplus crops are available both within the market and within economically feasible proximity. The fact that increased production will generate employment opportunities, and through increased production of DDGS decrease the price of animal feed, makes maize an opportune feedstock.[17] Such logic sounds extremely familiar considering the international perspectives but elides the more complicated nuances. Finally, prices within the grain market fluctuate considerably on the South African Futures Exchange, making it difficult for emerging farmers (and their existing commercial counterparts) to compete. When surpluses are produced, falling price signals reduce planting in the next season, which in turn leads to production deficits and rising prices, creating a continuous boom-and-bust cycle. In addition, year-on-year surplus production cannot be maintained, as local farmers cannot easily compete in the international market, which is a sign of both a potentially inefficient domestic agricultural sector and a skewed market internationally. The uptake of these surpluses for biofuels would stabilize this volatility internally, ostensibly stabilizing domestic food prices as well. Securing demand is, however, not the same as agricultural reform. It may well reinvigorate a declining commercial agriculture sector but any potential impacts on food security are neither immediate nor guaranteed.

Those supportive of biofuels argue that excluding maize will result in a lost opportunity for increased employment and in support provide sound economic reasons as to why. In much the same way, they have latched onto this employment-cum-food-security narrative to pursue their own interests, rather than considering the potential to change existing production practices[18] and transform the rural economy. It should be noted that not all agronomists come to such conclusions; for example, Ngepah (2010) from the Council for Scientific and Industrial Research shows that the impacts of using maize on the rural poor are actually negative but insignificantly so. Although it is disadvantageous to overlook basic market principles and economic rationality, and these agroeconomist arguments are theoretically legitimate, it is equally important not to rely solely on their predictions, especially when considering implications for food security.

One of the most important criticisms of relying purely on an agricultural economics perspective is the highly skewed political economy of the agricultural sector in South Africa. On the one hand, there is a highly capitalized commercial sector dominated largely by white farmers, who were highly subsidized prior to market liberalization (beginning in the 1980s), own a majority of the agricultural land, and produce upward of 90 percent of the country's produce (Vink and Kirsten 2003). In contrast there is an underdeveloped, small-scale subsistence-orientated sector dominated by black farmers, which was largely neglected during apartheid and has enjoyed only limited support thereafter (Vink and Kirsten 2003).

This persistent duality is a legacy of intra-agricultural biases introduced during apartheid in which incentives, laws, and institutions favored large farms and discriminated against smaller and labor-intensive farming (Lipton et al. 1996; see also Ortmann and Machethe 2003; Pieterse and Van Wyk 2005; Vink et al. 2000). Support within the agricultural sector remains biased toward the privileged more than toward the poor and liberalization has not benefited small-scale farmers and farm workers. Where support has been attempted, for example by the Comprehensive Agricultural Support Programme, the comprehensiveness is questionable (Greenberg 2010; Hall 2009). By and large, however, there has been a near official refrain from critical discussion of the economic and policy environment into which new farmers enter. Greenberg (2003) argues that continued agricultural restructuring programs, initiated largely by the apartheid government, have skewed the concentration of resources in the agricultural and agroprocessing industries toward the segment of the population with relatively greater wealth and resources at their disposal. Concentration within all links of the agrofood market, especially of maize (Chabane 2002), has further reduced the ability of the sector to provide food and employment to all those in need. The restructuring has also resulted in unstable and rising food prices (Greenberg, 2003: 96).

To be sure, agriculture, especially the commercial sector, remains important, in fact central to the rural economy, the wider agrofood industry, and national food security. Primary agriculture accounts for approximately 3 percent of South Africa's GDP (National Department of Agriculture 2009), while the agro-food industry accounts for approximately another 9 percent (NDA 2001; Vink and Kirsten 2003). Over a longer period agriculture's contribution to GDP has, however, declined from approximately 20 percent in 1951 to 6 percent in 1990 (Lipton and Simkins 1993: 361). This decline has been matched by falling employment but not productivity. Commercial agriculture represented almost one-third of formal sector employment in rural areas in 1998 even though regular employment on farms had declined by about 20 percent during the preceding decade (Aliber 2003). Steady declines in overall employment in commercial agriculture, which are expected to continue (Aliber et al. 2009), have also been marked by a shift toward smaller more skilled (and male) labor forces (Vink and Kirsten 2003), accompanied by a rise in casualization and labor broking (Greenberg 2010). Even the South African National Department of Agriculture has acknowledged that "it is doubtful whether agriculture can be regarded as a priority sector when interventions to create decent employment are implemented, unless solutions outside the commercial sector are explored" (Department of Agriculture, Forestry and Fisheries 2010: 27).

Commercial agriculture is fundamental for growth in the national economy, but it is socially dysfunctional in the way it is providing fewer livelihoods and exacerbating poverty and inequality in the population (FAO 2003, cited in Hall 2007). The economic challenge is to broaden the *distribution of incomes* from agriculture (Hall 2009: 129; emphasis in original). In addition, whereas expansion of labor-intensive agricultural output—as might be expected by an increasing demand for biofuels—expands opportunities for surplus or unemployed rural workers with low skills (Bardhan 2005, cited in Jacobs 2009), the alleviation of poverty depends on the actual wages. As Jacobs (2009) indicates in his review of South Africa's agro-food system, even by widening income distribution through a growing sector, the agricultural wages might be too low to lift wage-dependent rural households permanently above a socially acceptable deprivation threshold (Jacobs 2009).

The highly skewed nature of the agro-food system presents a dilemma. How would the uptake of biofuels reform or reinforce such inequalities? Within the Feasibility Report and Draft Industrial Biofuels Strategy, there is, for example, a strong reliance on the existing commercial sector and surplus production, which was based on more technical (quantitative) modeling processes within the feasibility studies. This large-scale, industrial, modern rationale permeates into the final strategy, although political influences shifted its geographical focus, rather ambitiously, toward development of the former homelands and the so-called second economy.[19] Existing commercial farming areas outside of former homelands were thereby excluded. It is questionable, however, whether such practices are consistent with the reality of these areas[20] and makes doubtful the coherence of the final strategy. In addition, without consideration for the broader political and economic contexts, any questionable nature of these approaches is likely to be perpetuated if not exacerbated when transposing them.

The centralization of agrarian-based (and notably the biofuels) policy has also meant little dynamic consideration of alternatives, especially at the local level. The overall imperative is that rural economies that remain dominated by the so-called backward subsistence farmers need to undergo transition to the more favorable agribusiness- based approaches, with current proposals reflecting the large-scale industrial rationale. This is important, as there is still a clear bias toward a transition or progress between the two, rather than transformation of one or the other or more important both. As Hall and Cliffe (2009: 17) lament in terms of the land reform process, "one of the main concerns … is the absence of, and need for, a wider vision of agrarian restructuring, that land reform is to bring about." There has therefore been limited space for counterpoints and alternatives (for example food sovereignty) to gain a footing within policy discourses around agrarian reform (transformation) in South Africa (Greenberg 2010).

Even the Feasibility Report put forward some recommendations of sustainable agriculture, yet these seem to be token gestures and have been marginalized through the further development of the policy. It also explicitly states that emerging farmers are unlikely to benefit and a separate strategy would need to be developed for them (BTT 2006: ix). This has neither happened nor does it seem likely.

Concluding Discussion

From the above discussion, it is questionable how fully the consequences of biofuels production on food security have been considered in South Africa. It is not about food prices or rural development, it is about both, and the links are more complicated and context-specific than ardent proponents and antagonists would have us believe. Underlying much of the above review is also a criticism of the inadequacies of neoclassical economics (free-market reforms) in planning for the impacts on food security, which are spread unevenly between the rich and the poor through complex interactions within the agro-food system. The outcome of adding biofuels to the mix is likely to be no different without a fundamental re-evaluation. Much of the debate that has taken place is documented above, but this only touches on the debate that needs to occur. Much of the attention that should be given to this complex issue is occluded by ideology and extractions of a much wider and more complicated assemblage and food system.

Although employment and rural development are important, food systems transcend both production systems imagined within econometric models and the essential needs of accessibility and include the importance of wider systems that interact with one another over the long and short term. From a systems perspective, if food security is to improve, there is a need to review not only production-based alternatives (including that related to the small-scale) but market and food distribution (infrastructural) changes and consumption patterns as well.

Many of the current arguments for biofuels in South Africa, though cloaked in food security rhetoric, are equally dismissive of the broader food security theories described above, eschewing critical perspectives of how the biofuels agenda may reinforce inequalities within the existing political economy of the agro-food system. Focusing on prices and the use of maize when drawing on limited domains of the food security framework elides attention being given to how biofuels can transform the agricultural sector (or production element of the agro-food system) to improve food security holistically, as well as interrogate the complex interlinkages between the two. Although looking more broadly or finding a middle ground lacks ideological purity, there is pragmatic value to taking, or at least including system-based and critical approaches to the food-versus-fuel debate. The answers (and questions) are also not easy ones. Simply excluding maize from a biofuels strategy paints all agricultures and all opportunities with the same brush, masking the finer details. This reifies the current theoretical impasse, dislocating holistic approaches from practical realities in which vested interests currently dictate practice and praxis substitutes for theory.

At its most basic, this article demonstrates that the skewed nature of the food production component (the agrarian sector) appears to be at odds with increasing food security and symptomatic of other components of the agro-food system in South Africa, characterized by concentration, industrialization, deregulation, and growing inequalities. In the past, centralized mills based on large-scale industrial agriculture, which was heavily subsidized and protected, have skewed the rural landscape, while concentration within the sector continues to do so. Although stifled by a moratorium on maize, the major maize-to-ethanol projects proposed within South Africa follow similar patterns (African Centre for Biosafety 2008). While the moratorium stands, there is a chance to critically re-evaluate the situation. At the heart of considering biofuels production should be reform of the existing agricultural political economy that remains largely based on substituting capital for labor, (super)exploitation, and dispossession (Greenberg 2010).

Finally, if biofuels are to share a space in the agro-food system, there is a need for institutional arrangements to ensure food and fuel do not compete in ways that are detrimental to food security, as is recognized internationally (Koning and Mol 2009). These require more than a market basis, as reliance on the market to ensure food security has been harmful to South Africa in the past, especially considering that many of the poor operate within a highly skewed and previously highly segregated market system (May 1999). Further, there are limited institutional arrangements capable of coordinating major food security interventions in times of crisis in South Africa (Drimie and Ruysenaar 2010), let alone situations where food competes with fuel. Getting the institutions right will only occur when a greater understanding the impact biofuels will have throughout the agro-food system is acquired.

ACKNOWLEDGMENTS

I thank Caryn Abrahams, Scott Drimie, Laura Pereira, and Peter Ranby and Hans Ruysenaar for their helpful comments and criticisms as well as the editorial assistance provided by the special edition organizers. I also want to thank the anonymous reviewers for their perceptive critiques. These have helped reshape and galvanize the article, although constrained space and time has limited the inclusion of all the valuable insights. The usual disclaimer applies.

SHAUN RUYSENAAR is pursuing a PhD in African Studies at the University of Edinburgh. Ruysenaar also works on the Policy Innovation Systems for Clean Energy Security project, a DFID-funded partnership of institutions in Kenya, India, Sri Lanka, Tanzania, and the United Kingdom, providing policymakers with new information and approaches to unlock the potential of bioenergy to improve energy access and livelihoods in poor communities. The views expressed in this review are entirely those of the author and do not necessarily represent DFID's policies or views.

NOTES

1. Ong and Collier (2006) use the construct of an assemblage as an analytical response to the more abstract notion of the global and distinctive specificity of the local when conceptualizing globalization and its effects. Assemblages differ in that they are a specific range of phenomena that articulate various transformations distinguished by a particular global quality; "in the space of assemblage, a global [and local] form is simply one among a range of concrete elements" (Collier 2006: 400). Assemblages are also a "tool for the production of global knowledge that strives to replace socially, politically and context bound forms of knowledge" (Smith 2010: 18).
2. A recent special edition of *Nature* (issue 474, 2011) provides a useful (and optimistic) update of the second-generation biofuels.
3. Although food security theoreticians began looking at access issues after the world food crises of the 1970s, those arguing from a regime perspective viewed the crisis as systemic from early on (Friedmann 1982). It is only recently that the two approaches have begun to combine. Goodman and Watts (1997) provide a useful synopsis, including theoretical insights and empirical examples.
4. "Neoliberal ideology believes that all types of social and political issues can more effectively and efficiently be 'managed' by introducing market-based governance mechanisms such as commercialization, competition and 'free' trade" (Büscher 2009: 3954). Further, specifically in neoclassical models, the state is exogenous to the economic reform processes. As Castree's (2006, 2010) provocative work describes, the true nature of neoliberalism remains fuzzy, representative of a grand narrative more than clearly articulated theoretical (or even empirical) concept. For this reason I refer to neoclassical principles where relevant, especially the marketization and liberalization that dominate the current agro-food regime in South Africa, and globally.
5. There are also counterpoints that are currently jostling for position within the overall hegemonic discourses of the global corporate regime (McMichael 2000; see also Mares and Alkon, in this volume). The most vocal and popularized movement embodies food sovereignty, which includes a host of alternative food networks to the corporate/industrial regime and embraces organic, small-scale productivity and repeasentization of the agrarian sector. Much of the focus is within local spheres and the grassroots level (the romanticized moral economy); however, these should not be considered antithetical to the global level. In terms of sovereignty, countermovements are increasingly becoming expressions of the global counterpoint, as is the case with the international La Via Campesina move-

ment. Similarly, one can see this in the rise of a global yet organic agro-food network as Raynolds (2004) describes.

6. Any declines in reserves will mean less cushioning in the markets, making prices even more volatile. Whether financial assistance or fair trade will replace the decline in food aid remains unclear (Clements 2008; Peskett et al. 2007). It is also questionable whether any associated agricultural reforms or turn-arounds will be quick or radical enough to ensure long-term food security and sovereignty (Clements 2008).
7. Global-oriented perspectives and in many cases country case studies have shown that enough land is available for both of food and fuels and new lands can be brought into production with little impact on food crops. Whether such optimism is justifiable is a matter for debate and depends on far more grounded research and full disclosure of the assumptions made toward the types of agriculture to be pursued, actual output potentials within environmental constraints and social and environmental externalities incurred by such development.
8. These frameworks do exist, such as those developed by the Food and Nutrition Technical Assistance project, systems approaches developed by Ingram and colleagues (e.g. Ingram and Ericksen 2010) and the Roundtable on Sustainable Biofuels toolkits (http://rsb.epfl.ch/) among others.
9. Inputs required for modern agriculture (fertilisers, pesticides, machinery) have created a supply-side linkage between food and fuel prices. Through knock-on effects, changing fuel prices are transferred to food prices. With biofuels now being produced from the same resource base, fluctuations of the oil prices more directly affect food prices through demand-side competition with fuel prices. The supply-side relationship is likely to remain more dominant (Pfuderer et al. 2009).
10. The debate is much wider than this and though state failures actually spurred on free market-based approaches, it is more realistic that the solutions reside somewhere between the two (Dorward et al. 2005).
11. For example, the International Assessment of Agricultural Knowledge, Science and Technology for Development (2009b, 3) insists that "there is growing consensus that the current model of agricultural knowledge, science and technology (AKST) requires revision. Business as usual is no longer an option." This is not to deny the important role science and technology plays but more so acknowledging caution in its application.
12. A dialectic compromise here is considered perhaps too simplistically in the Hegelian/Kantian sense of thesis, antithesis, and synthesis. The polemic approach is one in which the neoclassical thesis of the agronomist perspective is looking to gain greater currency than that of the politically orientated antithesis of access and externalities. Here, the compromise and tension is one in which both are considered correct, the synthesis is that neither are looking at food security aspects holistically.
13. Other food crops such as sugar cane and sugar beet were not considered as detrimental because they are not the region's staple.
14. Jatropha was also excluded due to environmental concerns in which the Department of Agriculture named it an invasive species and therefore detrimental to biodiversity; the decision was met with disappointment by a few groups already growing trials in the North West and the Eastern Cape Provinces. As Jatropha was not widely grown nor gained sufficient momentum as a feedstock there has been little debate around reincorporating it into the biofuels strategy (quite unlike maize). There is little evidence that the Department of Agriculture rigorously investigated Jatropha's noxious effects (a water use analysis is provided by Holl et al. 2007), although with other research showing that it is not the wonder tree originally thought, this might not be an issue (Ribeiro and Matavel 2009).
15. South Africa has an erratic climate with drought being a common problem, even though the maize farmers fraternity seem to consider it more of an aberration than a given and therefore do not plan around it (Marcus et al. 1996). Climate change predictions are that rainfall in South Africa will become more erratic and droughts more likely, putting further pressure on any (surplus) production. Significant reductions in maize production in southern Africa have been projected (Jones and Thornton 2003) with South Africa suffering the highest gross loss of 871,500 tons of maize in 2055. Although maize (–28.5%) and wheat (–15.7%) are the two crops that are hardest hit in southern

Africa, decreases in sugar cane (−5.5%) and soybean (−8.0%) production are also predicted to occur by 2030 (Lobell et al. 2008).

16. This calculation takes into account the predicted job losses from the existing refining sector and the new ones created in the biofuels sector (approximately 11,800 jobs).
17. Energy security is seemingly considered unimportant in South Africa, as the main objectives are geared toward rural development and employment creation. Given that South Africa produces fewer crops (output) per hectare than other countries, it probably has worse energy-in–energy-out efficiencies.
18. Van Zyl and Kirsten (1992) considered the existing types of agricultural production as being detrimental to a holistic approach to food security in South Africa (see also Greenberg 2010). That domestic production of maize now contains a large portion of genetically modified crops further inhibits export opportunities.
19. This is a metaphor that dominates public policy discourse on poverty in South Africa, in which the notion that poor people stay poor because they are trapped in a second economy disconnected from the mainstream first world economy is espoused (Du Toit and Neves 2007).
20. Small-scale subsistence farmers, some of which rely on communal land for grazing, dominate the former homelands. Snippets of the potential destruction large-scale biofuels present to these areas have been documented by land grab style takeovers witnessed in the Eastern Cape (African Centre for Biosafety 2008; Earthlife Africa et al. 2007). However, these should be contrasted to some of the more beneficial programs seen in Limpopo and the North West (Haywood et al. 2009).

REFERENCES

Abbott, Phillip, Christopher Hurt, and Wallace Tyner. 2008. *What's Driving Up Food Prices?* Illinois: Farm Foundation.

Adeyemo, Oyenike, and Russell Wise. 2009. "Bio-Fuels and Food Security: A Case Study of South Africa." *Africa Insight* 39 (3): 94–107.

African Centre for Biosafety. 2008. *Agrofuels in South Africa: Project, Players and Poverty.* Vol. 7. Biosafety, Biopiracy and Biopolitics Series. Johannesburg: PressPrint.

Aliber, Michael. 2003. "Chronic Poverty in South Africa: Incidence, Causes and Policies." *World Development* 31, 473–490.

Aliber, Michael, Mompathi Baiphethi, and Peter Jacobs. 2009. "Agriculture Employment Scenarios." Pp. 133–161 in *Another Countryside? Policy Options for Land and Agrarian Reform in South Africa.* Pietermaritzburg: University of Natal Press.

Altman, Miriam, Tim Hart, and Peter Jacobs. 2009. "Household Food Security Status in South Africa." *Agrekon* 48 (4), 345–361.

Ariza-Montobbio, Pere, Sharachchandra Lele, Giorgos Kallis, and Joan Martinez-Alier. 2010. "The Political Ecology of Jatropha Plantations for Biodiesel in Tamil Nadu, India." *Journal of Peasant Studies* 37 (4): 875–897.

Bardhan, Pranab, 2005. *Globalization, Inequality and Poverty: An Overview.* Berkeley: University of California.

Bernstein, Henry. 1992. "Poverty and the poor, in: Rural Livelihoods". Pp. 13–26 in *Rural Livelihoods: Crisis and Responses.* Oxford: Oxford University Press.

Biofuels Task Team (BTT). 2006. "An Investigation into the Feasibility of Establishing a Biofuels Industry in the Republic of South Africa." http://www.dme.gov.za/pdfs/energy/bio_feasible_study.pdf (accessed 23 August 2009).

Blanchard, Ryan, David Bek, Patrick O'Farrell, and Graham Von Maltitz. 2011. "Biofuels and Biodiversity in South Africa." *South African Journal of Science* 107 (5/6): 1–8.

Boamah, Prosper. 2010. "Competition Between Biofuel and Food? Re-Thinking Biofuel Narratives, Evidence from a Jatropha Biodiesel Project in Northern Ghana." Pp. 129–144 in *Biofuels, Land Grabbing and Food Security in Africa.* London: Zed.

Brown, Lester. 2008. "Why Ethanol Production Will Drive World Food Prices Even Higher in 2008." http://cleantech.com/news/2360/why-ethanol-production-will-drive-world-food-prices-even-higher-in-2008 (accessed 12 May 2009).

Bureau for Food and Agricultural Policy (BFAP). 2005. *Bioethanol Production in South Africa: An Objective Analysis.* Pretoria: Bureau for Food and Agricultural Policy.

———. 2007. *Modelling the Impacts of Macroeconomic Variables on the South African Biofuels Industry.* Pretoria: Bureau for Food and Agricultural Policy.

Büscher, Bram. 2009. "Connecting Political Economies of Energy in South Africa." *Energy Policy* 37 (10): 3951–3958.

Cadenas, Alfredo, and Sara Cabezudo. 1998. "Biofuels as Sustainable Technologies: Perspectives for Less Developed Countries." *Technological forecasting and Social Change* 58 (1–2): 83–103.

Castree, Noel. 2006. "From Neoliberalism to Neoliberalisation: Consolations, Confusions, and Necessary Illusions." *Environment and Planning A* 38 (1): 1–6.

———. 2010. "Neoliberalism and the Biophysical Environment 1: What 'Neoliberalism' Is, and What Difference Nature Makes to It." *Geography Compass* 4 (12): 1725–1733.

Chabane, Neo. 2002. *An Evaluation of the Influences on Price and Production in the Maize Market following Liberalisation.* Pretoria: Trade and Industrial Policy Strategies.

Chambers, Robert. 1992. *Sustainable Rural Livelihoods: Practical Concepts for the 21st Century.* Brighton England: Institute of Development Studies.

Charles, M, R. Ryan, N. Ryan, and R. Oloruntoba. 2007. "Public Policy and Biofuels: The Way Forward?" *Energy Policy* 35 (11): 5737–5746.

Clements, Rebecca. 2008. *Scoping Study into the Impacts of Bioenergy Development on Food Security.* Rugby: Practical Action Consulting.

Collier, Stephen. 2006. "Global Assemblages." *Theory, Culture & Society* 23 (2–3): 399–401.

Conway, Gordon. 1998. *The Doubly Green Revolution: Food for All in the Twenty-First Century.* Ithaca, NY: Comstock.

Cotula, Lorenzo., Nat Dyer, and Sonja Vermeulen. 2008. *Fuelling exclusion? The Biofuels Boom and Poor People's Access to Land.* London: International Institute for Environment and Development.

Coyle, William. 2007. "The Future of Biofuels: A Global Perspective." *Amber Waves* 5: 24–29.

Currie, Jeffrey, Allison Nathan, James Gutman, and Ruifang Zhang. 2007. *Food, Feed and Fuel. An Agriculture, Livestock and Biofuel Primer.* New York: Goldman Sachs.

Dauvergne, Peter, and Kate Neville. 2010. "Forests, Food, and Fuel in the Tropics: The Uneven Social and Ecological Consequences of the Emerging Political Economy of Biofuels." *Journal of Peasant Studies* 37 (4): 631–660.

Davison, Sue Canney, Francis Xavier Ochieng, and Tameezan wa Gathui. 2010. *Liquid Biofuels Strategies and Policies in Selected African Countries.* Rugby: Practical Action Consulting.

De Klerk, Mike, Scott Drimie, S Mini, R Mokoena, P. Randela, S. Modiselle, B. Roberts, Coleen Vogel, C. de Swardt, and Johann Kirsten. 2004. *Food Security in South Africa: Key Policy Issues for the Medium Term. Integrated Rural and Regional Development.* Pretoria: Human Sciences Research Council.

Department of Agriculture, Forestry and Fisheries. 2010. *Estimate of the Contribution of the Agriculture Sector to Employment in the South African Economy.* Pretoria: Economic Services, DAFF.

Devereux, Stephen, and Simon Maxwell. 2001. *Food Security in Sub-Saharan Africa.* Pietermaritzburg: University of Natal Press.

Diaz-Chavez, Rocio, Stephen Mutimba, Helen Watson, Sebastian Rodriguez-Sanchez, and Massaër Nguer. 2010. *Mapping Food and Bioenergy in Africa.* Ghana: Forum for Agricultural Research in Africa.

Doornbosch, Richard, and Ronald Steenblik. 2007. *Biofuels: Is the Cure Worse Than the Disease?* Paris: OECD.

Dorward, Andrew, Jonathan Kydd, and Colin Poulton. 2005. "Beyond Liberalisation: 'Developmental Coordination' Policies for African Smallholder Agriculture." *IDS Bulletin* 36 (2): 80–85.

Drimie, Scott and Shaun Ruysenaar. 2010. "The Integrated Food Security Strategy of South Africa: An Institutional Analysis." *Agrekon* 49 (3): 316-337.

Du Toit, Andries, and David Neves. 2007. *In Search of South Africa's Second Economy: Chronic Poverty, Economic Marginalisation and Adverse Incorporation in Mt Frere and Khayelitsha.* Cape Town: Programme for Land and Agrarian Studies.
Earthlife Africa, African Centre for Biosafety, GRAIN, SAFeAGE, Ekogaia, and The Thirld World Investment Gateway Trust. 2007. Submission to the Department of Minerals and Energy on the Draft Biofuels Industrial Strategy. Unpublished manuscript.
Eicher, C. K. Maredia, and I. Sitholeniang. 2006. "Crop Biotechnology and the African farmer." *Food Policy* 31 (6): 504–527.
Eide, Asbjørn. 2008. *Right to Food and the Impact of Biofuels (Agrofuels). The Right to Food.* Rome: Food and Agriculture Organization. http://www.fao.org/righttofood/publi08/Right_to_Food_and_Biofuels.pdf (accessed 14 July 2010).
Ericksen, Polly. 2007. "Conceptualizing Food Systems for Global Environmental Change Research." *Global Environmental Change* 18 (1): 234–245.
Ericksen, Polly, John Ingram, and Diana Liverman. 2009. "Food Security and Global Environmental Change: Emerging Challenges." *Environmental Science & Policy* 12 (4): 373–377.
Esterhuizen, Dirk. 2009. *South Africa Biofuels Annual.* Pretoria: Global Agricultural Information Network.
Evans, Alex. 2011. "Governance for a Resilient Food System." Oxfam Discussion Papers, Center on International Cooperation, New York University.
FAO. 2002. *The State of Food Insecurity in the World 2001.* Rome: Food and Agricultural Organisation.
———. 2008. *Soaring Food Prices: Facts, Perspectives, Impacts and Actions Required.* Rome: FAO.
Friedmann, Harriet. 1982. "The Political Economy of Food: The Rise and Fall of the Postwar International Food Order." *American Journal of Sociology* 88 (1): S248–S286.
———. 1990. "The Origins of Third World Food Dependence." Pp. 13–31 in *The Food Question: Profits versus People?* London: Earthscan.
———. 1993. "After Midas' Feast: Alternative Food Regimes for the Future." Pp. 213–233 in *Food for the Future: Conditions and Contradictions of Sustainability,* ed. P. Allen. New York: Wiley & Sons.
Friedmann, Harriet, and Philip McMichael. 1989. "Agriculture and the State System: The Rise and Decline of National Agricultures, 1870 to the Present." *Sociologia Ruralis* 29 (2): 93–117.
Funke, Thomas, Peter Klein, and Ferdie Meyer. 2011. "Biofuels Production in South Africa." *Biofuels* 2 (2): 209–220.
Funke, Thomas, P. Strauss, and F. Meyer. 2009. "Modelling the Impacts of the Industrial Biofuels Strategy on the South African Agricultural and Biofuel Subsectors." *Agrekon* 48 (3): 223–244.
Gasparatos, Alexandros, Per Stromberg, and Kazuhiko Takeuchi. 2011. "Biofuels, Ecosystem Services and Human Wellbeing: Putting Biofuels in the Ecosystem Services Narrative." *Agriculture, Ecosystems & Environment* 142 (3–4): 111–128.
Gerber, Nicolas, Manfred Eckert, and Thomas Breuer. 2008. "The Impacts of Biofuel Production on Food Prices: A Review." Zentrum für Entwicklungsforschung Discussion Papers on Development Policy, Bonn.
Godfray, H., J. R. Beddington, I. R. Crute, L. Haddad, D. Lawrence, J. F. Muir, J. Pretty, S. Robinson, S. M. Thomas, and C. Toulmin. 2010. "Food Security: The Challenge of Feeding 9 Billion People." *Science* 327 (5967): 812–818.
Goodman, David, and Michael Watts, eds. 1997. *Globalising Food Agrarian Questions and Global Restructuring.* London and New York: Routledge.
Greenberg, Stephen. 2003. "Political Stabilisation and Market Extension: Restructuring of Agriculture and its Impact on Food Security", in: *Interfund Development Update* 4 (2): pages unavailable.
———. 2010. "Status Report on Land and Agricultural Policy in South Africa, 2010." PLAAS Research Report, Institute for Poverty, Land and Agrarian Studies, Bellville.
Hall, Ruth. 2007. *Land Use and Livelihoods in South Africa's Land Reform.* Paper presented at the Conference on Another Countryside? Policy Options for Land and Agrarian Reform in South Africa Lord Charles Hotel, Somerset West: Programme for Land and Agrarian Studies.
———. 2009. "Dynamics in the Commercial Farming Sector." Pp. 121–131 in *Another Countryside?*

Policy Options for Land and Agrarian Reform in South Africa. Cape Town: Institute for Poverty, Land and Agrarian Studies.

Hall, Ruth, and Lionel Cliffe. 2009. "Introduction." Pp. 1–17 in *Another Countryside?: Policy Options for Land and Agrarian Reform in South Africa.* Cape Town: Institute for Poverty, Land and Agrarian Studies.

Haywood, Loren, Graham Von Moltitz, Kevin Setzkorn, and Nicholas Ngepah. 2009. *Biofuel Production in South Africa, Mozambique, Malawi and Zambia.* Pretoria: Council for Scientific and Industrial Research.

Hazell, Peter, and Rajendra Pachauri. 2006. "Bioenergy and Agriculture: Promises and Challenges." IFPRI Brief, Washington DC.

Hertel, T., and A. Winters, Eds. 2006. *Poverty and the WTO: Impacts of the Doha Development Agenda.* Washington DC: The World Bank.

Holl, M, M. Gush, J. Hallowes, and D. Versfeld. 2007. "Jatropha Curcas in South Africa: An Assessment of Its Water Use and Bio-Physical Potential." Water Research Commission Report, Pretoria.

Ingram, John and Polly Ericksen, 2010. *Food Security and the Vulnerability of Food Systems to Global Environmental Change.* Paper presented at RGS-IBG Annual Conference, 2 September 2010, London, Imperial College.

International Assessment of Agricultural Knowledge, Science, and Technology for Development. 2009a. *Global Report.* Washington, DC: Island Press.

———. 2009b. *Agriculture at a Crossroads: Executive Summary of the Synthesis Report.* Washington DC: IAASTD.

Jacobs, Peter. 2009. "Agricultural Market Reforms and the Rural Poor in South Africa" presented at PLAAS Poverty Workshop 2009, Cape Town, South Africa.

Jones, Peter and Phillip. Thornton. 2003. "The Potential Impacts of Climate Change on Maize Production in Africa and Latin America in 2055." *Global Environmental Change* 13 (1): 51–59.

Kent, George. 1999. "Globalization and Food Security in Africa." Pp. 17–35 in *Not by Bread Alone: Food Security and Governance in Africa.* Johannesburg: Wits University Press/Toda Institute for Global Peace and Policy Research.

Koning, Niek, and Arthur P. J. Mol. 2009. "Wanted: Institutions for Balancing Global Food and Energy markets." *Food Security* 1 (3): 291–303.

Kupka, Julia, and Wessel Lemmer. 2007. *Bio-Ethanol from Grains Will Increase Food Security.* Pretoria: SA Grain.

Labadarios, Demetre, Yul Davids, Zandile Mchiza, and Gina Weir-Smith. 2009. *The Assessment of Food Insecurity in South Africa.* Pretoria: Human Sciences Research Council.

Lipsky, John. 2008. "Commodity Prices and Global Inflation." http://www.imf.org/external/np/speeches/2008/050808.htm (accessed 14 July 2010).

Lipton, Merle, and Charle Simkins, eds. 1993. *State and Market in Post-Apartheid South Africa.* Johannesburg: Witwatersrand University Press.

Lipton, Michael, Frank Ellis, Merle Lipton. 1996. "Introduction." Pp. v–xvii in *Land, Labour and Livelihoods in Rural South Africa.* Durban: Indicator Press.

Liverman, Diana, Polly Ericksen, and John Ingram. 2009. "Governing Food Systems in the Context of Global Environmental Change." *IHDP Update* 3: 59–64.

Lobell, D. B., M. B. Burke, C. Tebaldi, M. D. Mastrandrea, W. P. Falcon, and R. L. Naylor. 2008. "Prioritizing Climate Change Adaptation Needs for Food Security in 2030." *Science* 319 (5863): 607–610.

Makenete, Andrew, Wessel Lemmer, and Julia Kupka. 2007. *The Impact of Biofuel Production on Food Security: A Briefing Paper with a Particular Emphasis on Maize-to-ethanol Production.* Johannesburg: South African Biofuels Association.

Marcus, Tessa, Kathy Eales, and Adele Wildschut. 1996. *Down to Earth.* Durban: Indicator Press.

Mathews, J.A. 2007. "Biofuels: What a Biopact between North and South Could Achieve." *Energy Policy* 35, 3550–3570.

Matondi, Prosper, Kjell Havnevik, and Atakilte Beyene, eds. 2010. *Biofuels, Land Grabbing and Food Security in Africa.* London: Zed.

Maxwell, Simon. 1990. "Food Security in Developing Countries: Issues and Options for the 1990s." *IDS Bulletin* 21 (3): 2–13.

Maxwell, Simon, and Caroline Ashley. 2001. "Rethinking Rural Development." *Development Policy Review* 19 (4): 395–425.

Maxwell, Simon, and Rachel Slater. 2003. "Food Policy Old and New." *Development Policy Review* 21 (5–6): 531–553.

May, Julian. 1999. "Food Security, Livelihoods and the State: The South African Experience." Pp. 85–101 in *Not by Bread Alone: Food Security and Governance in Africa.* Johannesburg: Wits University Press/Toda Institute for Global Peace and Policy Research.

McDaid, Liz. 2007. Comments on the Draft Biofuels Industrial Strategy Released by DME, November 2006. Cape Town: African Sustainable Fuel Centre.

McMichael, Phillip. 2000. "The Power of Food." *Agriculture and Human Values* 17: 21–33.

———. 2010. "Agrofuels in the Food Regime." *Journal of Peasant Studies* 37 (4): 609–629.

Meyer, Ferdie, P. Strauss, and Thomas Funke. 2008. "Modelling the Impacts of Macro-Economic Variables on the South African Biofuels Industry." *Agrekon* 47 (3): 327–345.

Mitchell, Donald. 2008. "A Note on Rising Food Prices." World Bank (Development Prospects Group) Policy Research Working Paper, New York.

Mohamed-Salih, M. A. 2009. "Governance of Food Security in the 21st Century." *Facing Global Environmental Change* 4 (5): 501–507.

Mol, Arthur. 2007. "Boundless Biofuels? Between Environmental Sustainability and Vulnerability." *Sociologia Ruralis* 47 (4): 297–313.

Molony, Thomas and James Smith. 2010. "Biofuels Food Security and Africa." *African Affairs* 109 (436): 489–498.

Moore, A. 2008. "Biofuels are Dead: Long Live Biofuels(?)—Part one." *New Biotechnology* 25 (1) : 6–12.

Mosley, Paul. 2002. "The African Green Revolution as a Pro-Poor Policy Instrument." *Journal of International Development* 14 (6): 695–724.

Naik, S.N., Vaibhav V. Goud, Prasant K. Rout, and Ajay K. Dalai. 2010. "Production of First and Second Generation Biofuels: A Comprehensive Review." *Renewable and Sustainable Energy Reviews* 14 (2): 578–597.

National Department of Agriculture (NDA). 2001. *The Strategic Plan for South African Agriculture.* Pretoria: National Department of Agriculture.

———. 2002. *The Integrated Food Security Strategy for South Africa.* Pretoria: National Department of Agriculture.

———. 2009. *Abstract of Agricultural Statistics 2009.* Pretoria: National Department of Agriculture (Directorate Agricultural Statistics).

Ngepah, Nicholas. 2010. "Socioeconomic Impacts of Biofuels: Methodologies and Case Study Examples." Pp. 149–174 in *Assessing the Sustainability of Bioenergy Projects in Developing Countries: A Framework for Policy Evaluation.* Newcastle upon Tyne: Newcastle University.

Ong, Aihwa, and Stephen Collier. 2006. *Global Assemblages: Technology, Politics and Ethics as Anthropological Problems.* Oxford: Blackwell.

Oosterveer, Peter, and Arthur Mol. 2010. "Biofuels, Trade and Sustainability: A Review of Perspectives for Developing Countries." *Biofuels, Bioproducts and Biorefining* 4 (1): 66–76.

Ortmann, Gerald and Charles Machethe. 2003. "Problems and Opportunities in South African Agriculture," Pp. 47–60 in *The Challenge of Change.* Scottsville: University of Natal Press.

Oxfam International. 2008. *Another Inconvenient Truth: How Biofuel Policies are Deepening Poverty and Accelerating Climate Change.* Oxfam Briefing Paper, London.

Paarlberg, Robert. 2002. *Governance and Food Security in an Age of Globalization.* Washington, DC: International Food Policy Research Institute.

Peters, Jorg, and Sascha Thielmann. 2008. "Promoting Biofuels: Implications for Developing Countries." *Energy Policy* 36 (4): 1538–1544.

Peskett, Leo, Rachel Slater, Chris Stevens, and Annie Dufey 2007. "Biofuels, Agriculture and Poverty Reduction". *Natural Resource Perspectives* (No. 107). London: Overseas Development Institute.

Pieterse, Jimmy and Barry van Wyk. 2005. *What's Cooking: AIDS Review 2005.* Pretoria: University of Pretoria: Centre for the study of AIDS,

Pfuderer, Simone, Grant Davies, and Ian Mitchell. 2009. "The Role of Demand for Biofuel in the Agricultural Commodity Price Spikes of 2007/08." http://www.defra.gov.uk/foodfarm/food/pdf/ag-price-annex%205.pdf (accessed 2 December 2009).

Ravallion, Martin. 1989. *Do Price Increases for Staple Foods Help or Hurt the Rural Poor?* Washington DC: World Bank.

Raynolds, L. 2004. "The Globalization of Organic Agro-Food Networks." *World Development* 32 (5): 725–743.

Reutlinger, Shlomo, and M. Selowski. 1976. *Malnutrition and Poverty: Magnitude and Policy Options.* Baltimore: Published for the World Bank by Johns Hopkins University Press.

Ribeiro, Daniel, and Nilza Matavel. 2009. *Jatropha! A Socio-Economic Pitfall for Mozambique.* Maputo: Justica Ambiental and Uniao Nacional de Camponeses.

Richardson, Ben. 2010. "Big Sugar in Southern Africa: Rural Development and the Perverted Potential of Sugar/Ethanol Exports." *Journal of Peasant Studies* 37 (4): 917–938.

Robertson, R. 2001. *Globalization: Social Theory and Global Culture.* London: Sage.

Roe, Emery. 1991. "Development Narratives, or Making the Best of Blue Print Development." *World Development* 19 (4): 287–300.

Rosegrant, M. 2008. "Biofuels and Grain Prices: Impacts and Policy Responses. IFPRI." Testimony before the U.S. Senate Committee on Homeland Security and Governmental Affairs. http://www.ifpri.org/publication/biofuels-and-grain-prices (accessed 8 November 2011).

Rosillo Callé, Francisco, and Francis Johnson, eds. 2010. *Food Versus Fuel: An Informed Introduction to Biofuels.* London and New York: Zed.

Royal Society. 2008. *Sustainable Biofuels: Prospects and Challenges.* London: Royal Society.

Sah, R, and J. Stiglitz. 1992. *Peasants versus City-Dwellers: Taxation and the Burden of Economic Development.* Oxford: Clarendon Press.

Sanchez, Pedro A., Glenn L. Denning, and Generose Nziguheba. 2009. "The African Green Revolution moves forward." *Food Security* 1 (1): 37–44.

Scoones, Ian, Stephen Devereux, and Lawrence Haddad. 2005. "Introduction: New Directions for African Agriculture." *IDS Bulletin* 36 (2): 1–12.

Sen, Amartya. 1981. *Poverty and Famines: An Essay on Entitlement and Deprivation.* Oxford: Clarendon Press.

Sharman, Amelia, and John Holmes. 2010. "Evidence-Based Policy or Policy-Based Evidence Gathering? Biofuels, the EU and the 10% Target." *Environmental Policy and Governance* 20 (5): 309–321.

Smith, James. 2010. *Biofuels and the Globalization of Risk: The Biggest Change in North-South Relationships since Colonialism?* London: Zed.

Strydom, Dirk. 2009. "The Economic Impact of Maize-Based Ethanol Production on the South African Animal Feed Industry." M.Com diss., University of Free State, Bloemfontein.

Strydom, Dirk, P. Taljaard, and B. Willemse. 2010. "Ethanol Blending Policies and the South African Animal Feed Industry." *Agrekon* 49 (2): 255–265.

Stuart, Tristram. 2009. *Waste: Uncovering the Global Food Scandal.* New York: W.W. Norton.

Sugrue, Annie, and Richard Douthwaite. 2007. *Biofuel Production and the Threat to South Africa's Food Security.* Wahenga Brief. Johannesburg: Regional Hunger & Vulnerability Programme (RHVP).

Swanepoel, Esmarie. 2007. "Food versus Fuel Debate Escalates." *Engineering News.* http://www.engineeringnews.co.za/article/food-versus-fuel-debate-escalates-2007-11-02 (accessed 18 November 2009).

Timmer, C., W. Falcon, and S. Pearson. 1983. *Food Policy Analysis.* Baltimore: Johns Hopkins University Press.

Tomlinson, Isabel. 2010. "Doubling Yields to Feed the 9 Billion: A Critical Perspective on a Dominant Discourse of Food Security." Paper presented at the RGS-IBG Annual Conference, Imperial College, London, 2 September.

Van Zyl, Johan, and Johann Kirsten. 1992. "Food Security in South Africa." *Agrekon* 31 (4): 170–184.

Vink, Nick, Johann Kirsten, and Johan Van Zyl. 2000. "Agricultural Policy: Undoing the Legacy of the Past." Pp. 22–44 in: *South African Agriculture at the Crossroads.* Basingstoke: Macmillan.

Vermeulen, Sonja, Emmanuel Sulle, and Swan Fauveaud. 2009. "Biofuels in Africa: Growing Small-Scale Opportunities." International Institute for Environment and Development Briefing paper, London.

Vink, Nick, and Johann Kirsten. 2003. "Agriculture in the National Economy." Pp. 3–21 in *The Challenge of Change. Agriculture, Land and the National Economy.* Pietermaritzburg: University of Natal Press.

Von Braun, Joachim. 2008. "When Food Makes Fuel: The Promises and Challenges of Biofuels." Pp. 5–13 in *Biofuels, Energy and Agriculture: Powering Towards or Away from Food Security? Proceedings of the Thirteenth Annual Development Conference held on 15 August 2007.* Canberra: ATSE Crawford Fund.

Von Braun, Joachim, and Diaz-Bonilla, Eugenio. 2008. "Globalization of food and agriculture and the poor." International Food Policy Research Institute Briefs, Washington DC.

Von Maltitz, Graham, and Alan Brent. 2008. *Assessing the Biofuel Options for Southern Africa.* Pretoria: CSIR.

Von Maltitz, Graham, Alan Brent, Maxwell Mapako, and Lorren Haywood. 2009. "Analysis of Opportunities for Biofuel Production in Sub-Saharan Africa." Center for International Forestry Research Environment Brief, Bogor, Indonesia.

Walker, Brian, Scott Barrett, Stephen Polasky, Victor Galaz, Carl Folke, Gustav Engström, Frank Ackerman, Ken Arrow, Stephen Carpenter, Kanchan Chopra, Gretchen Daily, Paul Ehrlich, Terry Hughes, Nils Kautsky, Simon Levin, Karl-Göran Mäler, Jason Shogren, Jeff Vincent, Tasos Xepapadeas, and Aart de Zeeuw. 2009. "Looming Global-Scale Failures and Missing Institutions." *Science* 325 (5946): 1345–1346.

Webb, Patrick, and Beatrice Rogers. 2003. "Putting the 'in' Back in Food Insecurity." USAID Occasional Paper.

Weber, Michael, John Staatz, John Holtzman, Eric Crawford, and Richard Bernsten. 1988. "Informing Food Security Decisions in Africa: Empirical Analysis and Policy Dialogue." *American Journal of Agricultural Economics* 70 (5): 1044–1052.

White, Ben, and Anirban Dasgupta. 2010. "Agrofuels Capitalism: A View from Political Economy." *Journal of Peasant Studies* 37 (4): 593–607.

Wiggins, Steve, E. Fioretti, J. Keane, Y. Khwaja, S. McDonald, Sarah Levy, and C. S. Srinivasan. 2008. *Review of the Indirect Effects of Biofuels: Economic Benefits and Food Insecurity.* London: Overseas Development Institute.

Wolde-Georgis, Tsegay, and Michael Glantz. 2009. "Biofuels in Africa: A Pathway to Development." International Research Centre for Energy and Economic Development Occasional Paper, Boulder, CO.

Woods, Jeremy. 2006. *Bioenergy and Agriculture: Promises and Challenges. Science and Technology Options for Harnessing Bioenergy's Potential.* IFPRI Brief, Washington DC.

World Bank. 1986. *Poverty and Hunger: Issues and Options for Food Security in Developing Countries.* Washington DC: World Bank.

———. 2008. *Biofuels: The Promise and the Risks.* Policy Brief of the World Bank, Washington, DC.

Young, E. 2004. "Globalization and Food Security: Novel Questions in a Novel Context?" *Progress in Development Studies* 4 (1): 1–21.

Ziegler, Jean. 2007. *The Right to Food: Note by the Secretary-General.* New York: United Nations General Assembly.

From Biotech to Nanotech

Public Debates about Technological Modification of Food

Jennifer B. Rogers-Brown, Christine Shearer, and Barbara Herr Harthorn

ABSTRACT: Technological modifications of food are being marketed as novel products that will enhance consumer choice and nutritional value. A recent manifestation is nanotechnology, entering the global food chain through food production, pesticides, vitamins, and food packaging. This article presents a detailed literature review on risk and benefit perceptions of technological developments for food and agriculture, including our own research from US deliberative workshops on nanotechnologies. The article suggests that many of the public concerns discussed in the literature on biotechnology in food are being raised in qualitative and quantitative studies on nanotechnologies for food: although nanotechnologies are generally perceived to be beneficial, many people express particular uneasiness about nanotechnological modifications of food. The article argues that these concerns represent material examples of unresolved social issues involving technologies and the food industry, including questions about the benefits of nanotechnology for food, and the heightened values attached to food as a cultural domain.

KEYWORDS: biotechnology, food and food safety, nanotechnology, public deliberation, risk perception, sustainability

This article reviews the literature on public perceptions of the technological alteration of food, particularly how past studies concerning biotechnology and food compare with upstream public assessments of nanotechnologies and food, including the Center for Nanotechnology in Society at the University of California, Santa Barbara (CNS-UCSB) public deliberations. Quantitative studies allow for gauging widespread public opinions, while qualitative studies allow for exploring the social hopes and concerns embedded in the discussion of the food industry and food as a cultural domain. Analysis of these studies suggests that many of the issues raised by participants in studies of biotechnology and food are also being activated by the use of nanotechnologies in food. Specifically, our deliberation work supports the literature on food, technology, and society studies by suggesting a widespread uneasiness about the alteration of food. For example, participants generally perceive greater risk with regard to nanotechnological alterations to food than most other nanotechnology applications. We argue this uneasiness reflects cultural values rooted deeply in ongoing individual and social concerns over technology and the food industry,

Environment and Society: Advances in Research 2 (2011): 149–169 © Berghahn Books
doi:10.3167/ares.2011.020109

including the symbolic value placed on food as a cultural domain. These issues and concerns are in turn informing and shaping assessments of nanofoods and food products.

From the picker, grower, transporter, corporation, and lawmakers, to the retailers, corporate owners, grocery stores, fast food restaurants, and your table, the trail of food production and consumption brings together a picture of a global world where food is central to societal and cultural reproduction. For most people in the world, food is not only consumed for sustenance, it embodies cultural rituals and plays a central role in symbolic interactions in everyday life (e.g., see Mintz and Du Bois 2002). In one case study of genetic modification and maize in Mexico, Rogers (2008) argues that maize symbolizes religious heritage and tradition for Mayan peoples, and that technological modification (in a laboratory, rather than in the field) is seen as a threat to economic stability and indigenous practices such as seed saving. Therefore, the following literature review of public debates on technological innovations for food is grounded in an understanding of food and its many meanings as a cultural domain. This distinction is important because the cultural meanings of food (e.g., as sacred, sustenance, and a source of income) are embedded in public debates of biotechnology and nanotechnology in food production and packaging.

Early Technological Advancements in the Lab: Biotechnology and Public Debate

We begin with a short overview of biotechnology and risk perception research for food production because the innovations and cultural implications of biotech provide a compelling historical parallel to emergent and different technological advancements through nanotechnologies. Despite key physical differences in the two types of technologies and the relative novelty of nanotechnologies, the cultural concerns and issues raised in deliberation and public debate highlight similarities in public sentiment regarding innovations in the food domain.

Biotechnology refers to many uses and applications and is an "overarching term for a wide variety of new technologies, of which genetic engineering and the cloning of animals are only two of the best-known examples" (Tokar 2001: 3). The technologies included under biotechnology all constitute some form of simulation and manipulation "of fundamental life processes at the cellular and molecular levels" (Tokar 2001: 3), such as recombinant DNA (rDNA), or genetic engineering (GE) (David and Thompson 2008). Genetic engineering of seeds often involves modifying or inserting genes into a plant in order to enhance resistance to pests, shelf life, and nutritional quality. The changes at the cellular and molecular levels are made in order to enhance particular physical or biochemical traits. With genetic engineering, the seed is altered quickly and at the cellular level, rather than the natural hybridization of improving seed quality through generations of breeding seeds (David and Thompson 2008; Kuzma and VerHage 2006). China grew the first genetically modified (GM) crop—tobacco—in 1988, and in 1994 the US followed with Calgene's "Flavr Savr" tomato (Stone 2010). Calgene altered a gene in the tomato to delay rotting, but the "Flavr Savr" failed to produce a significant profit and Monsanto, an agricultural biotechnology company, purchased Calgene in 1996.

Genetically engineered and "genetically modified crops and foods were claimed by their proponents to offer a range of benefits to a variety of beneficiaries, including higher productivity and lower pesticide costs for producers; less environmental pollution from pesticides and herbicides; and new crop varieties to ameliorate hunger in developing countries" (Gaskell et al. 2004: 185). A review of studies conducted by the Food Policy Research Institute on the savings versus costs of Bt cotton in developing countries highlights the reduced labor that results from reduced pesticide use, the differing reports of yield production, and the negative impact of intellectual property rights (IPR) on farmer's willingness to pay (WTP) for Bt cotton (Smale et al. 2009).[1] Recent news reports highlighted the ability of GM corn to lessen worm infestation in modified plant varieties and in nearby conventional seed fields (Charles 2010). Conversely, GM crops have some unanticipated consequences, such as harmful effects on useful bugs (ladybugs, lacewings, and monarch butterflies), an increased rate of cross-pollination, and a higher susceptibility to environmental problems (Tokar 2001: 6). The consumption of GM crops has also been harmful in human use. It is known to cause allergic reactions, digestive problems, an unprecedented spread of antibiotic resistance, and possible mutations in the growth of human organs (Tokar 2001: 6). The cross-pollination of genetically modified organisms (GMOs) with non-GMOs may result in the reproduction of GMO seeds without the farmer's knowledge and place the farmer in danger of growing a seed incapable of reproduction, forcing the farmer to purchase new seeds, or potentially to be sued by a seed company over patent claims.

Public backlash and the greatest activism against the genetic engineering of food were particularly pronounced in the UK and EU (Gaskell et al. 2004; Schurman and Munro 2010), with large public pressure in the UK and other northern European countries for public disclosure of information about the effects of GE for the public and for the environment. Gaskell et al. summed up some of the various concerns by stating that, "While environmentalists raise questions about gene drift, super-weeds, biodiversity, and the unknown long-term consequences of GM crops, the public is concerned about the ethics of genetic modification, the labeling of foods with GM ingredients, and the possible health effects of consuming GM foods" (2004: 185). Public concerns focused on human health risks, environmental risks, and distrust in the food industry to self-regulate and protect against possible risks. As stated by Schurman and Munro: "Wresting agricultural biotechnology into the public sphere, these critics transformed it from an elite technological development into a highly contentious social problem" (2010: xiii). In response to anti-GM sentiment and activism, policymakers in the UK and EU adopted restrictive regulation against the importation and growth of GM crops, including an "informal" moratorium on GM crop approvals after 1998 by the European Parliament (Schurman and Munro 2010). Public backlash against the commercialization of GE and GM foods in Europe led to calls for reconsideration of how such new technologies are introduced into society (Gaskell et al. 2004; Poortinga and Pidgeon 2004). Upstream public engagement rather than downstream engagement (after the technology or products are already in the marketplace, often without consumer awareness) offers the opportunity for anticipating public response and encouraging the public to actively participate in the research and development of emerging technologies (Horlick-Jones et al. 2007; Pidgeon and Rogers-Hayden 2007; Poortinga and Pidgeon 2004).

Kearnes et al. (2006) noted that researchers initially viewed negative public opinions of genetic engineering during the 1990s and early 2000s as a lack of public awareness about the "real" risks. This "deficit model" blamed the public for a lack of understanding of science, as well as sensationalist media and nongovernmental organizations (NGOs) for hyping the risks of new technologies like genetic engineering (Wynne 2006). Indeed, risk communication by trusted experts has long been argued by many policymakers and scientists as the solution to public skepticism over new technologies, with the idea that increased knowledge leads to increased acceptance.[2]

Studies of public perceptions of biotechnology and genetic engineering portrayed a different story. These studies suggested that confidence in experts is but one among many factors shaping public perceptions of risk of food technologies (Gaskell and Allum 2001; Siegrist 2008; Slovic et al. 1980; Thompson and Wesley 1996). Further, Freudenburg and Rursch's study (1994) found that providing additional information, such as through statistics about the risk issue, does not necessarily lead to increased acceptance. Other surveyed concerns about biotechnology for food involved not just potential dangers, but also moral concerns, usefulness, and benefits of the technological application, democratic considerations around funding and control, and uncertainties, such as unknown adverse consequences (Gaskell et al. 2003; Siegrist 2008). Classical risk analysis, with its focus on experts and technical knowledge, therefore falls short of addressing these social concerns, which in turn erodes trust in experts and institutions, one of the most important factors for public acceptance of food policies (Sapp et al. 2009; Siegrist et al. 2007).

Further, many public concerns cannot be fully assuaged by risk communication from experts, as such messages, if done transparently, include uncertainty and the limits of scientific knowledge. The new perceived risks associated with GM food were typically complex and uncertain (Savadori et al. 2004; Vlek 2010), for both the human body and the environment (Gaskell et al. 2003; Renn 2005). Perceived risks can also encompass moral, ethical, and religious concerns, such as possible violation of the order of nature, power relations, democratic rights, and the possible application of the technology to address poverty in developing countries (Lassen and Jamison 2006; Tveit et al. 2003).

Other research has broadened the concept of risk: in a qualitative study of ten countries (Austria, Denmark, Finland, France, Germany, the United Kingdom, Italy, Portugal, Sweden, and Switzerland), Gaskell et al. (2004) found that participants did not speak of "risk," but rather discussed "dangers," and this was a broad category that included moral and democratic hazards (see also Douglas and Wildavsky 1983), with participants seeing more danger associated with GM food than with biotechnology used in medical research. The researchers also noted the importance of benefit perceptions: participants were found to be more critical of and less risk tolerant toward GM food, due in part to the absence of perceived benefits and the possibilities of alternatives to achieve the same goals. Participants in their study "question[ed] the point of genetic modification of food. Was it necessary when there is plenty of food in the shops? Why change the character of food when it is already good and wholesome?" (Gaskell et al. 2004: 187). In an additional quantitative study of the Eurobarometer conducted by the same authors, they found that 60 percent of respondents believed GM foods offer no benefits and carry risks (Gaskell et al. 2003), while a representative subsample of a 1999 Eurobarometer survey by Costa-Font and Mossialos (2005a) found more support for biotechnology in health applications than food.

Scholars noted that public concerns over GM in Europe and elsewhere, including the lack of perceived benefits for the public, reflected broader, underlying sociocultural concerns of the time:

> Discussion of [GM] technology also reflected a broader set of tensions: global drives towards new forms of proprietary knowledge; shifting patterns of ownership and control in the food chain; issues of corporate responsibility and corporate closeness to governments; intensifying relationships of science and scientists to the worlds of power and commerce; unease about hubristic approaches to limits in human understanding; conflicting interpretations of what might be meant by sustainable development. These and numerous other 'nonscientific' issues condensed onto GM crops because of a particular range of institutional and cultural contingencies shaping the technology and its development. (Kearnes et al. 2006: 301)

In short, GM opponents were concerned not just with risks, which they saw as potentially long-term and uncertain; they were also skeptical about tangible social benefits, they suspected greed for profit at the expense of nature or the general public, and they were worried about the concentration of power in large industrial corporations (Hampel et al. 2000). This risk frame of corporate control over the food supply is already being applied to nanotechnology, which some scholars call the "nano-corporate paradigm" (Scrinis and Lyons 2007).

Policymakers and risk researchers are interested in possible connections between public perceptions of GM and emergent views on nanotechnologies. Public and private investors in nanotechnology development have expressed fear of nanotechnology becoming "the next GM" or reminding the world of the "GM experience" (Schurman and Munro 2010). These phrases invoke in the nanotech context the strong public resistance to GM in the UK and other parts of Europe, including the informal moratorium on GM crop approvals by the European Parliament in 1998. Anti-GM movements were more effective in Europe than in the US, due to the strong push for a moratorium, which resulted in labeling laws and a switch from governmental support of GM to resistance and regulation (Bray 2003; Schurman and Munro 2010). However, researchers have noted ongoing public response to nanotechnologies is minimal thus far, at least in part because public awareness or familiarity continues at very low levels (Kearnes et al. 2006; "Nanofood for Thought" 2010; Pidgeon et al. 2008; Satterfield et al. 2009). Kearnes et al. (2006) argue that at this upstream moment of nanotechnological innovation, the scientific and government communities have the novel opportunity to enhance public participation and open up nanotechnology R&D to public discussion.

Nanotechnology

Genetic engineering and nanotechnology are both considered foundational platforms where technological innovations can occur in areas such as medicine, food, and the environment. Nanotechnology, or nanotechnologies—given the diversity of the field—allow scientists to create, explore, and manipulate materials measured in nanometers (billionths of a meter). Such materials can have chemical, physical, optical, and biological properties that differ from those of their larger bulk material counterparts. At the nanoscale, 1–100 nanometers (nm), "unique phenomena create novel applications," as stated by the US National Nanotechnology Initiative (NNI). Some nanoscale materials at the molecular level, for example, show greater chemical reactivity or greater catalytic behavior compared to classical molecules (Barlow et al. 2009; Renn and Roco 2006). Indeed, the fields of nanotechnology and genetic engineering are both involved in molecular and genetic manipulation of food such as in the transfer of genes in plants and animals, however their processes and scale are different (Kuzma et al. 2008).

According to a definition in a 2006 European Nanoforum (a European Economic Interest Grouping)[3] report "Nanotechnology in Agriculture and Food," food is "nanofood" when synthesized nanoparticles, or nanotechnology techniques or tools, are used during cultivation, production, processing, or packaging (Joseph and Morrison 2006). Projected and current nanotechnology applications for food include foods created with lower fats, sugars, and salt, sensors that detect food spoilage, foods with increased shelf life, and new forms of pesticides. Examples of nanoenhanced food products include canola oil that blocks the delivery of cholesterol to the blood stream, tastier and more nutritious milkshakes, and nanocapsules of tuna fish oil (Omega 3) that are designed to break open when the capsule reaches the stomach, to avoid the unpleasant taste of fish oil (Das et al. 2009; Kuzma and VerHage 2006). *Nature Nanotechnology* recently reported findings that Zimmerman and colleagues at the Swiss Federal Institute of Technology in Zurich were able to use nanotechnology to develop zinc and iron fortified food with increased bioavailability, but with no loss in flavor or change in color (Miller 2010). Nanoencapsulation of nutrients that reduces their size can increase bioavailability, which has strong potential for vitamin preservation and delivery for humans and animals. Also, "nanotubes (micrometre sized, long & hollow) made of milk protein by self-assembly have the potential to be used as novel ingredients for viscosifying, gelation, nanoencapsulation, and controlled release purposes" (Das et al. 2009: 12).

The food industry has also found uses for nanotechnologies in food packaging. Such applications include those intended to reduce UV-light exposure or microbial growth (Joseph and Morrison 2006). Nanosensors are also being proposed and used to detect pathogens or contaminants like E. coli (Fauchet 2009; Li et al. 2006).

It is difficult to identify all of the different applications of nanofood currently on the market because of a lack of reporting standards or guidelines by any government; there exists neither an official definition nor obligatory labeling requirements for nanomaterial use, making tracking difficult. In a 2008 report, the international NGO Friends of the Earth Australia (FOEA), listed over 100 food and agricultural products containing nanoparticles, while a 2006 paper by Kuzma and VerHage indicated that the number of available nanotechnology food products on the market at that time was actually very low. As of March 2009, the Woodrow Wilson's Project on Emerging Nanotechnologies (PEN)—the main extant online inventory of nanotech products on the market worldwide—catalogued eighty-four food related nanotechnology products globally, fewer than the FOEA in 2008. However, these figures are somewhat murky. The general PEN "food and beverage" category includes dietary supplements and food packaging. As of August 2011, PEN's general "food" category, which contains four subcategories reports only fifty-five food-related nanotechnology products. In addition, a 2010 study by Berube et al. asserts that many of PEN's products do not actually contain nanoparticles or were not in fact available on the market.[4] Thus the exact number of food products enhanced with nanotechnology remains unknown, as most food companies do not communicate about their use of nanotechnology (Greßler et al. 2008). Conversely, a broader understanding of nanotechnologies in food recognizes that their use is not a new phenomenon. Some of the processes of cheese and ice cream production involve changes to food structure at the nanoscale (House of Lords 2010). The historical nonlaboratory production of cheese, ice cream, ketchup, and chocolate involves nanoscale processes, which account for the thickening of the final product. However, these processes differ from the deliberate engineering of food at the nanoscale by scientists in laboratories, which is a more recent phenomenon. Novelty is not a precondition for public concern about social and ethical issues about emerging technologies; however, scientific manipulation of food at the nano level is both new and carries significant uncertainties because of the lack of risk research that seems likely to drive amplified public concern.

Nanotechnologies have also been developed and proposed for use in agriculture. The European Nanoforum report argues that these nanotech applications are being hyped as being able to improve efficiency and sustainability, echoing some of the promises of the green revolution and biotechnology:

> Precision farming has been a long-desired goal to maximise output (i.e. crop yields) while minimising input (i.e. fertilisers, pesticides, herbicides, etc) through monitoring environmental variables and applying targeted action. ... One of the major roles for nanotechnology-enabled devices will be the increased use of autonomous sensors linked into a GPS system for real time monitoring. These nanosensors could be distributed throughout the field where they can monitor soil conditions and crop growth. (Joseph and Morrison 2006: 4)

The report goes on to note that nanodevices could be used to identify plant health issues before they become visible to the farmer, and may be capable of taking "appropriate remedial action" to help create a "healthy food culture" (Joseph and Morrison 2006: 5). Nanotechnologies have also been developed for use in pesticides, such as nanoencapsualtion to slowly release insecticides in a host plant and act as insect pest control. The agribusiness company Syngenta is looking into nanoemulsions in its pesticides. Such innovations are being marketed as "ecofriendly" because they diminish the need for larger scale pesticide spraying, although the full toxicological and ecotoxicological risks of the new materials remain unknown (Bhattacharyya et al. 2010).

The use of nanotechnologies is also being explored for use with livestock, including pathogen detection and removal; veterinary medicine; feed improvement and waste remediation; animal breeding and genetics; and identity preservation and tracking (Kuzma 2010). Finally, nanotechnology has also been proposed for water, including treatment and remediation, sensing and detection, and pollution prevention (Karn et al. 2009; Stilgoe 2007).

Scientific studies have pointed to possible risks related to the production of engineered nanomaterials (Dreher 2004; Hoet et al. 2004; Nel et al. 2006). In particular, some studies have shown nanoparticles to be toxic to animals and human tissues (e.g., Li et al. 2009; Oberdörster et al. 2005; Wang et al. 2006); however, other reports show uncertainty around the potential toxicity of nanoparticles (Behra and Krug 2008; Scheringer 2008). Nanotechnologically manipulated and enhanced materials also have the potential to result in unknown risks to the human body, such as the ability to pass through cell walls and the blood brain barrier (Oberdörster et al. 2005). Moreover, a scientific committee reported to the European Commission that, during production or consumption, nanoparticles will most likely be released into the environment and could pose risks such as toxic effects on aquatic organisms (Barlow et al. 2009). In the US, governmental concern about potential toxicities has resulted in the funding of over $575 million since 2008 in research on health, environment, and safety issues of nanotechnologies (Nano.gov).

Nanotechnology and Regulation

In September 2009, the Environmental Protection Agency (EPA) issued a final Significant New Use Rules (SNUR) on Multi-walled and Single-walled Carbon Nanotubes (CNTs), which would be the first "nanospecific" regulation of a nanoengineered material. There are also some voluntary and soft-law regulations that apply to nanotechnologies, such as the 2006 Berkeley, CA city ordinance requiring all facilities within the city limits that manufacture or utilize manufactured nanomaterials to submit toxicology reports. Certain regulations also apply to or cover nanotechnologies, although there can be gaps. For example, regulatory standards in the US for existing materials are mass- or concentration-based, grandfathering in many of their smaller, nanotech counterparts (Kuzma 2010); in Europe, REACH—a precautionary instrument in the EU—regulates chemicals and includes nanomaterials under its broader umbrella (European Commission—REACH; see also Bowman and Calster 2007 for critical review of REACH). Kuzma et al. (2008) argue that gaps in oversight in the US may be most prominent for convergent nanotechnology products that span various technologies, uses, and sectors, as they often do not fall neatly under the arm of any one law, agency, or industry (Kuzma et al. 2008). In other cases, the laws and institutions responsible for oversight may be clear, but nanomaterials may not trigger premarket review if the macroscale version of the product has already been approved (Davies 2007; Maynard 2007). Given the differing properties of nanoparticles at lower doses as compared to larger or bulk versions of products, such exemptions may merit reconsideration (Maynard 2007). In addition, REACH has recently been argued to fall far short of expectations in achieving compliance for nanomaterials (SCENIHR 2006).

Due in part to unknown risks and gaps in regulations, some NGOs are carefully following agri-food nanotechnology developments and raising concerns about oversight (ETC Group 2003; Friends of the Earth 2008). One of the groups raising the most concerns (and advocating a moratorium) is the Canada-based ETC Group (Action Group on Erosion, Technology and

Concentration). The group is widely viewed as the one that took the lead in the 1980s to raise health and environmental concerns about genetically modified food (Coolidge and Freedman 2004). Another group, Friends of the Earth Australia (FOEA), has been accused of confusing top-down nanotechnologies (using larger devices and components to produce smaller products or particles, which many chemical companies have long been engaged in) with bottom-up technologies (which encompass more concerns around novel techniques like molecular self-assembly and using small components to build bigger ones). It was also accused of drawing an artificial line at 100 nm to critique anything at or below that level (Berger 2010) (it should be noted that this is also the boundary set by the NNI to define the size scale for nanotechnologies). A number of NGOs have raised concerns over the unknown effects of engineered nanomaterials, with some advocating a precautionary approach, saying that neither they nor the technologies in which they are embedded should be on the market until their potential toxicological effects are known, and others calling for a full moratorium (ICTA 2007). Groups including the Center for Food Safety (US), Pesticide Action Network (North America), Greenpeace (UK), and the Institute for Agriculture and Trade Policy (US) have called for national and international regulation, labeling, and increased safety research (Lee and Kigali 2006).

Despite the efforts of these NGOs to raise concerns, surveys around the globe suggest that most people know little or nothing about nanotechnology (Satterfield et al. 2009), suggesting that concerns about nanotechnologies being raised by NGOs have not yet reached many people. Initial studies also suggest that although people may generally see more benefits than risks in nanotech development as a whole (Satterfield et al. 2009), they are more hesitant about nanotechnologies embedded in food, which shows the importance of considering specific technological application when looking at public perceptions of technologies like nanotechnology, particularly for food (Bauer 2005; Pidgeon et al. 2009; Rogers et al. forthcoming).

Nanofood and Public Perceptions

Although nanotechnology as a generic category was rated less risky and more beneficial than GM in a 2006 US survey (Currall et al. 2006), research suggests emerging public hesitation in acceptance of nanotechnology within commercial foods. According to survey data collected in the U.S. by Peter D. Hart Research Associates, Inc. (2007), on behalf of the Project on Emerging Nanotechnologies (PEN), when asked if they would use food storage products enhanced with nanotechnology, 12 percent said yes, 73 percent said that they need more information about the health risks and benefits, and 13 percent said no. When asked if they would purchase food enhanced with nanotechnology, a mere 7 percent affirmed, 62 percent said that they need more information about the health risks and benefits, and 29 percent said no.

Surveys on nanotechnologies and food in Switzerland also found public hesitation to buying nanotechnologies in food or food packaging. Although in this study, too, food packaging was assessed as less problematic than food itself (Siegrist et al. 2007; Siegrist et al. 2008). Trust and high perceived benefits (Siegrist et al. 2007), as well as naturalness and perceived control (Siegrist et al. 2008), were strong factors in the willingness of those surveyed to purchase nanofoods and food packaging. Siegrist et al. (2007) found that public reaction to nanoparticles for bacterial removal for food safety purposes will also be affected by the degree of trust in those responsible for oversight and the perceived benefits of reduced antibiotic use, as well as whether the public is consulted.

Why the seemingly low support for nanotechnology in food? One explanation is that GM constitutes some kind of template or mental model (cf. Morgan et al. 2002; Priest 2010) upon which people draw in making sense of largely unknown nanotechnologies. It has been shown in previous research that people often use availability heuristics, or cognitively derived examples, to process new (risk) information and to evaluate a new technology (Kahan et al. 2009), while affect heuristics have been shown to shape risk-benefit judgments (Finucane et al. 2000). Consumers' prior beliefs about GM food (which can be an availability heuristic or affect heuristic) have been found to be important for the evaluation of conflicting information in the EU (Costa-Font and Mossialos 2005b) and in the US, England, and France (Lusk et al. 2004). The studies found that the influence of benefit information on biotechnology acceptance was impacted by people's prior acceptance of GM food and biotechnology in general.

In a more explicit testing of this link, Vandermoere et al. (2009) tested the influence of GM views on perceptions of nanotechnologies in food by expanding on Siegrist's nanofood and food packaging study, through a study of how people react to "nano-outside" and "nano-inside" applications. In their 2008 Web survey of 752 participants in France, in which the majority (81.5%) had little to no knowledge of nanotechnology, Vandermoere's group analyzed trust in government agencies, familiarity with nanotechnology, views on genetic engineering, and trust in science and technology as effects on the participants' responses to two nanofood applications. In each description of the nanofood application (one for food packaging and one about nanofood emulsions for low-fat food alternatives and food coatings for preservation), the descriptions weighted heavily on the benefits and ended with one sentence about the unknown consequences (ignorance over risk). Most of the participants were ambivalent about nanofood packaging (44.1%) and nanofood (42.8%). Trust in governmental agencies as well as positive views on science and technology increased optimism about nanofood and food packaging. Overall, Vandermoere et al. (2009) found both positive and negative views of nanotechnology related more to social issues and views on science, technology, and nature than to people's lack of knowledge.

In a similar survey carried out in Munich, Germany, Marette et al. (2009) surveyed ninety-seven people on their WTP for orange juice fortified with nanotechnology, and a bottle fortified with nanotechnology to protect food nutrients from UV light. Provision of even general information on nanotechnologies and the products reduced the overall WTP for both products, particularly the juice. Respondents were then allowed to choose any additional information they wanted on the use of nanotechnologies in the products: health, environmental, or social risks and benefits. Most chose health and, despite discussion of the potential benefits, the information further reduced their WTP. The health information briefly mentioned potential toxicological risk and potential health benefits from greater absorption of nutrition. In a comparative survey of the WTP for nanofoods and food packaging of participants in France and Germany, Bieberstein et al. (2011) found that participants with a negative view of GM did bring WTP for nanofood down, but not as much as information on health, which brought down the WTP for nanofood and food packaging in both countries.

Some have suggested that people are hesitant about nanotech (and biotech) food because it raises issues of embodiment and ingestion. Indeed, a study on consumer perception of nanotechnology in Germany argued that the closer products were to the human body, the higher the consumer's reluctance to purchase or use them (René Zimmer et al. 2008). Participants in this study approved of improvements of surface treatments (e.g., paint, varnish and textiles) much more than applications for cosmetics and foods. Yet these results also clearly pertain to perceived benefits: much like biotechnology (Costa-Font and Mossialos 2005a), people often expressed more support for nanotechnologies in medical applications than food, even though medical applications are also near or within the body. A 2005 study of experimental groups at three different US sites asked the participants to identify up to five areas of highest interest regarding the benefits of nanotechnology, with 31 percent saying medical technologies, and only 6 percent saying food and nutrition (Macoubrie 2006). Similarly, a 2008 US telephone survey of 1,100 people found that participant support for nanopills and nanofuels varied by how information about the application was presented (in terms of risk, controllability, bodily invasion, and social distribution). However, nanofood was uniformly unsupported, regardless of how information about it was conveyed (Conti et al. 2011).

Yet some studies outside of the EU suggest that there is at least a modest level of support for nanotechnologies in food, particularly for achieving lower saturated fats and environmental sustainability. In a quantitative study in New Zealand of public perception of lamb or beef that has been genetically modified using nanotechnology to reduce saturated fat content, Cook and Fairweather (2007) found that the greatest concern (73.5%) by the 565 respondents was that "people [producers] may not comply with the rules and regulations that govern the processes used to make the product" (678). Sixty-one percent said it would feel unnatural to eat the modified lamb or beef. Almost half (46.6%) said that the development of the product is more about making money than making food; 29.6 percent said the product would be unethical; and only 16.8 percent were confident that there would not be any unexpected food risks from the lamb or beef. Despite some concerns by the respondents, 76.6 percent claimed they would definitely purchase the product, and the authors found this intention to buy correlated primarily with pro-technology attitudes, self-identity, and subjective norm (how they perceived others would respond to the purchasing of the product).

Similarly, a 2010 US survey of 750 people[5] on consumer perception of food technology by the International Food Information Council, a nonprofit communications organization supported by food industry companies, found that about half of those surveyed approved of nanotechnologies to improve the safety and quality of food, a finding that is much higher than most other studies. Along these lines, the survey found that support for plant biotechnology was somewhat low (32%), but increased greatly when it was suggested that the technology would be used for environmental sustainability (more food on less land with less water), lower fats, and reduced pesticide use. Although based on a relatively low sample size, the findings suggest that playing up the possible benefits of nanofood in areas that particularly appeal to the public, like improving environmental quality, can potentially increase public support; however perceived benefits may vary by country. Conti et al. (2011) also found that presenting any information, benefit- or risk-related, led to decreases in public acceptance, a likely upstream effect.

Public Deliberations on Nanotechnologies

To better assess and address these public responses, upstream oversight assessment through public deliberation has been proposed as an alternative to risk communication (Kuzma et al. 2008). As a set of emerging technologies, especially in the domain of food and agriculture, nanotechnologies offer an opportunity for upstream public engagement. Awareness and knowledge of nanotechnology is low in the general public. Though low awareness may impact public perception of nanotechnology (Kahan et al. 2009), studies show that risk and benefit perception of nanotechnology are also a result of public perception of GM, more general attitudes in science and technology, and trust in government, industry, and scientists, as well as situational and demographic variables. The goal of public deliberation is for experts and policymakers to engage with, listen to, and incorporate views of various members of the public into the development and oversight of new technologies. Such deliberations developed, in part, out of the public backlash against GM foods. In response to public backlash, the UK government directed the

Royal Society and Royal Academy of Engineering to carry out research on public perceptions of nanotech developments, leading to two UK public discussion workshops in 2003. In response to the workshops, the Royal Society wrote a report arguing that public dialogue should be part of early UK nanotechnology development (Royal Society and the Royal Academy of Engineering 2004).

Lancaster University and the UK NGO DEMOS, conducted a series of public deliberation experiments called Nanodialogues in England, with two of them focusing on specific application domains: one on nanotech for cleaning up contaminated land, and one on nanotech for cleaning water. Deliberating the use of nanotechnology for the treatment of contaminated land were London residents and a group of various experts (NGOs, government, and University researchers), with many of the residents determining that such uses should not be applied until the long-term effects are studied and known and the results made public (Stilgoe 2007). Several themes emerged from the focus groups: a precautionary approach to regulation, the need for more public involvement, the reporting of all toxicology findings, and the importance of application context for discussions about science, technology, and risk. In 2006, Zimbabwean residents participated in focus groups held by DEMOS, Practical Action, and Lancaster University to discuss nanotechnology for cleaning water with scientists from around the world. The residents raised distributive justice concerns around access and accessibility (Grimshaw et al. 2006).

Also in the UK and the US, from 2005 to 2007 the Nanotechnologies Engagement Group was a public engagement project on nanotechnology. The participants highlighted a series of social concerns related to nanotechnology, such as the social distribution of benefits and risk. The participants were concerned that nanotechnology may benefit certain sectors of society, or the producers themselves, and others would be left out. The UK participants referred to their experiences with GM food and bovine spongiform encephalopathy (BSE, or commonly known as mad-cow disease), whereas the US participants focused on asbestos, dioxin, polychlorinated biphenyls (PCBs), Agent Orange, nuclear waste, and the Gulf War Syndrome (Miller 2008). Gaskell et al. (2004) argued that in the GM case, European publics saw benefits accruing primarily upstream in the value chain—for example grocers benefit by longer product shelf life while consumers assumed the risks downstream; similar patterns appear to be emerging in the nanotechnology case with food applications. Similarly, indigenous rights activists and scholars are already raising concerns that nanotech developments in food may lead to more patents and technological innovations that can further accelerate social inequalities, particularly in developing nations (Wetter 2010).

In the 2010 UK House of Lords report on nanotechnology and food, Nick Pidgeon is cited as an expert for his work on public deliberation of technology, including nano and GM, and involvement with different stakeholder groups. Drawing on Pidgeon's testimony, the House of Lords reports that it "had been difficult to persuade food companies to fund any public engagement projects. The PEN [Project on Emerging Nanotechnologies] suggested that the same was true of the industry in the United States and the GMA [Grocery Manufacturers Association] also told us that companies had retreated from a public dialogue on the subject in recent years" (2010: 70). Survey data shows general ambivalence and some concern toward food packaging and nanofood (Hart Research Associates 2009; Siegrist et al. 2007), but qualitative study can help researchers examine how the public engages with nanotechnologies they know very little about. How do they discuss nanotechnology and nanofood? Do US participants draw on their experiences with and knowledge of GM food or other science and technology domains? How do risk and benefit perception relate to views on ethical development, distribution, and trust in government, scientists, and corporations?

Studies suggest that application matters in perceptions of risk and benefit of nanotechnologies (Pidgeon et al. 2009), particularly food, and that those in the food industry might benefit from exploring rather than dismissing these views. Specifically, quantitative and qualitative data show that publics are generally supportive or somewhat ambivalent about nanotechnology as a whole; however, nanofoods routinely elicit more concern than other applications about potential risks associated with trust, lack of benefits, unknown environmental and health effects, distributive and procedural justice issues, and privacy. More research is needed to fully understand the future of public responses to nanotechnology and whether or not they will be "the next GM." Drawing on our own set of deliberation workshops conducted by the CNS-UCSB with members of the California public, we briefly analyze how our findings related to public thoughts on nanotechnologies and food compare with and expand on previous research.[6]

Six deliberative workshops were run in the Santa Barbara, California area in the fall of 2009 by an interdisciplinary research team of social and physical scientists. The four and half hour long workshops focused on applications of nanotechnology for either health and human enhancement or energy and environment, with gender composition varied between the groups. The aim of these workshops was to facilitate a public discussion on different nanotechnology applications, and examine how social location shapes views on the risks and benefits of nanotechnologies. Nanotech food and agricultural applications were discussed in both sets of workshops, but primarily in the health and human enhancement workshops. Out of 67 participants in six workshops, 33 were men and 34 were women, and each workshop ranged between nine and thirteen participants.[7] After the audiotapes were transcribed, we conducted a content analysis using N*Vivo of the six workshop transcripts, as well as quantitative analysis of pre- and post-survey results.

Genetic Modification as a Reference

In all six deliberations many participants voiced reservations about nanotechnology and food. Many of their concerns echoed those expressed about GM: unknown environmental and health risks, lack of perceived tangible benefits, concerns about tampering with nature, distrust around corporate influence in food production, and perceived lack of effective government oversight. Kearnes et al. (2006) point out that such concerns often encompass what can be seen as social risks, as unknown health risks raise issues around trust and regulation. We therefore analyze the common (although not ubiquitous) reference to GM by these nanotech deliberation participants, to see how it colored and informed their upstream assessments and discussion of nanotech for food. Out of six workshops, 85 sections of dialogue were coded with references to food, of which about 30 percent mentioned genetic modification of food. Participants expressed concerns around nanofood in the health/human enhancement workshops as well as the energy/environment workshops, despite greater discussion about food in the health workshops and a higher overall benefit-focus in the energy workshops. We argue that references to GM stand as an example of unresolved social issues around food that concern the participants, which they believe may continue in the case of nanotechnologies.

We found a high use of GM as a reference in deliberations about nanotechnology food applications, often drawn upon as an example—or mental model—of ongoing social concerns with technology and the food industry. In this first example, the participant discusses GM as a way to raise questions about who benefits and profits from nanofood:

> Gary: Where I see the food industry—again the one which is run by on one purpose only, and that is profit. … If you actually are able to get it on your crops then Monsanto sues you for stealing their products.

> Facilitator: For GM—for genetically modified—
>
> Gary: Genetically modified materials.
>
> Facilitator: So some of the nan—
>
> Gary: I worry about that—is this going to be true with nanotechnology as well as applied agriculture? (Health/HE, male only, white, 66+ years old)[8]

Here, Gary brings up Monsanto and genetic modification to raise more general issues about the privatization of food, the patenting of nature, and distrust of the food industry.

When asked which nanotech application could be the most controversial, food came up in every workshop, possibly because some nanofoods are already on the market and there is not yet any mandatory labeling. The uninformed and unconsented exposures represented by the dissemination of nanofood products raised issues concerning informed consent, in patterns very similar to those of genetically modified foods:

> Chad: I think food [will be most controversial], because I think that's where materials we put in that we won't be told about or that [it is] just like genetic engineered food today. Where there's no list, there's no labeling and ...
>
> Facilitator: And people will care about it more?
>
> Chad: Yeah, yeah because it will be, like with medicines, you're gonna know. It's more, there's more of a chance of informed consent, you know what you're gonna get in this medicine, you're gonna know, you know. With foods you're gonna be less likely to know, and the lack of knowledge of whether that's there or not, it's more pressure. (Health/HE, male only, white, 36–45 years old)

Chad is concerned that, like GM foods, nanotech foods will continue not to be labeled, a concern shared by many of the participants. The general public will not be aware of what they are buying, thus limiting their ability to make informed choices.

Other participants saw nanofoods as the next step in an ongoing trend of unnecessary and potentially unhealthy food modification:

> Caitlynn (Health/HE, female only, mixed race, 18–25 years old): It frightens me too. I know that some of the food that you find in the store, actually most of it is genetically altered as it is. It's been tampered. It's not what it used to be, and that as it is, is scary. To add to it and to make it even more, I don't know how my body would react to it. I have a whole slew of stomach problems, and foods do not react well with it. Let's just add more things to it, and I may not be able to eat at all. So—
>
> Facilitator: That could be a bad outcome. Patty.
>
> Patty (white, 46–55 years old): Yeah I am concerned about genetically modified foods, and I lately have been changing my diet, and I'm losing weight and just, you know if comes out of the ground, and it doesn't have process—if you can't understand or read the ingredients on the label—it is too over processed. I mean it is obviously not going to be that good for your body so why add more to that formula?
>
> Facilitator: Okay.
>
> Patty: Or to eat as naturally as possible.

These exchanges show a general concern with nanofood, and part of that concern parallels feelings about GM as involving "tampering" with nature, with possible risks and a lack of perceived benefits for the public.

A preference for natural and less-processed foods was particularly vocalized in the all-female health/human enhancement group. A longitudinal study measuring perceptions of food risks and social location in Canada found gender to be the only robust variable across the classes of issues and time periods—with more women concerned about food risks than men (Dosman et al. 2001). The following dialogue highlights preference for natural foods, but also shows the key points of contention about nanofood and the use of GM as a reference for discussing broader social issues and trust in the food industry. This passage is a go-around of the participants as they answer a question about how they feel about nanofood:

> Dora (Health/HE, female only, Latina, 26–35 years old): I agree I wouldn't feel comfortable having food *manipulated or injected* with chemicals. It is just … you know?
>
> Facilitator: Okay.
>
> Dora: Very natural. It should be *natural.*
>
> Facilitator: India.
>
> India: (Latina, 56–65 years old) I agree. I don't feel comfortable with it, and I don't think that it would necessarily be done to make it better for me.
>
> Facilitator: Okay so the beneficiary would not necessarily be you? The consumer, okay. Deborah.
>
> Deborah (white, 18–25 years old): I believe that it is completely wrong. I don't really want my zucchini *injected* with more vitamins I just—
>
> Facilitator: Do you want the soil it is grown in injected with more?
>
> Deborah: No. I don't. I think that the world, the earth is here, you know, and I *don't think that we need to* genetically modify it. …
>
> Sadie (white, 66+ years old): I think we need more education and help for the public to obtain real food. Food that has not been *processed.*
>
> Facilitator: Okay. Jaquelyn
>
> Jaquelyn (African American, 26–35 years old): I would say depends.
>
> Facilitator: Here's our scientist.
>
> Jaquelyn: No. Because when I think of things on the molecular level, I just sort of think of it in terms of molecules. So like you grow a banana from the ground, and it's composed of certain compounds like just natural elements from the periodic table, and so essentially they're just sort of saying instead of it growing it from the earth why not just kind of grow it in a lab?
>
> Facilitator: Put it together from the atomic level.
>
> Jaquelyn: Put it together, put it together from the atomic level yourself instead of sort of, instead of doing the same thing just growing from the earth. So and essentially with *genetically modified* food, it's kind of like if you think of the evolutionary process plants kind of adapted themselves over time because maybe there was a predator that used to eat it and so it maybe grew thorns so it wouldn't be eaten, so it would survive. So it sort of yeah or something happened and sort of genetically modified itself, but instead we are just kind of doing that in the lab. So I would say it depends. It just needs more research. Right now people—we don't fully understand how maybe things grow or you know how things change over time but if you figure that out, I think, I think it would be safe.

This passage is striking due to the use of the terms "manipulated," "injected," "should be natural," "wrong," "we don't need to," and "processed." Although the group was generally apprehensive about nanofood, Jaquelyn—a graduate student in the physical sciences—offered a more proscience and technology viewpoint. She did not appear to change the minds of her fellow workshop attendees, but she did suggest that if nanotech is similar to GM, it does not necessarily signify an "unnatural" process. Her final words, however, do show some concern about the lack of risk research and it shows that she, too, uses GM as a template for making sense of nanofood. An analysis of the use of GM as a reference for understanding nanotech food applications highlights the concerns that our participants expressed about nanofoods, and shows that many of these concerns center on social risks such as trust, consent, embodiment, naturalness, monopolization, and control.

Although Vandermoere et al. (2009) did not find views on GM to be the strongest predictor of public views on nanotechnology in food in France, we found numerous references to it in our workshops. As they deliberated to make sense of nanotechnologies of food, participants referred to GM and "alterations/modifications" of food, and these references related to overall discussions of uneasiness about technology in the food industry. Participant statements suggest that their views on GM are not simply affective processing shortcuts, and are not mutually exclusive with concerns like perceived benefits and trust, but are rather bound up in them. Such assumptions, concerns, and discussions centered on issues of health risks (including the unknown risks of ingesting nanomaterials), lack of informed consent (nanofood is not labeled and participants do not have faith that it will ever be labeled, as GM foods are still not labeled, in contrast to medicines that carry detailed labeling), lack of trust in industry and government to safely regulate and develop nanofood, and a concern about who benefits from nanofood. All of these concerns for nanotechnology and biotechnology in food products are tied up in perceptions and social connections to food as a cultural domain.

Conclusions

A brief look at the CNS-UCSB study shows support for many of the key findings in the literature on risk perceptions of nanofood, particularly:

1. Quantitative and qualitative studies suggest perceived risks of nanofood are greater than risk perceptions of nanotechnologies in other applications (e.g., health or energy).
2. Discussion of nanotechnologies for food often raise broader concerns with food and the food industry, and highlight the particular salience of food as an important cultural domain.
3. A majority of people sampled in the US, like their counterparts in the UK (see Pidgeon et al. 2009), question the benefits of nanotechnology in food, and would like some regulation of nanofood, such as labeling, which mirrors unaddressed concerns with GM foods.

In short, our literature review and deliberative research suggest that many of the social concerns discussed in the literature on biotechnology and food—such as control over food, unknown risks, and potential changes to food and ecology—are being raised around the use of nanotechnology in food, highlighting a host of social concerns that many people feel have not been adequately addressed. Such concerns are sometimes tempered, however, by the perceived potential for nanotechnologies to address wider social problems like food shortages, hunger, and environmental degradation. In addition, there is distrust that such technologies will actually reach the people and countries that might need them. Thus while scholars debate whether nanotechnology will lead to the "next GM," the public wonders more specifically about the future of nanofood and whether it is being developed thoughtfully, carefully, and to the benefit of the world's people.

Quantitative and qualitative research on technologies for food illustrate several common concerns such as variations in public perception research cross-nationally and regionally, food as a sacred and significant cultural domain, which reflects concerns about altering nature and food through technologies, and a perceived lack of benefit for consumers. Our research and the research of those who have engaged in qualitative research of nanofood risk and benefit perceptions also suggest that it is necessary for scholars to supplement or go beyond survey research to understand how people discuss an unfamiliar new technology. Public engagements can take advantage of this upstream context in the history of nanotechnology to explore public perception and offer a forum for people to affect decisions that will determine the future of food. At this point, public participation can be key for exploring and thus addressing ongoing social hopes and concerns around new technologies in food.

ACKNOWLEDGMENTS

This work is supported by the National Science Foundation (NSF) through grant #0824042 to Barbara Herr Harthorn and Coop. Agreement No. SES 0531184 to the Center for Nanotechnology in Society at UCSB. Any opinions, findings, conclusions, or recommendations expressed in this material are those of the authors and do not necessarily reflect the views of the NSF. The authors thank the following people for their contributions to the research on which this article is based: 2007 workshops that developed the protocol on which this study is based—Karl Bryant (SUNY New Paltz), Nick Pidgeon (Cardiff Univ, UK), Tee Rogers-Hayden (Univ of East Anglia, UK), Hillary Haldane (Quinnipiac University), Joseph Summers (MIT), and Terre Satterfield and Milind Kandlikar (Univ of British Columbia); 2009 workshops—Rachel Cranfill (UCSB), Amanda Denes (UCSB), Shannon Hanna (UCSB), Indy Hurt (UCSB), Tyronne Martin (UCSB), Julie Whirlow (UCSB). The authors also thank the four anonymous reviewers, whose helpful comments have contributed to an improved final version of this article.

JENNIFER B. ROGERS-BROWN, PhD, is an assistant professor of sociology at Long Island University, C.W. Post. Her areas of research include intersections of gender, race, and class, environment and technology, and globalization. Her dissertation analyzes maize in Mexico, free trade, bioengineering, gender, and indigenous cultures. As a postdoctoral scholar at the Center for Nanotechnology in Society at UCSB from 2008 to 2010 she studied gender, public deliberation, and public perceptions of nanotechnology.

CHRISTINE SHEARER, PhD, is a postdoctoral scholar at the Center for Nanotechnology in Society at the University of California, Santa Barbara (UCSB). Her areas of research include environmental sociology and science and technology studies, with a focus on policy and regulations. Her dissertation looked at vulnerability and adaptation for Alaska Natives facing permanent displacement from environmental change, and was published under the title *Kivalina: A Climate Change Story* (2011).

BARBARA HERR HARTHORN is a cultural anthropologist who is associate professor of feminist studies, anthropology and sociology and Director of the NSF Center for Nanotechnology in Society at the University of California, Santa Barbara (CNS-UCSB). She is also a research group leader and executive committee member in the UC Center for Environmental Implications of Nanotechnology. In CNS-UCSB she leads an interdisciplinary group studying nanotechnology risk perception and public deliberation. She is editor (with Laury Oaks) of *Risk, Culture, and Health Inequality: Shifting Perceptions of Danger and Blame* (2003) and a fellow of the American Association for the Advancement of Science.

NOTES

1. For additional information on IPR and geographical indications, a form of IPR, and their impact on indigeous populations and rural communities in emerging economies, see Parasecoli and Tasaki's article in this volume.
2. For a discussion of the importance of risk communication and discussion of the deficit model see Slovic (1999), Gaskell et al. (2004), and Siegrist (2008).
3. Nanoforum is a news organization with member organizations. It was originally funded by the European Commission.
4. PEN notes that it used self-reporting and publicly available information to identify products that utilize nanotechnology. Therefore, the project did not check the validity of each product to verify if it contains nanotechnology.
5. The survey was a representative national sample and conducted by the Cogent Research of Cambridge, MA (Food Insight 2010). It was weighted on maritial status and education.
6. The 2009 workshops built on CNS-UCSB's 2007 cross-national (US-UK) study of nanotech deliberation (Pidgeon et al. 2009).
7. Recruitment and screening were performed by a professional third party (social survey center), without disclosing the nanotechnology topic for the sessions. With our aim of democratic inclusiveness, recruitment was designed to produce a diverse quasi-representative quota sample of participants designed to match as closely as possible the demographics of the central coast California area along race/ethnicity, occupation/education, age, and income criteria. The participants' incomes reflect the relative affluence of the community and are not reflective of national demographics.
8. Notes describing a participant's social location and workshop participation are provided after each quote in the following format: pseudonym, workshop topic, gender composition of group, gender of participant (if not clear from gender composition), race/ethnicity of participant, and age of participant.

REFERENCES

Barlow, Sue, Andrew Chesson, John D. Collins, Albert Flynn, Anthony Hardy, Klaus-Dieter Jany, Ada Knaap, Harry Kuiper, John Christian Larsen, Pierre Le Neindre, Jan Schans, Josef Schlatter, Vittorio Silano, Staffan Skerfving and Philippe Vannier. 2009. "Scientific Opinion of the Scientific Committee on a Request from the European Commission on the Potential Risks Arising from Nanoscience and Nanotechnologies on Food and Feed Safety." *The EFSA Journal* 958: 1–39.

Bauer, Martin. 2005 "Distinguishing Red and Green Biotechnology: Cultivation Effects of the Elite Press." *International Journal of Public Opinion Research* 17 (1): 63–89.

Behra, Renata, and Harald Krug, H. 2008. "Nanoparticles at Large." *Nature Nanotechnology* 3: 253–254.

Berger, Michael. 2010. "Blowing Hot Air—How Not to Criticize Nanotechnology." *Nanowerk.* http://www.nanowerk.com/spotlight/spotid=19060.php (accessed 27 November 2011).

Berube, David M., Eileen M. Searson, Timothy S. Morton, and Christopher L. Cummings. 2010. "Project on Emerging Nanotechnologies–Consumer Product Inventory Evaluated." *Nanotechnology Law & Business* 7 (2): 152.

Bieberstein, Andrea, Jutta Roosena, Stéphan Maretteb, Sandrine Blanchemanchec, and Frederic Vandermoere. 2011. "Consumers between Indifference and Mistrust: French and German Behaviour regarding the Potential Introduction of Nanotechnologies in Food." *INRA Sciences Sociales* 1. http://www.inra.fr/sae2/publications/iss/eng/ (accessed 27 November 2011).

René Zimmer, Herausgegeben von, Rolf Hertel, Gaby-Fleur Böl. 2008. *Wahrnehmung der Nanotechnology in der Bevölkerung: Repräsentativerhebung und morphologisch-psychologische Grundlagenstudie.* Berlin, Germany: BfR Wissenschaft. www.bfr.bund.de/cm/350/wahrnehmung_der_nanotechnologie_in_der_bevoelkerung.pdf (accessed 28 November 2011).

Bhattacharyya, Atanu, Asim Bhaumik, Pathipati Rani, Suvra Usha Mandal, and Timothy T. Epidi. 2010. "Nano-Particles—A Recent Approach to Insect Pest Control." *African Journal of Biotechnology* 9 (24): 3489–3493.

Bowman, Diana M., and Geert Van Calster. 2007. "Does REACH Go Too Far?" *Nature Nanotechnology* 2: 525–526.

Bray, Francesca. 2003. "Genetically Modified Foods: Shared Risk and Global Action." Pp. 185–208 in *Risk, Culture, and Health Inequality: Shifting Perceptions of Danger and Blame,* ed. B. H. Harthorn and L. Oaks, Westport, CT: Praeger.

Charles, Dan. 2010. "Genetically Modified Corn helps Common Kind, Too." NPR News. http://www.npr.org/templates/story/story.php?storyId=130405227 (accessed 10 August 2011).

Conti, Joe, Terre Satterfield, Barbara Herr Harthorn, and Nick Pidgeon. 2011. "Vulnerability and Social Justice as Factors in Emergent US Nanotechnology Risk Perceptions." *Risk Analysis,* doi:10.1111/j.1539-6924.2011.01608.x.

Cook, Andrew J., and John R. Fairweather. 2007. "Intentions of New Zealanders to Purchase Lamb or Beef made using Nanotechnology." *British Food Journal* 109 (9): 675–688.

Coolidge, Carrie, and Michael Freedman. 2004. "Nanofear." *Forbes* 174 (4): 168–170.

Costa-Font, Joan, and Elias Mossialos. 2005a. "Ambivalent Individual Preferences Towards Biotechnology in the European Union: Products or Processes?" *Journal of Risk Research* 8 (4): 341–354.

———. 2005b. "Is Dread of Genetically Modified Food Associated with the Consumers' Demand for Information?" *Applied Economics Letters,* 12 (14): 859–863.

Currall, Stephen C., Eden B. King, Neal Lane, Juan Madera, and Stacey Turner. 2006. "What Drives Public Acceptance of Nanotechnology?" *Nature Nanotechnology* 1 (3): 153–155.

Das, Mukul, Neha Saxena, and Premendra D. Dwivedi. 2009. "Emerging Trends of Nanoparticles Application in Food Technology: Safety Paradigms." *Nanotoxicology* 3 (1): 10–18.

David, Kenneth, and Paul B. Thompson. 2008. *What Can Nanotechnology Learn from Biotechnology? Social and Ethical Lessons for Nanoscience from the Debate over Agrifood Biotechnology and GMOs.* Amsterdam: Elsevier.

Davies, J. Clarence. 2007. "EPA and Nanotechnology: Oversight for the 21st Century." Project on Emerging Technologies. http://www.nanotechproject.org/file_download/197 (accessed 28 November 2011).

Dosman, Donna M., Wicktor L. Adamowicz, and Steve E. Hrudey. 2001. "Socioeconomic Determinants of Health-and Food Safety-Related Risk Perceptions." *Risk Analysis* 21 (2): 307–318.

Douglas, Mary, and Aaron Wildavsky. 1983. *Risk and Culture: An Essay on the Selection of Technological and Environmental Dangers.* Berkeley: University of California Press.

Dreher, Kevin L. 2004. "Health and Environmental Impact of Nanotechnology: Toxicological Assessment of Manufactured Nanoparticles." *Toxicological Sciences* 77 (1): 3–5.

ETC Group. 2003. "Down on the Farm: The Impact of Nanoscale Technologies on Food and Agriculture." http://www.etcgroup.org/en/materials/publications.html?pub_id=80 (accessed 27 November 2011).

Fauchet, Philippe M, ed. 2009. *Frontiers in Pathogen Detection: From Nanosensors to Systems (Proceedings Volume).* Proceedings of SPIE Volume 7167.

Finucane Melissa, Ali Alhakami, Paul Slovic, and Stephen Johnson. 2000. "The Affect Heuristic in Judgments of Risks and Benefits." *Journal of Behavioral Decision Making* 13 (1): 1–17.

Food Insight. 2010. "2010 Consumer Perceptions of Food Technology Survey." www.foodinsight.org/Resources/Detail.aspx?topic=2010_Consumer_Perceptions_of_Food_Technology_Survey (accessed 10 August 2011).

Freudenburg, William R., and Julie A. Rursch. 1994. "The Risks of 'Putting the Numbers in Context': A Cautionary Tale." *Risk Analysis* 14 (6): 949–958.

Friends of the Earth. 2008. "Out of the Laboratory and onto Our Plates: Nanotechnology in Food & Agriculture." Report.

Gaskell, George, and Nick C. Allum. 2001. "Sound Science, Problematic Publics? Contrasting Representations of Risk and Uncertainty." *Politeia* 17 (63): 13–25.

Gaskell, George, Nick Allum, Martin Bauer, Jonathan Jackson, Susan Howard, and Nicola Lindsey. 2003. "Climate Change for Biotechnology?" UK Public Opinion 1991–2002. *Journal of Agrobiotechnology Management and Economics* 6 (1 and 2): Article 12.

Gaskell, George, Nick Allum, Wolfgang Wagner, Nicole Kronberger, Helge Torgersen, Juergen Hampel, and Julie Bardes. 2004. "GM Foods and the Misperception of Risk Perception." *Risk Analysis* 24 (1): 185–194.

Greßler, Susanne, André Gazsó, Myrtill Simkó, Michael Nentwich, and Ulrich Fiedeler. 2008. "Nanopartikel und Nanostrukturierte Materialien in der Lebensmittelindustrie." *Nano-Trust-Dossier* 4.

Grimshaw, David J., Jack Stilgoe, and Lawrence Gudza. 2006. "The Role of New Technologies in Potable Water Provision: A Stakeholder Workshop Approach." Report on the Nano-Dialogues Held in Harare, Zimbabwe.

Hampel, J., A. Klinke, and O. Renn,. 2000. "Beyond 'Red' Hope and 'Green' Distrust: Public Perception of Genetic Engineering in Germany." *Politeia* 16 (60): 68–82.

Hart Research Associates. 2007. "Awareness of and Attitudes toward Nanotechnology and Federal Regulatory Agencies." Project on Emerging Nanotechnologies, Washington, DC.

———. 2009. "Nanotechnology, Synthetic Biology, & Public Opinion." Project on Emerging Nanotechnologies, Washington, DC.

Hoet, Peter H. M., Irene Bruske-Hohlfeld, and Oleg V. Salata. 2004. "Nanoparticles: Known and Unknown Health Risks." *Journal of Nanobiotechnology* 2 (1): 12–26.

Horlick-Jones, Tom, John Walls, Gene Rowe, Nick Pidgeon, Wouter Poortinga, Graham Murdock, and Tim O'Riordan. 2007. *The GM Debate: Risk, Politics and Public Engagement.* New York: Routledge.

House of Lords. 2010. "Nanotechnologies and Food." Science and Technology Committee: 1st Report of Session 2009–10.

International Center for Technology Assessment (ICTA). 2007. *Principles for the Oversight of Nanotechnologies and Nanomaterials.* Washington, DC: International Center for Technology Assesment. http://www.icta.org/doc/Principles%20for%20the%20Oversight%20of%20Nanotechnologies%20and%20Nanomaterials_final_with_post-release%20signatories.pdf (accessed 28 November 2011).

Joseph, Tiju And Mark Morrison. 2006. "Nanotechnology in Agriculture and Food." Nanoforum Report, Institute of Nanotechnology, Reino Unido.

Kahan, Dan M., Donald Braman, Paul Slovic, John Gastil, and Geoffrey Cohen. 2009. "Cultural Cognition of the Risks and Benefits of Nanotechnology." *Nature Nanotechnology* 4 (2): 87–91.

Karn, Barbara, Todd Kuiken, and Martha Otto. 2009. "Nanotechnology and *in situ* Remediation: A Review of the Benefits and Potential Risks." *Environmental Health Perspectives* 117 (12): 1813–1831.

Kearnes, Matthew, Robin Grove-White, Phil Macnaghten, James Wilsdon, and Brian Wynne. 2006. "From Bio to Nano: Learning Lessons from the UK Agricultural Biotechnology Controversy." *Science as Culture* 15 (4): 291–307.

Kuzma, Jennifer. 2010. "Nanotechnology in Animal Production: Upstream Assessment of Applications." *Livestock Science* 130 (1–3): 14–24.

Kuzma, Jennifer, Jordan Paradise, Gurumurthy Ramachandran, Jee-Ae Kim, Adam Kokotovich, and Susan M. Wolf. 2008. "An Integrated Approach to Oversight Assessment for Emerging Technologies." *Risk Analysis* 28 (5): 1197–1220.

Kuzma, Jennifer, and Peter VerHage. 2006. *Nanotechnology in Agriculture and Food Production: Anticipated Applications.* Woodrow Wilson International Center for Scholars: Project on Emerging Technologies. http://www.nanotechproject.org/news/archive/new_report_on_nanotechnology_in/ (accessed 27 November 2011).

Lassen, Jesper, and Andrew Jamison. 2006. "Genetic Technologies Meet the Public: The Discourses of Concern." *Science, Technology & Human Values* 31 (1): 8–28.

Lee, Julian, and Rwanda Kigali. 2006. "Comparing NGO Influence in the EU and the US." In *Programme on NGOs and Civil Society.* Geneva, Switzerland: Centre for Applied Studies in International Negotiation.

Li, J. G., Q. N. Li, J. Y. Xu, X. Q. Cai, R. L. Liu, Y. J. Li. 2009. "The Pulmonary Toxicity of Multi-Wall Carbon Nanotubes in Mice 30 and 60 Days after Inhalation Exposure." *Journal of Nanoscience and Nanotechnology* 9 (2): 1384–1387.

Li, B., G. Sauve, M. C. Iovu, M. Jeffries-El, R. Zhang, J. Cooper, S. Santhanam, L. Schultz, J. C. Revelli, and A. G. Kusne. 2006. "Volatile Organic Compound Detection using Nanostructured Copolymers." *Nano Lett* 6 (8): 1598–1602.

Lusk, Jayson L., Lisa O. House, Carlotta Valli, Sara R. Jaeger, Melissa Moore, J.L. Morrow, and W. Bruce Traill. 2004. "Effect of Information about Benefits of Biotechnology on Consumer Acceptance of Genetically Modified Food: Evidence from Experimental Auctions in the United States, England, and France." *European Review of Agricultural Economics* 31 (2): 179–204.

Macoubrie, Jane. 2006. "Nanotechnology: Public Concerns, Reasoning and Trust in Government." *Public Understanding of Science* 15 (2): 221–241.

Marette, Stephan, Jutta Roosen, Andrea Bieberstein, Sandrine Blanchemanche, and Frederic Vandermoere. 2009. "Impact of Environmental, Societal and Health Information on Consumers' Choices for Nanofood." *Journal of Agricultural & Food Industrial Organization* 7 (2): Article 11.

Maynard, Andrew. 2007. "Nanotechnology: The Next Big Thing, or Much Ado about Nothing?" *Annals Occupational Hygiene.* 51 (1): 1–12.

Miller, Dennis D. 2010. "Food Nanotechnology: New Leverage against Iron Deficiency." *Nature Nanotechnology* 5 (5): 318–319.

Miller, Georgia. 2008. "Nanotechnology and the Public Interest: Repeating the Mistakes of GM Foods?" *International Journal Technology Transfer and Commercialisation* 7 (2/3): 274-280.

Mintz, Sidney W., and Christine M. Du Bois. 2002. "The Anthropology of Food and Eating." *Annual Review of Anthropology* 31: 99–119.

Morgan, Granger M., Baruch Fischoff, Ann Bostrom, and Cynthia J. Atman. 2002. *Risk Communication: A Mental Models Approach.* Cambridge, UK: Cambridge University Press.

"Nanofood for Thought." 2010. *Nature Nanotechnology* 5: 89.

Nano.gov. 2011. "NNI Budget." Nano.gov National Nanotechnology Initiative. http://www.nano.gov/about-nni/what/funding (accessed 10 August 2011).

Nel, André, Tian Xia, Lutz Mädler, and Ning Li. 2006. "Toxic Potential of Materials at the Nanolevel." *Science* 311 (5761): 622–627.

Oberdörster, Günter, Eva Oberdörster, and Jan Oberdörster. 2005. "Nanotoxicology: An Emerging Discipline Evolving from Studies of Ultrafine Particles." *Environmental Health Perspectives* 113 (7): 823–839.

Pidgeon, Nick, Barbara Herr Harthorn, Karl Bryant, and Tee Rogers-Hayden. 2008. "Deliberating the Risks of Nanotechnologies for Energy and Health Applications in the United States and United Kingdom." *Nature Nanotechnology* 4 (2): 95–98.

Pidgeon, Nick, and Tee Rogers-Hayden. 2007. "Opening up Nanotechnology Dialogue with the Publics: Risk Communication or 'Upstream Management'?" *Health, Risk & Society* 9 (2): 191–210.

Poortinga, Wouter, and Nick F. Pidgeon. 2004. "Trust, the Asymmetry Principle, and the Role of Prior Beliefs." *Risk Analysis* 24 (6): 1475–1486.

Priest, Susanna, Ted Greenhalgh, and Victoria Kramer. 2010. "Risk Perceptions Starting to Shift? US Citizens are Forming Opinions about Nanotechnology." *Journal of Nanoparticle Research* 12 (1): 11–20.

Renn, Ortwin. 2005. "Risk Perception and Communication: Lessons for the Food and Food Packaging Industry." *Food Additives & Contaminants: Part A* 22 (10): 1061–1071.

Renn, Ortwin, and Mihail Roco. 2006. "Nanotechnology and the Need for Risk Governance." *Journal of Nanoparticle Research* 8 (2–3): 1–42.

Rogers, Jennifer. 2008. "The Ma(i)ze of Globalization: Free Trade, Gender, and Resistance in Oaxaca." PhD diss., University of California, Santa Barbara.

Rogers, Jennifer, Christine Shearer, and Barbara Herr Harthorn. Forthcoming. "How Application Domain Affects Risk and Benefit Perception in Public Deliberation on Nanotechnologies." In *The Social Life of Nanotechnology,* ed. B. H. Harthorn and J. Mohr. New York: Routledge.

Royal Society and the Royal Academy of Engineering. 2004. "*Nanoscience and Nanotechnologies: Opportunities and Uncertainties.*" Report.

Sapp, Stephen G., Charlie Arnot, James Fallon, Terry Fleck, David Soorholtz, Matt Sutton-Vermeulen, and Jannette J. H. Wilson. 2009. "Consumer Trust in the US Food System: An Examination of the Recreancy Theorem." *Rural Sociology* 74 (4): 525–545.

Satterfield, Terre, Milind Kandlikar, Christian E.H. Beaudrie, Joseph Conti, and Barbara Herr Harthorn. 2009. "Anticipating the Perceived Risk of Nanotechnologies." *Nature Nanotechnology* 4 (11): 752–758.

Savadori, Lucia, Stefania Savio, Eraldo Nicotra, Rino Rumiati, Melissa Finucane, and Paul Slovic. 2004. "Expert and Public Perception of Risk from Biotechnology." *Risk Analysis* 24 (5): 1289–1299.

Scientific Committee on Emerging and Newly Identified Health Risks (SCENIHR). 2006. "Modified Opinion (after Public Consultation) on the Appropriateness of Existing Methodologies to Assess the Potential Risks Associated with Engineered Adventitious Products of Nanotechnologies." European Commission. Report

Scheringer, Martin. 2008. "Nanoecotoxicology: Environmental Risks of Nanomaterials." *Nature Nanotechnology* 3 (6): 322–323.

Scrinis, Gyorgy, and Kristen Lyons. 2007. "The Emerging Nano-Corporate Paradigm: Nanotechnology and the Transformation of Nature, Food and Agri-food Systems." *International Journal of Sociology of Food and Agriculture* 15 (2): 22–44.

Schurman, Rachel, and William A. Munro. 2010. *Fighting for the Future of Food: Activists Versus Agribusiness in the Struggle over Biotechnology.* Minneapolis: University of Minnesota Press.

Siegrist, Michael. 2008. "Factors Influencing Public Acceptance of Innovative Food Technologies and Products." *Trends in Food Science & Technology* 19 (11): 603–608.

Siegrist, Michael, Marie-Eve Cousin, Hans Kastenholz, and Arnim Wiek. 2007. "Public Acceptance of Nanotechnology Foods and Food Packaging: The Influence of Affect and Trust." *Appetite* 49 (2): 459–466.

Siegrist, Michael, Stampfli Nathalie, Kastenholz Hans, and Keller Carmen. 2008. "Perceived Risks and Perceived Benefits of Different Nanotechnology Foods and Nanotechnology Food Packaging." *Appetite* 51 (2): 283–290.

Slovic, Paul. 1999. "Trust, Emotion, Sex, Politics, and Science: Surveying the Risk-Assessment Battlefield." *Risk Analysis* 19 (4): 689–701.

Slovic, Paul, Baruch Fischhoff, and Sara Lichtenstein. 1980. "Facts and Fears: Understanding Perceived Risk. Pp. 181–216 in *Societal Risk Assessment: How Safe Is Safe Enough?*, ed. Richard C. Schwing and Walter A. Albers. New York: Plenum.

Smale, Melinda, Patricia Zambrano, Guillaume Gruère, José Falck-Zepeda, Ira Matuschke, Daniela Horna, Latha Nagarajan, Indira Yerramareddy, and Hannah Jones. 2009. "Measuring the Economic Impacts of Transgenic Crops in Developing Agriculture during the First Decade: Approaches, Findings, and Future Directions." *Food Policy Review* 10, doi: http://dx.doi.org/10.2499/0896295117FPRev10.

Stilgoe, Jack. 2007. *Nanodialogues: Experiments in Public Engagement with Science.* London: DEMOS.

Stone, Glenn Davis. 2010. "The Anthropology of Genetically Modified Crops." *Annual Review of Anthropology* 29: 381–400.

Thompson, Paul B., and Dean Wesley. 1996. "Competing Conceptions of Risk." *Risk: Health, Safety and Environment* 7: 361–374.

Tokar, Brian, ed. 2001. *Redesigning Life: The Worldwide Challenge to Genetic Engineering.* London and New York: Zed Books.

Tveit, G., Madsen, K. H., and Sandoe, P, eds. 2003. "Biotechnology and the Public." *Danish Centre for Bioethics and Risk Assessment.*

Vandermoere, Frederic, Sandrine Blanchemanche, Andrea Bieberstein, Stephan Marette, and Jutta Roosen. 2009. "The Public Understanding of Nanotechnology in the Food Domain: The Hidden Role of Views on Science, Technology, and Nature." *Public Understanding of Science.* 20 (2): 195–206.

Vlek, Charles. 2010. "Judicious Management of Uncertain Risks: Developments and Criticisms of Risk Analysis and Precautionary Reasoning." *Journal of Risk Research* 13 (4): 517–543.

Wang, B., W. Y. Feng, T. C. Wang, G. Jia, M. Wang, J. W. Shi. 2006. "Acute Toxicity of Nano-and Micro-Scale Zinc Powder in Healthy Adult Mice." *Toxicology Letters* 161 (2): 115–123.

Wetter, Kathy Jo. 2010. "Big Continent and Tiny Technology: Nanotechnology and Africa." *Foreign Policy in Focus.* http://www.pambazuka.org/en/category/features/67525 (accessed 10 August 2011).

Wynne, B. 2006. "Public Engagement as a Means of Restoring Public Trust in Science: Hitting the Notes, but Missing the Music?" *Community Genetics* 9 (3): 211–220.

The "State" of Ecological Thinking in the Political Science Classroom

Joanna L. Mosser

ABSTRACT: Scholars identify the classical and neoliberal commitment to consumption, production, and self-directing individualism as a cultural barrier to ecological thinking and action. The state's complicity in the production of market-based norms and practices hostile to ecological thinking is widely acknowledged. Some solutions, in turn, advocate the liberating force of critical pedagogies that cultivate alternative conceptions of the individual, place, production, consumption, and environment. Missing in this literature is a consideration of the implications of state-based instructional methods for the pursuit of such critical, liberating pedagogies. This article revisits the sovereign territorial state as a modern form of political authority and explores the implications of the state's project of self-authoring standardization and consolidation for the development of ecological thinking and action. The epistemology and ontology of the modern state is rooted in a praxis of subject-hood that dismisses, and constructs as dangerous, the anarchic, self-authoring tendencies of the everyday. Recovering the everyday as a site of authorship, agency, and choice is a first step to creating individuals who take seriously the demands of ecological thinking and action.

KEYWORDS: critical pedagogy, ecological, everyday, neoliberalism, praxis, state

Introduction: Bringing the State Back In

The classical liberal commitment to "self-directing" individualism (Bowers 2009) and capitalist cultural-economic practices that license production and consumption (Spretnak 1999) often are credited with the fragile hold of ecological thinking and action. As a corrective, scholars call upon the liberating force of a "pedagogy of the earth" (Gadotti 2008) to contest the alliance of market and individualism and militate against the captured and co-opted nature of rhetorical gestures to—and practices of—sustainable development, which serve merely to feed the economic aims of the neoliberal state (Banerjee 2003; Kahn 2008). Such literatures seek to cultivate, in neoliberalism's stead, forms of reflective practice that uproot "capitalist cultural spaces" (Kahn 2008: 5), contest the conception of the autonomous, self-directing individual they feed, and expose the oppressive enviro-cidal, classist, racist, and sexist practices they cultivate (Gruenewald 2003).

Environment and Society: Advances in Research 2 (2011): 170–181 © Berghahn Books
doi:10.3167/ares.2011.020110

Missing in this literature, however, is an exploration of the full extent to which the state shapes and constrains the pursuit of critical, liberating pedagogies and epistemologies. In a modern neoliberal democracy like the United States, it is difficult to disentangle philosophical commitments to neoliberal market individualism and the state as a modern form of political authority. In part, with good cause: The state and neoliberalism's parallel consolidation has been mutually reinforcing, with the modern state providing deference to and active endorsement of neoliberal commitments through hollowed-out citizenships that trade political authorship for passive client-consumership (Katz 2002; Ozkan and Foster 2005; Scammell 2000). The mutual affection, consolidation, and reinforcement of state and neoliberalism is taken so much for granted that the modern state's unique way of knowing the world has been given little independent attention in environmental studies, and more specifically ecopedagogical circles. Indeed, what is the peculiar epistemology of stateness, and what are the implications of an exclusively state-centered pedagogy for the cultivation of reflective ecological thinking and practice in the political science classroom?

In this article, I revisit the sovereign territorial state as a modern and contingent form of political authority and explore the implications of a state-centered pedagogy for the cultivation of ecological thinking and practice in the political science classroom. I begin with a brief discussion of the development of the state, focusing in particular on the project of self-authorship and consolidation that distinguishes the state from rival historical (or future) approaches to political authority and the set of constitutive practices in which the modern state engages. I then turn to the co-optation of political thinking by the state in the political science classroom, and suggest ways to render this classroom more hospitable to and accommodative of approaches that question the managerial-organizational commitments of state building, state development, and statist consolidation. Finally, I consider the requirements of ecopraxis, unpacking the statist roots of neoliberal greenwashing and exploring the challenge of fostering ecological praxis vis-à-vis a state-centric pedagogy. I argue, throughout, that we cannot credit an enviro-cidal alliance of public policy and philosophical liberalism—neoliberal praxis—for the failures of environmental thinking and a commitment to co-opted practice (as sustainable development). Instead, we must look deeper, inside the state, to consider the ways in which the modern state's way-of-knowing constitutes, or makes, the individual who studies it. The ecocrisis is not merely a "crisis in cultural ways of knowing" (Bowers 2009: 49); rather, it is a paradigmatic crisis of statist epistemology.

The State's Ways of Knowing

The co-optation of political authority by the territorial state began with the 1648 Peace of Westphalia. The Westphalian model, as outlined by Held (1995: 78), established the governing principle of a world order of sovereign territorial states that "recognize no superior authority" and resort to force to maintain the integrity of their territorial borders. Supranational legal norms, in turn, must provide for "minimal forms of co-existence" and guard against "impediments on state freedom." The Westphalian thesis itself has been contested as a causal "fallacy" based on a "teleological interpretation of what happened in the subsequent three hundred years"—indeed, as Keating (2006) indicates, the word sovereignty appears nowhere in the treaty. Nonetheless, this historical moment is still widely regarded in political science, and in international relations circles, as the first gesture toward the normative organizing logic of the bounded territorial state as a self-regulating and self-directing system (Keating 2006: 195).

It is important to acknowledge the trumping claim of the modern territorial state as a contingent fact of political practice with a justification rooted merely in a—perhaps reflective[1]—habit of convention, norm, and institutionalized practice. Indeed, the normative push and pull of statist self-direction marked a significant, if gradual, turn in political practice. Historically, political authority had been lodged in tribe, empire, feudal systems, estate, and absolutist state[2] (Held 1995)—forms of authority that turned on more or less fractious social, political, and economic relationships (Pierson 2004: 41). What separates the modern state from such premodern forms of political authority is the modern state's ability to convert a pretention to authority into an administrative, military, and legal capacity to deliver it (Dryzek and Dunleavy 2009; Pierson 2004)—a capacity dependent on the state's ability to extract resources and repackage them as public goods that feed the stability and legitimacy aims of the state (Offe 1984; Piven and Cloward 1983). Indeed, the modern state's project is one of legitimate and credible consolidation of power over societies of territorially proximal strangers.

The territorial state's co-optation of imagination and legal and administrative practice is deeply consequential. Most strikingly, the modern territorial state's trumping claim to sovereignty troubles the pursuit of human social, economic, and political rights (DeLaet 2005). Rights are conferred and enforced—or not—at the will of a granting state, and typically in ways that comport with a governed community's local understandings of the relationship between state and the governed. Individuals without a state—those who, for fear of persecution on the basis of race, religion, nationality, social membership, or political belief, cannot claim or call on the protective arm of their state of nationality—find themselves in a particularly vulnerable position. International law confers no right to asylum but, instead, allows states to define the criteria of admission and regulate and refuse entry, reducing the asylum-seeker, refugee, or stateless person to a state of profound—and profoundly insecure—social, political, and legal liminality (Mosser forthcoming). Indeed, there is no formal, codified recognition in law of the accident of birth upon which membership is conferred and social, economic, civil, and political rights distributed.

Instead, state making and statist consolidation are the governing principles of the global order. Groups of people that explicitly reject or resist this project—historically, indigenous tribes, vagrants, the homeless, the Romani of Southeast Europe, and the self-governing hill peoples of Southeast Asia, for example—are dismissed as barbaric, uncivilized, backward, and a threat to security that must be disciplined, tamed, and contained (Scott 2009). The mobility and fluidity of these peoples serve as constant evidence of the state's failures and constitute an act of profound cultural and economic resistance to the state's "legibility" project (Scott 2009); indeed, history, guided by the grand narrative of the development of the modern state, is a story marked by the civilizing, disciplinary, centralizing hand of the territorial state and its technologies of administrative stabilization and standardization (Scott 1999). The state's job is one of territorial expansion and full social and economic mobilization. That which resists the "planning mentality" of the state is dismissed as inefficient (at best) or a threat to civilization itself (Scott 2009: 6).

The State in/of the Political Science Classroom

The state has virtually co-opted and captured the political (Keating 2006). The job of the American political scientist, in particular, is so strongly bounded by the state that the discipline only very recently acknowledged a need for a more self-conscious theory of the "special meanings and activities" of the state (Skocpol 1985: 3–4). The formative period of the 1950s, 1960s, and

1970s in American political science, in which behavioral-pluralist perspectives on politics carved out a state in permanent "passive receivership" among dueling social interests (Schneider and Ingram 1997), had a constraining effect on the discipline's reflexivity. Government "was not taken very seriously as an independent actor" (Skocpol 1985) but, rather, understood as a set of norms, rules, practices, and institutions charged with placating, accommodating, and distributing benefits among a sufficiently broad cross-section of vocal social groups to ensure systemic stability. Still, efforts to bring the state back in (including, most notably, a volume of the same name [Skocpol 1985]) were limited; while an effort was made to carve out a role for the state-as-political-actor—and an understanding of the state's institutions as constitutive of social practices—analyses focused primarily on the viability of conceptualizing the state as actor and the how of the state's constitutive impact on social and political action. Less attention was given to the for-granted nature of the state in the political scientist's thinking. Indeed, remarkable was the resistance—evident in the introductory chapter of *Bringing the State Back In*—to "offer[ing] any new general theory of the state or of states and social structures" (Skocpol 1985: 8). The state is a fixture of the political scientist's working environment, in short, but the political scientist has not only "simply taken it for granted" but also "tended to celebrate it" (Keating 2006).

The state plays an equally central, and equally given, role in the political science classroom. By passive, ready, or benign commission or neglect, politics tends to be framed by teachers—and understood by students—as that which implicates, relates to, is rooted in, or is bounded by the state. Thus, students tend to expect—and instructors, to deliver—a set of themes and topics that turn on the agents, institutions, processes, and programs of government and the formal and informal processes by which government "choose[s] [among] a variety of collective goals for society and seek[s] to implement them" (Peters 2007: 25). For the student or scholar of international relations, the discipline virtually necessitates the state (Keating 2006). The "political" is that which implicates, relates to, implies, is rooted in, or bumps up against the agents and institutions of the final trumping, or final enforcement, authority in a particular place (Nicholson 2007). For most of the world today, this is the state.

This conception of politics and the political, to be sure, is not universally shared among political scientists. Feminist-minded conceptions of politics, for example, take a radically decentered view of politics, finding the political in domains of social life, and among persons, typically described as private (Squires 2007). Other, Marxist-minded approaches push for a similarly decentered view of the political but argue that politics is a by-product—and a contingent one—of systems of class domination institutionalized and maintained by the state (Callinicos 2007). Other perspectives, finally, embrace a process-based conception of politics, departing from the "arena"- or "site"-based approach to politics to argue that politics is a constant feature of all human societies—indeed, because all human societies engage in questions and decisions about the "use, production, and distribution of [material and non-material] resources" (Leftwich 2007: 106). These perspectives do not exhaust the range of conceptions available, but they serve to illustrate that, indeed, the discipline does engage in self-reflexive conversation about the inner and outer boundaries of the political (Nicholson 2007).

Nonetheless, the state is a (generally) nonoptional category of—or background constraint on—analysis, with the result that politics becomes the province of "the rather narrow band of usually formal institutions to do with states and governments and their more or less complex relations with their people, whether as voters, citizens, or subjects" (Leftwich 2007: 102). Textbooks on American government are particularly symptomatic of this constraint, with most containing introductory chapters on the politics of incentive and strategic accommodation at the Constitutional Convention, chapters on the political linkages—parties, interest groups, voting and elections, media—that facilitate principal-agent relationships between citizen and state, chapters on the key political institutions that translate demands into binding patterns of rule, process, and program, and chapters that explore the troubled pursuit of civil rights and liberties within the neopluralist constraints of the American political system. Upper-level courses in politics, certainly, can (but often do not) contest this state-based approach to the subject, but for students who take introductory politics courses in high school or college, politics becomes that which implicates the activities, personnel, and institutions of the state in its many, varied relationships with a bounded citizenry. Instructors are not fully responsible for this construction of politics—indeed, local and national news outlets privilege the agents, processes, and institutions of government in their delivery of political news—but these casual constructions of politics tend to get the informal stamp of legitimacy in classroom materials, lectures, and conversation.

The State and "Real" Authorship

This approach to politics takes much for granted—not least, the state as trumping form of legitimate political authority, the bounded community of members it governs, and the rules and practices that reinforce the commands of one and the compliance of the other. Left unchallenged, an exclusively state-bounded view of politics and the political can lead to a "degenerative" construction of politics (Schneider and Ingram 1997) that implies or insists on analyses rooted in or filtered through the institutions of the state. During a class session on constructions of nature in my environmental politics and policy course, an upper-level seminar that enrolls politics, sociolegal studies, philosophy, English, and environmental science majors, I engaged students in a discussion of the political construction of the Arctic National Wildlife Refuge (ANWR) as a desolate, uninhabited space (Schlosser 2006). We also explored an argument for the discursive roots of international environmental treaties signed between 1870 and1990 (Frank 1997). The students, overall, were quick to note the political expediency of ANWR's construction as desolate—and its political exploitation to justify oil exploration for national security purposes. They also were intrigued to discover that global environmental agreements spiked on the emergence of ecosystem (vis-à-vis feral or resource-based) models of nature that emphasized the intimate life-sustaining connection between the human-social and natural. In both cases, the conversation strayed from talk of policy-as-text, regulation, and political action as filtered through the institutions of government. Near the end of our conversation, a political science student politely, even sheepishly, raised her hand and asked if we could talk about "real" policy—presumably, government action vis-à-vis the dispute over ANWR.

Well-meaning early- and mid-semester requests for "real" policy—as formal, codified text, doctrine, regulation, or pattern of enforcement—happen often in my policy classes, which are rooted in a claim that policy consists in, and begins with, social and political constructions of subjects, authors, problems, solutions, and targets. Moreover, these constructions are not simply the background or context within and on which formal, codified text is formulated; to the extent these constructions are constitutive of patterned relationships of power—whether in the halls of Congress or in daily life—they *are* policy. Indeed, policy involves an authoritative attempt to stabilize meanings—in short, to "put an end to politics."[3] I convey to students that a study of ANWR policy does not, and cannot, begin in the halls of Congress or among US Fish and Wildlife Service or US Geological Survey personnel; instead, it begins with the peculiar way we humans constitute—imagine, create, and construct—the natural (see, e.g., Cronon 1996: 69–90). Any formal, codified text speaks in, by, and through these constructions. As a result, when we get back to "real" policy in subsequent weeks, students find themselves not only speaking a language of strategy, negotiation, and choice among governmental actors but also asking questions about the social and political constructions that structure preference, interest, choice, thought, and action. Still, the striking absence of the state in the middle third of the course causes no small amount of discomfort. Students expect, but do not immediately get, a set of themes and topics that turn on the final enforcement authority of the state.

Instead, students get a radically nonstate, decentered conception of the domain and essence of the political that invites them to find politics in the nonstate civic spaces that define their everyday lives—grocery, mall, classroom, coffee shop, and workplace. They also hear a constant invitation to see these ordinary spaces as constitutive of the stuff that real policymakers do. Yet for many of my students politics is not transparently present or evident in these experiences. Politics is the domain of the state, the province of backroom pluralistic wheelings-and-dealings, and the product of demands of the unusually engaged and resourced that produce knotty, inconsistent, and toothless regulations. More specifically, they tend not to view the "practice of everyday life" (de Certeau 1984) as a site of authorship. The mundane everyday is life, certainly, but life is something different from what one studies in the classroom. It also is different from what political actors do as they produce text, process, and program in places, and for persons and entities we do not know or see. Authorship is the province of agents, processes, and institutions of the state, and life is a site of receivership in which we find ourselves subject to the commands of statist authors.

Indeed, the political science classroom can "exaggerate the importance of state law," to the detriment of "society's nonlegal normative resources" (Sarat and Kearns 1993: 22). It is necessary but insufficient to acknowledge the project of legibility and consolidation in which the modern state engages. It is equally necessary and equally insufficient to acknowledge that the state, pursuant to this effort, engages in a postindustrial project of governance that extends the reach of surveillance and discipline to home, work, play, and civic life and enlists the subject in its own subjugation (Foucault [1977] 1995; Simon 2007). What is missing is an acknowledgement and exploration of the creative, authoring capacities of those parts of everyday life that "seem on the face of things, removed from law, or at least not dominated by law from the outset" (Sarat and Kearns 1993: 55). Sarat and Kearns offer several illustrative examples:

> Everyday life ... is a mélange of social practices and normative commitments that stand, no doubt, in a wide variety of relations to law and legal materials. But to call a practice a social practice is to direct attention to its underlying intelligibility, even if its connection with law and legal concepts were nonexistent. Thus, for example, it might be argued that the institution of giving and accepting promises is a social practice no matter how much it may, as a purely contingent matter, be shaped by contract doctrine. The crucial point is this: the institution of promising does not conceptually depend on anything drawn from the law of contracts. Similarly, though more contentiously, the institution of marriage as a wide-ranging commitment to another for a lifetime might be extricated from its legal accompaniments and still retain its nonlegal significance as a powerful normative enterprise. Finally, there is little reason to suppose that the belief that one should respond to beneficent acts in kind is a conviction that owes its existence to law. (1993: 56)

The political science classroom, in particular, can embrace instrumental and constitutive conceptions of law, politics, and policy that start first with the agents, institutions, and activities of the state. From an instrumentalist perspective, the state is the agent-author that attempts, effectively or not, to lend order and structure to social life by reproducing and enforcing rules targeted toward society's well-being. From a constitutive perspective, the state is an agent-author whose activities make us—not by shaping our otherwise independent, authentic activities, but by shaping us internally, constructing our wants, needs, preferences, and other and self-understandings (Foucault [1977] 1995). What is different and innovative about an everyday approach to law, politics, and policy is its effort to recover everyday life and reclaim the individual as a person and author, not wholly or primarily a subject or target (Sarat and Kearns 1993). Similarly, the state-centric focus of the political science classroom conveys the idea that law does, should, and may dominate the everyday and one's sense of self and self-in-politics, often in undetected ways, and that one's proper disposition toward the law is one of subject and receivership.

Depoliticizing the Everyday

The political rationality of neoliberalism cultivates a similar brand of subjectship that naturalizes a depoliticized everyday—in ways that render problematic any critical pedagogical effort to recover the everyday as a site of agency. Scholars of critical pedagogy, including most notably Paulo Freire and Henry Giroux, demonstrate the ways in which constructions of an apolitical, objective, and neutral classroom serve, politically, to reinforce status quo relationships of domination and oppression. Such relationships of domination and oppression continue largely because individuals do not *see* them (Freire [1968] 2000); in particular, students tend not see how their own lives are structured through socially and economically oppressive systems of meaning and practice. Their activities, behaviors, habits, and practices seem, to them, normatively natural, authentic, and self-directed, and stories of oppression are engaged as distant, if compelling and moving, abstractions.

Failures to see, in large part, are rooted in depoliticized constructions of the work of the classroom. Evans (2009: 4) notes how curriculum standardization, standardized testing, and attacks on academic tenure are suggestive of the "corporatization of education in the modern world." Cramer and Mosser (2011) similarly document the ways in which the classroom and the university are actively depoliticized by governing authorities. Public relations and marketing campaigns that carve out a university's distinctive market niche; neoconservative demands for "pure" universities and classrooms devoted to the pursuit of "truth" and "perfection of learning" (Oaks 1977, cited in Cramer and Mosser 2011: 7); Internal Revenue Code regulations that posit political inaction as the university's requisite disposition;[4] arguments for liberal education reforms that gesture to employer demands for their primary justification (Humphreys and Carnevale 2010); and efforts to import the claims and methods of outcomes assessment into higher education position the university—and the classroom—at the nexus of the consumerist neoliberal market. More important, such efforts serve to reinforce the seemingly neutral but politically powerful notion that it is the job of the university to participate in and foster the type of "prostrate" (Scott 2009: 4), disciplined—thus "legible"—civil society that fuels the commitments of competitive global production and consumption.

In late-modern neoliberal states, this legibility project involves the transformation of the citizen-agent into citizen-consumer. Indeed, what distinguishes the neoliberal state from the laissez-faire liberal state is the former's recognition that neither the market nor economic rationality are natural or normatively neutral (Brown 2006); rather, individuals must be invited—through elaborate systems of positive and negative incentives (Schram et al. 2010)—to actively embrace the aims of the market and integrate them into their everyday lives, replacing metrics of social justice with metrics of utilitarian efficiency, and displacing social/collective constructions of problems with constructions that foster the internalization of risk and embrace rhetorics and practices of personal responsibility (Schram 2000). Indeed, neoliberalism's goal is to "mobilize [us] as consumers" (Passavant 2005) who exercise choice and exit as an alternative to engagement, voice, and dissent. In the classroom, this can translate to student demands to get what they pay for, echoing recent Pew Research Center (2011) survey findings that adult Americans increasingly question the "monetary value" of a college degree.

It is in this context—the political rationality of the neoliberal state—that a critique of sustainable development and the greening of consumer products must be positioned and understood. The greening of the market is dismissed, most frequently, as a public relations ploy: a merely rhetorical gesture to discourses of social and environmental responsibility that appropriates ecological virtue for business-building ends. Sustainable development, similarly, is criticized, particularly among scholars of ecopedagogy, as a captured discourse that maintains the very paradigm of self-directing individualism that is hostile to ecological thinking and action (Kahn 2008).

However, a stronger critique is available to us. Passavant (2005: 28), for example, notes the ways in which data-tracking practices have fed the process of "neo-liberal state formation" by equipping governing authorities with information about consumer habits and preferences. Modern technologies of consumership also fuel a desire to be tracked: When I give my birthdate or email address to a company or participate in online product surveys, I do so in the explicit or implicit hope that the company will track my responses and award me with coupons and other incentives to buy. Important, then, rhetorics of sustainable development and efforts to green the market are not vectors in a Focauldian disciplinary project. The political rationality of the neoliberal state is rooted in control and a recognition that economic rationality and market-based norms are fragile and in need of constant cultivation (Brown 2006) through incentives that induce market-conforming behaviors by a "disciplined shopper" (Passavant 2005: 3). Greenwashing, then, is not objectionable merely because it involves the exploitation—for market purposes—of environmentalism's compelling moral claims. Sustainable development, similarly, is not objectionable merely because it demands a compromised brand of environmental action that is consistent with moderated production and consumption demands. Rather, both practices traffic uncritically in conceptions of engagement that push consumption as the requisite (dis)position of the individual. In the market, we are targets. Consumers, not producers. Policy-takers, not authors.

Political science classrooms and texts can similarly embrace a high-modernist vision of a state that knows, orders, and disciplines the anarchic tendencies of the social and economic world and a neoliberal vision of political engagement that curtails the domain of the political. State-centric pedagogies can feed the impression that political action is action that intersects with agents, processes, and institutions of the state. It is not local, it is not mundane, and it is not present in classroom conversation, critique, and analysis. Indeed, state-centric pedagogies can encourage teachers and students to cede the demands of legibility and authorship—of "naming the world" for Freire ([1968] 2000)—to the agents, activities, processes, and institutions of the state. The state's project of subject making is, and has been, so tremendously successful that we fail to regard the non- and parastate everyday as a site of political authorship and regard everyday practices that deviate from the norm as different, strange, inefficient, backward, traditional, or, in our most dismissive moods, dangerous and anarchic. Any farmer who embraces the true spirit (if not legal demands) of organic knows this well (Pollan 2006), as does any participant in community-supported agriculture. Such practices remain, in the public's mind, off the grid and, therefore, discomfort-inducing, and calls to ensure their safety have produced a sea of regulations that offer stable, routinized terms, meanings, and procedures governed by a centralized bureaucracy—terms, meanings, and procedures that can seem, at times, dismissive of, or downright hostile to, the local norms governing these practices.

Repoliticizing the Everyday

As a teacher my job is to destabilize the necessity and seeming inevitability of what is, to recover the possibility of alternative constructions, and to build a positive case for the freeing—indeed, liberating!—contingency of policy problems and solutions. My job also is to repoliticize—and render up-for-grabs—the mundane activities we do, and the things we think, feel, consume, and wear, often by unreflective habit, on a daily basis. I begin, often, with our considered or knee-jerk reactions to and understandings of these things, inviting students to unpack the norms, expectations, and constructions surrounding proper citizen-citizen and citizen-state relations, the nature of politics, their expectations of and for politics, their expectations of and for each other, and the mundane norms and expectations that govern their daily activities at work, in the classroom, at home, and at play. Real policy—written text, formal actors and institutions that process public demands—comes much later, after students have done the surprisingly difficult job of seeing contested meanings and understandings (and hence politics) at work in places they typically do not see it. I do these things in all of my classes (in American politics, the politics of crime and punishment, the public policy process, the politics of membership and citizenship in the US), not simply those devoted explicitly to environmental politics and policy. In my mind, ecological praxis begins with a perceptive eye that sees possibilities for (re)making, (re)naming, and (re)authoring the world.

In my course on environmental politics and policy, this commitment to (re)making, (re)naming, and (re)authoring means, in part, taking seriously the claims of anarchist and philosophical anarchist theorists like Peter Kropotkin and Robert Paul Wolff, despite the images and aversions the term anarchy initially elicits among students. Doing so allows us to understand why and how political constructions of the eco-terrorist—and, more mundanely, political constructions of the environmental activist as radical—are so powerfully constitutive of the (narrowed) boundaries of political engagement. Doing so allows us, in the final third of the course, to take seriously—and understand the primary justification for—nonstate and parastate solutions to environmental problems, including food co-ops, Slow Food movements, neighborhood gardens, democratized workplaces (Shrivastava 1995), and managerial forest councils comprised of nontechnocrat locals (Agrawal 2005). We render legible and rescue so-called (and so-dismissed) alternative solutions to environmental problems that promote democratic modes of engagement (Shrivastava 1995), (re)position and (re)imagine the individual as author and coparticipant in the process of meaning making (Agrawal 2005), empower decidedly local, everyday, and practical ways of knowing and being in the world (Scott 1999), and confer authority on local metrics and processes of standard-setting, legibility making, and trust-building whose authority is conferred and confirmed by (and in) local practice, not by the actors and institutions of the state. We confer political authority, not alterity, on a movement like Slow Food, whose motive force and contribution is its drive to denaturalize food: to render it, again, up for grabs, and a subject of imagination, choice, community, and connection in a market that otherwise privileges standardization and abstraction.

Cultivating ecological thinking and practice in the classroom involves harnessing the anarchic—and, by anarchic, I mean authoring—capacities of the everyday. We need not contest or overturn the state; indeed, if we read anarchic this way, we simply reinforce the dismissive construction of practices that resist or operate beyond the disciplining arm of the state. We must, however, be mindful that the peculiar epistemology, ontology, and pedagogy of the state is a praxis of subject making that regards the anarchy of the everyday—its resistance to discipline, its malleability, its hostility to classification and categorization—as dangerous. The everyday, however, must be recovered as a site of agency, choice, action, and authorship before we—and our students—can successfully embrace a liberating project of "naming our world" (Freire 2000). The biggest barrier, here, is not production, consumption, and liberal individualism but, rather, the surrender of a practice of reflective authorship and meaning making to the agents, institutions, and processes of the state and the cultivation of state-centric pedagogies that narrow the scope and domain of the political. Indeed, in recovering the everyday as a site of political authority, we must first invite students to see their everyday—too often normatively natural, neutral, and given—as a site of agency and contest. Why do I eat what I eat? Where does my food come from? What constitutive role does food play in my community? Such questions render contingent what is, too often, taken as given and, in so doing, gives students license to remake their everyday lives as sites of nonstate political reflection and action.

JOANNA L. MOSSER is assistant professor of politics at Drake University, Des Moines, Iowa. Her research and teaching interests include neoliberalism and the state, critical pedagogy, democratic theory, and the politics of risk in late modernity.

NOTES

1. There might be good reason to endorse the convention of territorial sovereignty. International legal norms award the final trumping claim to territorial states, troubling claims of human social, political, and economic rights (DeLaet 2005); however, according to Keating (2006), democratic practice may necessitate the kind of communal trust, affinity, and attachment territorial proximity can provide.
2. Inclusion of the absolutist state in forms of traditional, or premodern, political authority is somewhat controversial among scholars of the modern state. As Pierson (2004: 44–47) indicates, the absolutist state, a form of political authority characteristic throughout the sixteenth and seventeenth centuries, is probably "best seen as a *transitional* form" of authority, one that embodied a movement away from earlier feudal forms of authority and movement toward the several practices—standing army, bureaucratization, territorial-based systems of taxation, and a "formal diplomatic service"—characteristic of the modern territorial state.
3. My thanks to an anonymous reviewer for identifying and inviting this connection to Graeber's *Fragments of an Anarchist Anthropology* (2004: 9).
4. To qualify for tax-exempt, 501(c)(3), status, a nonprofit organization "may not be an action organization" that "attempt[s] to influence legislation" or "participate[s] in campaign activity for or against political candidates." http://www.irs.gov/charities/charitable/article/0,,id=96099,00.html (accessed 30 May 2011).

REFERENCES

Agrawal, Arun. 2005. *Environmentality: Technologies of Government and the Making of Environmental Subjects.* Durham, NC: Duke University Press.

Banerjee, Subhabrata Bobby. 2003. "Who Sustains Whose Development? Sustainable Development and the Reinvention of Nature." *Organization Studies* 24 (1): 143–180.

Bowers, C.A. 2009. "Educational Reforms that Foster Ecological Intelligence." *Green Theory & Praxis: The Journal of Ecopedagogy* 5 (1): 26–50.

Brown, Wendy. 2006. "American Nightmare: Neoliberalism, Neoconservatism, and De-Democratization." *Political Theory* 34 (6): 690–714.

Callinicos, Alex. 2007. "Marxism and Politics." Pp. 53–66 in *What is Politics?* ed. Adrian Leftwich. Cambridge: Polity Press.

Cramer, Renee, and Joanna Mosser. "Regulatory Policy in the (Academic) Everyday." Paper presented at the Annual Meeting of the Law and Society Association, San Francisco, CA, June 2011.
Cronon, William. 1996. "The Trouble with Wilderness; or, Getting Back to the Wrong Nature." Pp. 69–90 in *Uncommon Ground: Rethinking the Human Place in Nature*, ed. William Cronon. New York: W.W. Norton.
de Certeau, Michel. 1984. *The Practice of Everyday Life*. Trans. Steven Randall. Berkeley: University of California Press.
DeLaet, Debra. 2005. *The Global Struggle for Human Rights: Universal Principles in World Politics*. Belmont, CA: Wadsworth.
Dryzek, John, and Patrick Dunleavy. 2009. *Theories of the Democratic State*. New York: Palgrave Macmillan.
Evans, Tina Lynn. 2009. "Reflections in a Broken Mirror: Higher Education and the Challenges of Sustainability." *Green Theory & Praxis: The Journal of Ecopedagogy* 5 (1): 1–13.
Foucault, Michel. 1995 (1977). *Discipline and Punish: The Birth of the Prison*. New York: Vintage.
Frank, David John. 1997. "Science, Nature, and the Globalization of the Environment, 1870–1990." *Social Forces* 76 (2): 409–437.
Freire, Paulo. [1968] 2000. *Pedagogy of the Oppressed*. New York: Continuum.
Gadotti, Moacir. 2000. "Education for Sustainability: A Critical Contribution to the Decade of Education for Sustainable Development." *Green Theory & Praxis: The Journal of Ecopedagogy* 4 (1): 15–64.
Graeber, David. 2004. *Fragments of an Anarchist Anthropology*. Chicago, IL: Prickly Paradigm Press.
Gruenewald, David. 2003. "The Best of Both Worlds: A Critical Pedagogy of Place." *Educational Researcher* 32 (4): 3–12.
Held, David. 1995. *Democracy and the Global Order: From the Modern State to Cosmopolitan Governance*. Palo Alto, CA: Stanford University Press.
Humphreys, Debra, and Anthony Carnevale. 2010. "The Economic Value of Liberal Education." http://www.aacu.org/leap/index.cfm (accessed 30 May 2011).
Kahn, Richard. 2008. "From Education for Sustainable Development to Ecopedagogy: Sustaining Capitalism or Sustaining Life?" *Green Theory & Praxis: The Journal of Ecopedagogy* 4 (1): 1–11.
Katz, Michael. 2002. *The Price of Citizenship: Redefining the American Welfare State*. New York: Holt.
Keating, Michael. 2006. "Sovereignty and Plurinational Democracy: Problems in Political Science." Pp. 191–208 in *Sovereignty in Transition*, ed. Neil Walker. Oxford: Hart.
Leftwich, Adrian. 2007. "Thinking Politically: On the Politics of Politics." Pp. 1–22 in *What Is Politics?* ed. Adrian Leftwich. Cambridge: Polity Press.
Mosser, Joanna. Forthcoming. "In-Betweens: The Social, Political, and Legal Status of the Asylum Seeker in the U.S." In *The Making of Modern Immigration: An Encyclopedia of People and Ideas*, ed. Patrick J. Hayes. Santa Barbara, CA: ABC-CLIO.
Nicholson, Peter P. 2007. "Politics and the Exercise of Force." Pp. 41–52 in *What Is Politics?* ed. Adrian Leftwich. Cambridge: Polity Press.
Offe, Claus. 1984. *Contradictions of the Welfare State*. MIT Press.
Ozkan, Derya, and Robert John Foster. 2005. "Consumer Citizenship, Nationalism, and Neoliberal Globalization in Turkey: The Advertising Launch of Cola Turka." *Advertising and Society Review* 6 (3).
Passavant, Paul A. 2005. "The Strong Neo-Liberal State: Crime, Consumption, Governance." *Theory & Event* 8 (3).
Peters, B. Guy. 2007. "Politics Is about Governing." Pp. 23–37 in *What Is Politics?* ed. Adrian Leftwich. Cambridge: Polity Press.
Pew Research Center. 2011. "Is College Worth It? College Presidents, Public Assess Value, Quality, and Mission of Higher Education." http://pewsocialtrends.org/files/2011/05/higher-ed-report.pdf (accessed 30 May 2011).
Pierson, Christopher. 2004. *The Modern State*. New York: Routledge.
Piven, Frances Fox and Richard Cloward. 1983. *Regulating the Poor: The Functions of Public Welfare*. New York: Vintage Books.

Pollan, Michael. 2006. *The Omnivore's Dilemma: A Natural History of Four Meals.* New York: Penguin Press.

Sarat, Austin, and Thomas R. Kearns. 1993. "Beyond the Great Divide: Forms of Legal Scholarship and Everyday Life." Pp. 21–62 in *Law in Everyday Life,* ed. Austin Sarat and Thomas R. Kearns. Ann Arbor: University of Michigan Press.

Scammell, Margaret. 2000. "The Internet and Civic Engagement: The Age of the Citizen-Consumer." *Political Communication* 4 (17): 351–355.

Schlosser, Kolson. 2006. "U.S. National Security Discourse and the Political Construction of the Arctic National Wildlife Refuge." *Society and Natural Resources* 19 (1): 3–18.

Schneider, Anne Larason, and Helen Ingram. 1997. *Policy Design for Democracy.* Lawrence: University Press of Kansas.

Schram, Sanford F. 2000. "The Medicalization of Welfare." *Social Text* 18 (1): 81–107.

Schram, Sanford F., Joe Soss, Linda Houser, and Richard C. Fording. 2010. "The Third Level of U.S. Welfare Reform: Governmentality under Neoliberal Paternalism." *Citizenship Studies* 6 (14): 739–754.

Scott, James. 1999. *Seeing Like a State: How Certain Schemes to Improve the Human Condition Have Failed.* New Haven, CT: Yale University Press.

———. 2009. *The Art of Not Being Governed: An Anarchist History of Upland Southeast Asia.* New Haven, CT: Yale University Press.

Shrivastava, Paul. 1995. "Ecocentric Management for a Risk Society." *Academy of Management Review* 20 (1): 118–137.

Simon, Jonathan. 2007. *Governing through Crime: How the War on Crime Transformed American Democracy and Created a Culture of Fear.* New York: Oxford University Press.

Skocpol, Theda. 1985. "Bringing the State Back In: Strategies of Analysis in Current Research." Pp. 3–38 in *Bringing the State Back In,* ed. Peter B. Evans, Dietrich Rueschemeyer, and Theda Skocpol. Cambridge: Cambridge University Press.

Spretnak, Charlene. 1999. *The Resurgence of the Real: Body, Nature, and Place in a Hypermodern World.* New York: Routledge.

Squires, Judith. 2007. "Politics beyond Boundaries: A Feminist Perspective." Pp. 119–134 in *What Is Politics?* ed. Adrian Leftwich. Cambridge: Polity Press.

BOOK REVIEWS

BLASER, Mario. 2010. *Storytelling Globalization from the Chaco and Beyond,* 292 pp. Durham: Duke University Press. ISBN 978-0-8223-4545-9.

Mario Blaser's *Storytelling Globalization from the Chaco and Beyond* is part of the "New Ecologies for the Twenty-First Century" series published by Duke University Press under the editorship of Arturo Escobar and Dianne Rocheleau. An Argentine-born anthropologist who now holds a Canada Research Chair in the Department of Anthropology at Memorial University in Newfoundland, Blaser draws on nearly twenty years' field experience with the Yshiro people of the Paraguayan Chaco in the book. The Gran Chaco, for readers unfamiliar with the region, is a dry central plain—comparable in many respects to the North American Great Plains—that encompasses portions of northern Argentina, southeastern Bolivia, and northwestern Paraguay. The Yshiro of Paraguay (known in older ethnographic literature as the Chamococo) are divided into two related groups: the Ebitoso, many of whom during the early twentieth century began working for wages along the Paraguay River, and the Tomaraho, who retreated into the Paraguayan interior and avoided such contact until the 1940s. Both groups have continued to practice some hunting and fishing until the present day; the Ebitoso have been more extensively proselytized by both Catholic and Protestant evangelical missionaries while the Tomaraho have hewed more closely to traditional views and practices. In the 1990s their separate trajectories came together due in part to the end of the Stroessner regime and its specific forms of patronage and in part to the emergence of international indigenous peoples/environment/development funding mediated by nongovernmental organizations (NGOs). The story is of self-evident interest to anthropologists of Paraguay and the Chaco, especially because of Blaser's long engagement with the Yshiro and his thorough competence on the history and present prospects of indigenous politics in Paraguay. The time depth of his relationship with the Yshiro across a period of remarkable transformation, the extent of his knowledge of the relevant academic literature in the region, and his command of the ins and outs of the universe of NGOs, government, and indigenous activism in Paraguay are all nonpareil.

It is not just area specialists to whom the work will appeal, however. I will take one example of several to demonstrate how useful this text will be to anyone interested in indigenous peoples, development, and environment issues in comparative context. Blaser describes how during the 1990s Yshiro people hoped to access development resources putatively available to indigenous peoples of the Paraguayan Chaco through a project called Prodechaco, which was funded by the European Union. Although many Paraguayan NGOs working on indigenous issues insisted that resolving indigenous land claims ought to be a precondition of any such EU funding being disbursed in Paraguay, Paraguayan ranching and agribusiness interests successfully convinced evaluators that land claims were impossible to adjudicate because of the

Environment and Society: Advances in Research 2 (2011): 182–206 © Berghahn Books
doi:10.3167/ares.2011.020111

problem of "representation" in Paraguayan indigenous communities. Prodechaco redesigned its aims accordingly: "now a central objective of the project would be to develop the representational skills of the communities ... an objective that circumvented the land issue" (184). The tutelage of the Yshiro by outsiders in this process, which Prodechaco institutionalized as a step preparatory to addressing Yshiro land claims, predictably became the basis for charges that the Yshiro were not truly representing themselves. This new "problem" itself has become a major focus of NGO activity. This formula, and its temporality, will be recognizable to anyone who works in the field of indigenous peoples, environment, and development: a shift from the address of substantive demands to an obsession with proceduralism, in the guise of a continuation of "activism" and "collaboration" that in fact disguises the thoroughness with which many concrete indigenous demands have been shifted from center stage. Blaser's ethnography is unusual because rather than offering a circa 1990s snapshot of this process, it orients it in longer historical time and shows all the more effectively the starkness of its consequences for the people involved.

Blaser's own position nevertheless is an optimistic one. He argues that the present moment—not merely in Paraguay, but globally—is marked by what he calls "ontological conflicts" between a modernist worldview and associated set of practices and a coalescing set of challenges to that worldview and its practices. He says that while "modernist" knowledge practices seek to divide subjects from objects, are obsessed with distinguishing subjectivity from objectivity, and are ordered by "Cartesian moral logic," nonmodern (say, Yshiro) and postmodern (say, globalized) knowledge practices are connective, oriented toward networks and hybrids, and—as either a cause or a consequence—are not so governed but are instead structured by a relational moral logic. Narrative is important here; hence the title, "storytelling globalization." Blaser ends his book in the hope that the "pluriverse" (as opposed to a "universe") envisioned by both nonmodern cosmologies and postmodern globalization is on the verge of opening up.

The book, then, draws on and contributes to a burgeoning recent anthropological literature on ontology, which entertains the premise that modernity is a particular kind of thing, the very strangeness of which is best thrown into relief by combined reference to critique on the part of postmodernists (Bruno Latour—though he would not call himself a postmodernist—being the key figure here, and whose work is heavily referenced in Blaser's text) and ethnographic documentation of nonmodern difference. What is striking about this trend is the repetitive exactness with which indigenous cosmologies of diverse kinds are supposed to resemble a far more uniform set of postmodern critiques of modernity. The book succeeds in applying and expanding the Latourian critique of modernity, but it is less evaluable according to the kindred lights of the Yshiro cosmology by which it is ostensibly also guided. This is because the *yrmo* (Yshiro world, the dynamic continuum between *sherwo* and *om*, being and nonbeing), *wozosh* (something like transformation, related to decay), and the *puruhle* (mythic times, and stories about these times) are more often invoked (frequently accompanied by citations of older ethnographic literature the non-Chaco specialist will not have read) than described or brought into focus via ethnographic example.

This absence is quite marked in the chapter of the book that is likely to be of most interest to readers of *Environment and Society.* Blaser discusses the implementation and collapse of a sustainable hunting program carried out under the auspices (at least initially) of Prodechaco. The narrative is rather confusing, in the way that stories about development initiatives with multiple actors tend to be, but the upshot seems to be that a quota system to be brokered by the Yshiro came in for heavy criticism because Yshiro people sold some of their allotments to non-Yshiro hunters in order to generate more community income than otherwise would have been possible. Blaser emphasizes the lack of fit between the modernist view of ecosystem management, which focuses on human-to-nature dynamics, and the Yshiro view, which focuses instead on human-to-human dynamics. In the latter view, quantities of game animals will reflect the degree to which, for example, kinspeople are fulfilling their mutual responsibilities for mutual care. Two features of his discussion are to my mind unsatisfying. First, the modernist side of the equation comes in for one-size-fits-all condemnation. Indeed, the self-serving motives of the hunting industry in Paraguay are subjected to rather less scorn than are the concerns of academic ecologists to evaluate how regional populations of game animals are doing in scientific terms. Surely though both may be described as modern they apprehend the world rather differently, with rather distinct consequences.

This blanket treatment is emblematic of a real weakness of current discussions of ontology that treat modernity as if it were one unitary thing, rather than an internally complex phenomenon with self-critical capacities. By the same token, in Blaser's text the Yshiro side of the coin is similarly handled. Though it may be the case that the Yshiro do not treat nature as an external object, presumably there exists a range of perspectives among the Yshiro regarding how those human-to-human dynamics are going. If my own fieldwork experience in indigenous communities is any guide, it is certain that some Yshiro thought the quota income was distributed fairly while others disagreed, and both opinions referred to the perceived state of game animals to bolster their arguments. We do not hear about this, however; (nonmodern) Yshiro ontology is treated monolithically the better to employ it to critique a similarly homogenized modernist ontology. For my part, while I am sympathetic to this critical impulse I do not think it is the proper aim of ethnography, which is to document the way internally complex sets of social relations and cultural notions hang imperfectly together in particular times and places. Blaser's book does quite a lot of this and does it very well, as the example I offered previously demonstrates, but the ontology discussion as it is handled in contemporary anthropology does not. Blaser's very fine book is less good than it might otherwise be for having joined it.

Kathleen Lowrey
Department of Anthropology
University of Alberta

HALVERSON, Anders. 2010. *An Entirely Synthetic Fish: How Rainbow Trout Beguiled America and Overran the World,* 288 pp. New Haven, CT: Yale University Press. ISBN 978-0-300-14087-3.

An Entirely Synthetic Fish asks, "So just what is a rainbow trout?" And it answers: "That depends on whom you ask, and maybe even when." This study of historical archives charts how sportsmen and their political allies became fixated on this unassuming fish—originally found only in a few coastal rivers in the Pacific Rim—and began remaking the waterways of the United States to promote its well-being.

Departing from the personalities of nineteenth-century acclimatizers—who sought to populate the wildlands of America with the "best" species of fish—Halverson explores the ambivalent legacies of large-scale breeding and stocking operations centering on the rainbow trout. Having completed a PhD in biology and studied the decimation of frog populations by introduced fish, Halverson is keyed in to the impacts of rainbow trout on local flora and fauna. Ecological disasters, which arose as a result of trout management practices, figure prominently in the book. He describes how in 1962, for example, Wyoming officials killed all the native fish in a huge watershed so that introduced rainbow trout would face no competition.

This book is heavy with solid evidence from historical archives but light on theoretical analysis. Halverson missed an opportunity to engage with lively scholarship on related subjects. *An Entirely Synthetic Fish* is not in dialog with scholars who are contributing to the multispecies zeitgeist that is sweeping the humanities and the social sciences. It is, instead, well-researched historical nonfiction that will undoubtedly find a popular readership among outdoorsmen.

Formerly an avid angler himself, Halverson reports that earlier in his life he "got bored" catching "another ten-inch stocked rainbow." Now he appreciates trout anew. "Hold a rainbow in your hand, and you are holding a savior of democracy," he claims. "Look that fish in the eye, imagine all the effort that humans have put into helping the species achieve a nearly global conquest, and ask yourself which one of you is subordinate in the relationship." Still, Halverson prefers catching "natives" to stocked rainbow trout.

On this last point, Halverson sustains a sense of ambivalence in his concluding paragraphs: "Reading through the letters and public pronouncements of the men who were most responsible for spreading nonnative species like rainbow trout throughout the world in the nineteenth century, I have been struck by the similarity of the rhetoric to those who promote native species restoration today. They, too, were sure they were doing the right thing for the world."

Eben Kirksey
CUNY Graduate Center

HECKLER, Serena, ed. 2009. *Landscape, Process, and Power: Re-Evaluating Traditional Environmental Knowledge*, 289 pp. New York: Berghahn Books. ISBN 978-1-84545-549-1.

In this edited volume the authors take a broad look at the idea of traditional environmental knowledge (TEK) and its practice around the world. While fully embracing the idea that TEK has been "challenged, deconstructed, and reinvented" (p. 1), the volume's editor Serena Heckler notes that approaches to TEK seem to be converging around the interrelated themes of landscape, power, and process. In each chapter, the authors, the majority of whom presented on a panel at the 2004 International Congress of Ethnobiology, address at least one of those themes to provide a diverse contribution on TEK. *Landscape, Process, and Power* will appeal to audiences of anthropology, cultural geography, ethnoecology, and conservation and sustainable development studies.

The book is composed of eleven chapters contributed by anthropologists, ethnobotanists, ecologists, conservation and development practitioners, and an indigenous activist. The chapters cover a wide geographic range, including Europe, Ethiopia, Indonesia, Kenya, and Papua New Guinea. The volume's diversity is both its strength and weakness, as the contributions never quite seem to synthesize around its themes of landscape, power, and process. Yet the chapters' ethnographic richness, varied natural resources, and attention to landscape, power, or process makes this volume an uncommon and expanded treatment of TEK.

The book begins with three chapters that artfully describe traditional environmental knowledge. First, editor Heckler overviews TEK and change, illustrating how the volume's themes of landscape, process, and power are integral to newly expanding notions of TEK. Next, Stanford Zent explores scientific representations of indigenous knowledge, focusing on the fruitful genealogy of TEK and the power of scientific knowledge. In the third chapter, Miguel Alexiades addresses the globalization of traditional knowledge and its subsequent commoditization, politicization, and fragmentation. In these introductory chapters, the emphasis is on the process and power of traditional knowledge, as well as on its integration into scientific literature, itself a changing landscape.

The next four chapters place greater attention on power and landscapes. The contributions by David Carss, Sandra Bell, and Mariella Marzano explore decreasing habitat and fish stocks of Europe and the resultant conflicts among fishermen, the fish-eating Great Cormorant, and conservationists. This section's theme on the power of resource management is continued in Emma Gilberthorpe's chapter on cosmology and landscape in Papua New Guinea (PNG). She details how Fasu people's social networked landscape of paths is being de-emphasized in a material landscape of oil extraction. The emphasis on the material landscape is furthered in William Thomas contribution on Hewa peoples' traditional knowledge of birds in PNG. Thomas examines how ecology's embrace of disturbance facilitated Hewa and conservationists' comanagement of birds in the Hewa landscape mosaic of gardens and old-growth and secondary forests. In the following chapter, Takeshi Fujimoto addresses overlooked elements of landscape by examining how Ethiopia's Malo people use wild plants to indicate appropriateness for agriculture. He summarizes wild plant use as agricultural indicators worldwide and advocated for greater attention to such indirect uses of plants.

In the volume's next four chapters, the contributors detail the power and processes of TEK. In his chapter, Manuel Boissière discusses knowledge sharing among Indonesia's Yali and Hupla peoples, who live in the same village. He finds that the dominance of the more numerous Yali is apparent in some areas (e.g., naming landscape features, shamanism), but not in others (e.g., myths). Also in Indonesia, Daniel Vermonden studies how traditional knowledge of portable trapping, angling, and shark finning is learned. Taking on the Convention of Biological Diversity's assertion that TEK is orally transmitted, Vermonden found that fishing knowledge is learned through practice, through both one's own practice and others' practice as a resource for expertise. Paul Sillitoe's chapter similarly emphasizes the power of technoscience in the attempt to apply the concept of carrying capacity in the New Guinea highlands. He thoroughly explains why an assessment of a landscape's carrying capacity for people involves too much simplification—of population, land use, climate, arable land, and so on—to make it tenable. In the final chapter, Aneesa Kassam and Francis Chachu Ganya counter a discourse on the purported irrationality of nomads by illustrating how Glabra pastoral nomads use and respect customary laws to manage their water rights and pastoral commons in Kenya.

In its entirety, *Landscape, Process, and Power* broadens the traditional TEK literature by emphasizing amplified notions of time and space. In so doing, it offers something of a scattershot of traditional environmental knowledge, with contributions that narrowly integrate with one another through the volume's themes. However, to do otherwise would tame its diversity, to take away some of the volume's power to highlight the multiplicity that has become TEK in the twenty-first century.

Julie Velásquez Runk
Department of Anthropology
University of Georgia

HELMREICH, Stefan. 2009. *Alien Ocean: Anthropological Voyages in Microbial Seas,* 422 pp. Berkeley: University of California Press. ISBN: 978-0-52025-062-8.

In *Alien Ocean: Anthropological Voyages in Microbial Seas,* Stefan Helmreich adventurously tracks oceanographers conducting microbial research. In doing this, he opens up a large part of the world traditionally overlooked in anthropological study. The ocean is a seemingly unpeopled, neutral, blank, wild, and nationless. But by making the ocean an ethnographic site, Helmreich affords us a nuanced look at the ways in which the ocean is lived upon, studied, and immersed into by scientists. Helmreich writes elegantly about a vast space where blankness is made meaningful to human and microbial life as well as the future of the planet in its entirety. In a space typically imagined as unknown nature, what kinds of knowledges and pursuits of knowledges are unfolding to make the ocean known, if not comprehensible?

Helmreich's fieldsites are varied and mimic the partial and temporary sorts of work that microbial oceanographers do—Helmreich, his scientists, and various iterations of microbes and their attached meanings travel to conferences, workshops, laboratories, ships, and the deep sea. He expertly shows us what it means to follow scientific objects. He suggests that his imagined science studies exhortation "follow the microbes" would involve "living in a lab in a sleeping bag, killing time rereading Latour's *Science in Action*" (133). Instead, Helmreich's following of microbes resists the conception of stalking objects linearly: he sidles up to the organisms for brief stints in the ocean, in the lab, and in conference presentations. In each chapter, Helmreich analyzes the ocean, the life that composes it, and the possibilities within it.

The microbes advance and retreat in the projects and stories that marine biologists orchestrate, helping to make careers, tell vignettes about the changing environment, and become marketable and patentable subjects. In between these brief moments where scientists and oceanic microbes make sense to each other, the organisms do their thing, suspended in laboratory coolers and the dark and deep ocean. In the chapter titled "Message from the Mud," Helmreich describes how researchers gather microbes, how they sense the ocean environment, and how they recreate themselves as embodied scientists in relation with the organisms and environments they are interested in.

The chapter titled "Dissolving the Tree of Life" takes us to hydrothermal vents, where life makes sense in ways different to what we are accustomed. Upon their discovery, vents seemed to hold the promise of explaining primitive life: to contemporary vent researchers, the creatures found there are recapitulating the Darwinian, phylogenetic tree of life as well as our insistence that genes serve as boundaries between species at the same time that they create novel individuals. By making a mess out of what it means to be related to each other—kinship at its broadest—gene transfers between hyperthermophiles in the ocean point to the "net" of life, reminding us that the way we conceptualize nature allows and hinders specific representations of it.

In the chapter titled "Blue-Green Capitalism," marine microbes are made meaningful as bits and possibilities of biotechnology. Most marine biotech companies are at the stage where they are trying to articulate the promises of the extremophiles. Some make them out to be workhorses, "the blue-collar workers of the environment," able to be reconfigured and put to biological use outside of the deep ocean (125–126). The fetish of the microbe is also bound up in matters of ownership, discovery, marketing, and patents. Indigenous claims to microbes discovered or claimed by nonindigenous corporations, such as those made by Native Hawaiians, highlight the creative and contested geographies of the open ocean, a space continuously reterritorialized, in the name of knowledge, nativeness, and capital.

"Alien Species, Native Politics" is an important contribution to the anthropological study of nonnative species. In this chapter Helmreich explores how scientists use the tropes of native and alien contextually, in relation to the places where they are working, the cultural and historical dynamics of the places, people, and other species attended to, and their views on the fixity or fluidity of nature. Specifically, Hawaiian organisms count as endemic or invasive depending on which colonial timeline one uses (are canoe species brought to Hawaii from other parts of Polynesia native or introduced?), whether Hawaiian or Latin species names are used, and the acceptability or rejectability of DNA evidence of proof of relationship among subspecies.

At first glance, "Abducting the Atlantic" is a comprehensive accounting of the methods and motivations involved in the scientific gathering of oceanic DNA. At its heart, however, the chapter is a beautiful story about competition between a charismatic, media-savvy, mega-funded scientist and the humble, earnest, folksy, student and faculty team racing to collect these DNA samples. I love the social drama in this chapter—the save-the-world characterizations of finding the ocean's "genome," the likelihood that such a heroic mission could be contaminating the biota that it is collecting, and ways in which the university scientists reject the idea of collecting a stable, permanent ocean "genome" and instead speak elegantly about the dynamism of ocean genetics.

"Submarine Cyborgs" beautifully describes the phenomenological experience of riding a submarine to the ocean floor to take samples—the sociality of it, the fusion-dependence between human and machine, and the politics of national boundaries at the bottom of the sea. Helmreich describes the awe-inspired alienation that submerged scientists experiences, along with their physical "merging with their data" that placing oneself onto the ocean floor suggests. This chapter also tracks the legally ambiguous space of much of the sea floor, which raises questions about extranational space and claims to the organisms and the knowledge generated from them.

In my Antarctic work, I learned that the US Department of State legal counsel for Antarctica was also the lawyer for the open oceans and outer space. So I was pleased to see Helmreich make a similar kinship conceptually, taking us from the oceans to space. In "Extraterrestial Seas," he uses the oceans as a means for imagining "interplanetary ecological stewardship"—an environmentalism that extends the concept of the ecosystem beyond familiar boundaries (270). The oceans, their politics, and their scientifically defined knowledges reorganize what kinds of life may be possible on our planet and past it. Perhaps the extremophile microbes at the deep sea vents relate so differently to the rest of life on Earth because they relate better to microbes on other planets.

In sum, *Alien Ocean* opens up new spaces for what can be considered anthropological. By using the open and deep oceans as a cultural landscape, Helmreich provides us with a close look at how knowledge is formed in an international and interspecies way, and always in relation to scientific expertise. At its core, the book is about microbial politics—negotiations at the broadest sense of the term—between species, nations, personalities, profiteering, and conceptions of life itself. Through his beautiful accounting of the anthropology of microbial seas, and the scientific practices that inform it, Helmreich takes us on an adventure about the possibilities of life in otherworldly places.

Jessica O'Reilly
Department of Sociology
College of St. Benedict and St. John's University
Collegeville, Minnesota

HOLIFIELD, Ryan, Michael PORTER, and Gordon WALKER, eds. 2010. *Spaces of Environmental Justice*, 272 pp. New York: Wiley-Blackwell. ISBN 781-4-44332-452.

In the late 1980s, a growing awareness about the uneven distribution of environmental hazards in low-income communities and communities of color gave rise to the environmental justice movement. As that movement has matured over the past three decades, scholarship about it has also matured and flourished. Early research was often centered in the United States and largely devoted to empirically documenting inequities in the placement of toxic sites and the enforcement of environmental regulations (e.g., Brown 1995; Bryant and Mohai 1992; Wildavsky 1997; Zimmerman 1994). More qualitative studies concentrated on the ways in which race and class experiences inflected activists' expansive definitions of environmental justice (e.g., Checker 2005; Novotny 2000) and on possibilities for cross-class and cross-race coalitions (e.g., Alley et al. 1995; Checker 2002; Moberg 2001).

As the century turned, environmental justice activists extended their networks globally. In 2002, for instance, activists from around the world traveled to Johannesburg, South Africa to attend the Earth Summit for sustainable development. Similarly, environmental justice scholars began situating their research in global contexts and using broader and more sophisticated theoretical frameworks.

Spaces of Environmental Justice, a new edited volume published by Wiley-Blackwell, exemplifies the maturation of this field of research through eight chapters of cutting edge critical geography. Originally presented as papers at the Association of American Geographers annual conference in Chicago in 2007, the chapters complicate traditional questions of space and scale by drawing on Marxist urban political ecology and, in some cases, actor-network theory (ANT). In so doing, they strive to develop more critical analyses of the implications of environmental justice as a discursive framework for activism, policy, and research.

The book is organized into two parts. The first concerns the general theorization of environmental justice, with each chapter emphasizing a different theoretical thread (spatiality, the nonhuman, gender, and the state). The second part grounds those theoretical developments in specific case study material. In the first chapter, Gordon Walker calls for greater plurality in approaches to spatiality including a consideration of the distribution of environmental "goods" (i.e., green space and open space) as well as "bads" (i.e., environmental hazards). The second chapter, by Ryan Holifield, addresses heated debates over Marxist urban political ecology (UPE) and ANT (as defined by Bruno Latour). In brief, Holifield explains that critics of ANT claim that it does not go far enough in analyzing the political factors that drive environmental inequalities while UPE comes under fire for being overly structural and economically deterministic. After reviewing these debates, Holifield uses a case study from his own research in Minnesota to reconcile the two approaches. Although he successfully answers ANT's critics by demonstrating the degree to which that approach can engage politics, by the end of the chapter, Holifield's version of ANT is almost indistinguishable from UPE, except that the former includes considerations of nonhuman agency.

Although it addresses a less sexy topic, chapter 3 is perhaps one of the volume's most significant contributions. Here, authors Susan Buckingham and Rakibe Kulcur delve into the issue of gender—a surprisingly neglected topic in environmental justice literature. The authors argue passionately for a reconsideration of scale that includes the level of the body and the household, as well as the gendering of institutions. Also provocative is Hilda Kurtz's chapter, which reminds us not to neglect the analysis of state-generated definitions of racial categories, as even in a postracial era, they are central to the cause of environmental justice.

The second part of the book presents a diverse and interesting set of case studies. A chapter on gold mining rethinks conceptions of justice by foregrounding the agency of illegal gold miners in Ghana. Certainly the most methodologically interesting of the collection, this chapter also details a highly successful example of participatory research. Meletis and Campbell's chapter on ecotourism in Costa Rica serves as an important rejoinder to economically deterministic arguments. In this innovative study, the authors find that profiting from ecotourism did not inhibit the community from collective action when they traced a local solid waste crisis to the ecotourism industry.

The final two chapters feature work by leading environmental justice scholars. Karen Bickertstaff and Julian Ageyman discuss the importance of scale in a partnership between Friends of the Earth (FOE) and a local community in northeast England. Their analysis reveals how local engagement diminished as FOE shifted the focus of its campaign to global issues. The last chapter by Sze et al. draws on the case of a statewide network of grassroots organizations in California to discuss how an environmental justice framework can be applied to water issues. Here, the authors show that contrary to common scholarly assumptions about the environmental justice movement's provincialism, activists asserted a broad critique of the socioeconomic structures that perpetuate environmental injustice. At the same time, opportunities for activists to voice these critiques were preconfigured to emphasize reformist rather than radical actions. Ultimately, this chapter adds as much to environmental justice literature as it does to our understandings of social movement organizing under neoliberal regimes.

Despite the array of new insights offered by contributions to this volume, it is far more useful to experienced scholars than undergraduates or even graduate students being introduced to environmental justice. The collection has not successfully made the shift from a special issue loosely organized around a central theme (it first appeared as an issue of *Antipode*) to a collected volume. That is to say that the chapters speak to scholars, activists and professionals already well versed in the debates and issues surrounding environmental justice rather than to those new to the conversation. At the same time, those in the former category should take note as this volume clearly represents a crucial step toward the next phase of environmental justice activism and scholarship.

Melissa Checker
Department of Urban Studies
Queens College, New York
Departments of Anthropology and Environmental Psychology
The Graduate Center, City University of New York

References

Alley, Kelly, Charles Faupel, and Conner Bailey. 1995. "The Historical Transformation of a Grassroots Environmental Group." *Human Organization* 54(4): 410–416.

Brown, Phil. 1995. "Race, Class and Environmental Health: A Review and Systemization of the Literature." *Environmental Research* 69 (1): 15–30.

Bryant, Bunyan, and Paul Mohai, eds.1992. *Race and the Incidence of Environmental Hazards: A Time for Discourse.* Boulder, CO: Westview Press.

Checker, Melissa. 2002. "'It's in the Air': Redefining the Environment as a New Metaphor for Old Social Justice Struggles." *Human Organization* 61 (1): 94–105.

———. 2005. *Polluted Promises: Environmental Racism and the Search for Justice in a Southern Town.* New York: New York University Press.

Moberg, Mark. 2001. "Co-Opting Justice: Transformation of a Multiracial Environmental Coalition in Southern Alabama." *Human Organization* 61 (2): 377–389.

Novotny, Patrick. 2000. *Where We Live, Work and Play: The Environmental Justice Movement and the Struggle for a New Environmentalism.* Westport, CT: Greenwood.

Wildavsky, Aaron. 1997. *But Is It True? A Citizen's Guide to Environmental Health and Safety Issues.* Cambridge, MA: Harvard University Press.

Zimmerman, Rae. 1993. "Issues of Classification in Environmental Equity: How We Manage Is How We Measure." *Fordham Urban Law Journal* 21 (3): 633–669.

LANSING, J. Stephen 2006. *Perfect Order: Recognizing Complexity in Bali,* 240 pp. Princeton, NJ: Princeton University Press. ISBN 978-0-69102-727-2.

In *Perfect Order: Recognizing Complexity in Bali,* J. Stephen Lansing sets out to complicate the "conventional Western social science" view of how "traditional societies" are organized. The picture that emerges from his impressive research on Balinese water temples is certainly complex, filled with ample detail to quicken the heart of any red-blooded ethnographer. Although the theoretical contributions of the book may be questioned, the interdisciplinary methodology works to reassert ethnography as a truly holistic science. The underlying aim of the research is seemingly to critique and inform the Western development industry, which historically has neglected or rejected "traditional" ways of being as irrational and contrary to modernization. This is a common anthropological stance—perhaps all too common now. To move beyond this critique for a more substantial impact on the field of international development, anthropologists, Lansing included, must revamp their perception of and interactions with the field.

In the span of a decade, Lansing and his colleagues conducted a comprehensive study of the traditional irrigation system in Bali, including archeological investigation to better understand Bali's history of irrigation development, ecological modeling to conceptualize how the whole network functions to manage water and control pests, and survey research and participant observation to explicate the social and cultural dynamics that contribute to the success—or failure—of both individual *subaks* (groups that manage irrigation) and the water temple network as a whole. The vast empirical evidence produced is alone enough to declare the book a triumph. Nevertheless, theory, too, is expected. One argument Lansing makes is that "traditional" social organization is not as simple as our social science forefathers, such as Emile Durkheim, would suggest. He more specifically rejects French anthropologist Louis Dumont's 1979 thesis that *Homo hierarchicus* rules South Asia's caste system while *Homo aequalis* characterizes the more modern, democratic West. Lansing's study demonstrates how the water temple organization is in fact a bottom-up democracy built on rationalism—not magic, as it appears—and a different sense of self: an interconnected agent as opposed to an atomistic Western self.

A quick critique of Lansing's theoretical perspectives takes two tacks. The first has to do with Lansing's use of contemporary scholarship on Bali, which Bali scholar Howe (2006) argues is lacking. The other relates to Lansing's discussion of social theory, which is at worst a straw man and at best outdated. Though Durkheim is indeed part of our canon, it is more common to talk about culture as "fuzzy" (or some newer analogy) than as an "organism." Ethnographers, at least postmodern ones, do not try as hard anymore to put social life in neat boxes. Life is messy and we like it that way.

Still, Lansing has done incredible research. The interdisciplinary team employed diverse methods, many of which are not commonly used by ethnographers. As a supposedly holistic approach to investigating social life, ethnography could certainly benefit by incorporating more methodology (both analytical frameworks and data collection methods) from other relevant fields. The present study's ecological modeling as well as the complexity framework, for instance, might be added to the ethnographer's toolkit. I am not suggesting that the ethnographer be an expert in everything from computer programming to water sampling—just surf with someone who is (as Lansing did). This study illustrates the value of collaborative research, which has been steadily gaining attention in anthropology.

In the end, however, the book left me unsatisfied. Much of the introduction critiqued the development industry—the Green Revolution, five-year plans, and Lansing's own interactions with development workers who, though happy to visit water temples with him, did not take them seriously in their work. I expected Lansing to return to this theme in the end, but he did not. Besides the composition critique (the introduction should prepare the reader for the conclusion), *Perfect Order* raises the question of how anthropologists can best influence the development industry. Lansing states that early project findings prompted the Ministry of Agriculture to change their approach, but I would argue that development anthropologists can do much more.

First, we might check our usual criticism to make way for a little praise. Academic anthropologists are quick to vilify the development industry, or any power structure, in our efforts to valorize the underdog. There is often good reason for this, as I can attest after thorough review of the development literature. Yet there is also positive development work, as I witnessed during my experiences as a Peace Corps volunteer and independent consultant. Of course we know better than to characterize anything as monolithic, but at some point we started equating being critical with negativity and the opposite with cheerleading. So in deference to our mothers who advised, "If you don't have anything nice to say, don't say anything at all," allow me in this brief space to compliment the UK Department for International Development Guidance Sheets outlining livelihoods approaches to development. An anthropologist could have developed these guidelines, not least for their emphasis on context.

Second, we have to learn to speak with development practitioners and planners, which means adapting to their communication conventions. Despite its innovations, *Perfect Order* is a traditional ethnography. As such, it will not find a wide audience among practitioners in the development field. Certainly, ethnographers should continue to publish in the genre they know best, but we can simultaneously write for other audiences. In the case of Lansing's research, only one of the resulting publications was nonacademic (a technical report by the ecologist on the team). Some anthropologists have made the leap, such as the consultants at Technical Assistance for Non-Governmental Organizations (TANGO) International. They have produced scores of documents resulting from their conversations with the development industry, such as "Household Livelihood Security Assessments: A Toolkit for Practitioners" commissioned by CARE, International. A discussion on how ethnographers could make that communicative leap would advance works like *Perfect Order* to their fullest potential.

Juliana Essen
Soka University of America
Aliso Viejo, California

References

Howe, Leo. 2006. Review: Perfect Order: Recognizing Complexity in Bali. *Anthropological Quarterly,* 79 (4): 777–782.

TANGO International. "Household Livelihood Security Assessments: A Toolkit for Practitioners." http://www.tangointernational.com/index.php?mh=1&mi=101.

UK Department for International Development. Guidance Sheets. http://www.eldis.org/go/topics/dossiers/livelihoods-connect/what-are-livelihoodsapproaches/training-and-learning-materials.

LYON, Sarah, and Mark MOBERG, eds. 2010. *Fair Trade and Social Justice: Global Ethnographies,* 320 pp. New York: New York University Press. ISBN 978-0-81479-621-4.

This edited volume sets out to answer whether fair trade can use market mechanisms to transform the free market. The book explores the paradox of seeking social justice through market-based movements such as fair trade. It is organized into three sections, which cover a wide array of fair trade goods, including agricultural products such as coffee, tea, flowers, and bananas, artisan products, and an alternative currency system. This diversity of products provides a rich introduction to fair trade. Moreover, the volume contributes to a number of pertinent debates in the literature on fair trade, including critiques of increased labor, which is demanded by fair trade's quality and environmental standards placed on producers, the gendered impacts of fair trade, and how consumer actions and preferences influence fair trade standards and production practices.

The first section explores four different commodity systems and the disparities between the strong "social transformation" rhetoric used to promote fair trade and the modest on-the-ground effects fair trade certification has had on farmers' and farm workers' lives. Smith's chapter focuses on the increasing quality standards fair trade coffee farmers must meet, while Moberg explains how fair trade's environmental standards have led to increased labor burdens on banana farmers in St. Lucia. Smith argues that fair trade has increased its focus on quality and the specialty coffee market has sought out third party certification and other programs to offer more environmentally friendly and socially just coffee. In St. Lucia, banana farmers now face several problems associated with fair trade including higher labor costs and the spread of an invasive weed due to pesticide restrictions. Both chapters highlight the increased labor burdens producers face to meet the requirements of the fair trade system.

Ziegler and Besky's chapters also focus on labor and fair trade certification. Ziegler compares fair trade's labor and environmental standards for cut flowers to those of other sustainable or ethical labels for flowers in the US and European markets. Besky compares local labor standards in Darjeeling India to Fair Trade Labelling Organization (FLO) labor standards and finds that FLO standards may be undercutting stronger local labor standards on tea plantations there.

The volume also adds to a small but growing body of scholarship on the gendered impacts of fair trade certification. Moberg argues that female and male banana farmers in St. Lucia have been equally able to take advantage of the benefits of the fair trade market. However, Lyon and Dolan's chapters illustrate that entrenched patriarchies in Guatemala and Kenya, respectively, stand in the way of extending the benefits of fair trade to women.

The second section of the book explores gender and ethnic differences in the impacts of fair trade. Both Dolan and Lyon question fair trade's ability to secure gender equity in societies with longstanding patriarchies. Lyon writes about women struggling to find assistance for their weaving while their husbands and other male relatives are the priority of the local coffee cooperative, fair trade buyers, certifiers, and nongovernmental organizations. Coffee production itself is not a practical or acceptable activity for some women so they have sought alternative income earning activities such as weaving. Similarly, Dolan argues that the gendered inequities she observed among Kenyan small-scale tea producers are more the result of local patriarchies than the impact of transnational commodity chains. Alternatively, Wilson addresses the negotiation of ethnic identity among Ecuadorian artisans who are either permitted or denied access to fair trade depending on whether or not they "act" indigenous enough.

The final section is comprised of three chapters that explore fair trade consumption and the multiple and competing meanings associated with consuming fairly traded products. For example, Papavasiliou writes about the alternative currency in Ithaca, New York called HOURS and its implications for fair trade in this community. The HOURS system, like fair trade, emphasizes the significance of direct relationships between producers and consumers. Doane also explores the relationship between consumers and producers by comparing the perceptions and understandings of fair trade by Midwestern students and coffee roasters to those of Mexican coffee farmers.

Furthermore, M'Closkey explains how fair trade knock-offs of Navajo weavings have had a devastating effect on traditional Navajo weavers in the Southwest United States. Novica, an online fair trade store, has supported the reproduction of Navajo designs by Zapotec weavers in Oaxaca, Mexico. Knock-offs, fair trade or otherwise, are sold for much less than authentic Navajo weavings, thus further marginalizing Navajo artists.

In the conclusion, Jane Henrici reiterates the collective thesis of the volume that fair trade attempts to transform the free market from within by relying on the goodwill of consumers to buy sustainably produced products. The volume demonstrates a number of the challenges and tensions faced by fair trade producers, such as the need for additional labor to meet fair trade standards and the disconnect between fair trade's strong market transformation rhetoric and the actual social and economic benefits afforded to producers. Moreover, the volume highlights the importance of exploring differences within fair trade producer communities by exploring how gender and ethnic differences impact people's access to the fair trade system. In the end, Henrici calls for more transparency in the fair trade system to improve conditions for producers and provide consumers with accurate information.

This volume will be useful for scholars studying fair trade to compare and contrast different commodity systems against each other. It is also a welcome addition for scholars, like myself, examining the gendered implications of ethical trading systems such as fair trade. Moreover, it could be used in courses on social movements, alternative markets, or globalization.

Rebecca Mari Meuninck
Department of Anthropology
Michigan State University

MARSH, Kevin R. 2007. *Drawing Lines in the Forest: Creating Wilderness in the Pacific Northwest,* 192 pp. Seattle: University of Washington Press. ISBN: 978-0-29598-702-6.

American wilderness conservation has received great attention in the annals of popular and scholarly literature. Wilderness conservation, or rather preservation, has often been celebrated for protecting certain bucolic ideals on which the country was supposedly founded; presented as the lasting achievement of a forward-thinking populace within the framework of effective, participatory governance. Consequently, much academic work on the subject has tended to focus on the symbolic notions of wilderness motivating conservation efforts and/or the class-based interests often overlooked in more popular depictions of these movements. In *Drawing Lines in the Forest,* Kevin Marsh turns away from previous examinations of wilderness idealisms to an analysis of the more pragmatic aspects of wilderness designation in the Pacific Northwest. Marsh argues that the creation of wilderness in this area was foremost a material process of land use zoning based on the construction of contested boundaries between landscapes of resource extraction and preservation. Analyzing wilderness designation through the lens of land use zoning allows Marsh to make important contributions to environmental history by providing a means to move beyond discussions centered on problematic conceptions of purity that only serve to further solidify the ideological gap between nature and culture. As Marsh states, "when thinking historically, focusing on wilderness as a form of land use in specific places rather than as a vague and romantic ideology brings us back to the land and illuminates more constructively the historical and environmental significance of political disputes over wilderness areas" (7). Significantly, this approach helps us to understand the complexity of the roles and relationships of power swinging between the actors involved in wilderness debates in the region. Marsh's purposefully materialist examination of wilderness construction, however, does leave a few holes in the story that need to be addressed.

Drawing Lines in the Forest centers on six chapters that recount the changing political economic relationships of wilderness designation in Washington and Oregon from 1950 to 1984. The author begins most chapters with bucolic vignettes drawn from personal experiences of hiking or working as a US Forest Service ranger in the Skykomish district. The vignettes are important not just because they draw the reader into the various chapters, but also because they begin to reveal the theoretical foundations and personal values that are left inexplicit but, nevertheless, structure the arguments that follow. In the course of the reading, it becomes clear that Marsh is a bit more sympathetic to the conservationist cause than he is to that of the American timber industry or even the US Forest Service. Recognizing this philosophical leaning makes it easier to understand some of the author's analytical choices.

The wilderness conservation movement began in the Pacific Northwest as a reaction to increasing levels of Forest Service facilitated timber extraction on public lands in the midst of growing demand and reduced availability following World War II. Initial processes of wilderness designation hinged on the actions and influence of three players: the US Forest Service, the timber industry, and wilderness conservationists. The Forest Service and timber industry maintained the power to structure early wilderness debates while the fledgling conservation movement was originally led by "a loose-knit collection of hikers, scientists, and social liberals" that "would evolve over the next few decades into a powerful grassroots movement in Oregon and across the country" (27). The Forest Service's increased involvement in the timber economy, combined with popular visions of wilderness as an aesthetic setting of untouched land that spanned across interest groups, limited initial wilderness concessions to areas that lacked commercial timber or any semblance of human influence. This narrow view of wilderness and its appropriate values and uses changed as the conservationist movement grew, gaining power and legislative influence. Wilderness debates eventually encompassed the commercially valuable forests and de facto wildernesses of the region for reasons of science and economics as well as aesthetics and recreation as the increasingly powerful conservationist faction worked past the Forest Service and its commitment to multiple use, timber management. Conservationists appealed to Congress to make wilderness designation a legislative process, thus removing primary decision-making power from the Forest Service. These appeals culminated in the passage of the 1964 Wilderness Act, as the pendulum of power swung away from the Forest Service and the timber industry. Marsh argues, "passage of the Wilderness Act of 1964 opened the door for all citizens to get involved on all sides of the debates over where to draw the lines, and the process of defining perimeters of wilderness areas played a major role in expanding the participatory nature of American politics in the post-war era" (151). Consequently, Marsh's biggest contribution to the academy is his effort to highlight the work of conservationists, Congress, the changing conditions of the American timber economy, and the changing mission of the Forest Service in this lesser studied, post-Wilderness Act era.

In this analysis, Marsh tries to paint a complex picture of the give and take between these often competing interests to show how the work of wilderness conservation was never "one-sided, and it never will be" (15). Wilderness conservation was the result of difficult compromises made in the midst of oscillating material interests and power structures. Although Marsh's efforts are predominantly successful, he tends to present an uncomplicated vision of conservationists and their movement as well as an incomplete analysis of forestry and what it means to be a forester. Of course, this simplification is necessitated by the direction of the book. Yet, I do find myself wanting more insight into how the initial class make-up of the conservationist movement provided access to structures of influence that facilitated growth on a national scale. On page 81, Marsh quotes conservationist John Hazle who asks, "Wilderness is for everyone, not just a few?" This is a tremendously important question that is just as important to engage with today as it was in 1959. At its most symbolic level, is it possible for wilderness to be truly egalitarian or would that make it something quite different? It certainly does not seem that wilderness is an appropriate place for foresters to practice forestry. Marsh tends to present foresters as a nebulous group of hard individuals who do nothing more than clearcut trees to supply timber markets. But what is a forester, what are they trained to do, and why? Is a clearcut an intrinsically bad practice? Essentially, what does a forester see when he or she walks into the woods? Likewise, what does a conservationist see? The author argues, "Since World War II wilderness in the United States has been less an idealized abstraction than a set of very real, valued pockets of the American landscape" (143–144). However, it may be more appropriate to say that wilderness has been a set of very real, valued pockets of the American landscape precisely because it is based upon idealized abstractions. This difference in emphasis gets to the crux of my critique for this book. The material and the symbolic are always symbiotic. In his efforts to redress wilderness romanticisms through a purposefully materialist examination, Marsh does not fully analyze the gravity of the idealisms bound in these material processes of wilderness designation.

Jason Roberts
Department of Anthropology
University of Texas at San Antonio

MUSCOLINO, Micah S. 2009. *Fishing Wars and Environmental Change in Late Imperial and Modern China,* 286 pp. Cambridge: Harvard University Asia Center. ISBN: 0674035984.

In recent decades, Garret Hardin's (1968) concept of the "tragedy of the commons" has become an important element of many analyses of contemporary fisheries management. Inspired by the debate surrounding the idea, historian Micah Muscolino traces the dynamics of ocean fisheries' decline in southeast China. He explores controversies over proprietary claims to fishing territories and their entanglement with culture, economy, politics, and science. As "the first major study of Chinese fisheries from the perspective of environmental history" (2), the book offers a rich account of the disputes over fishing in modern China from the late Qing Dynasty in 1800 to the end of Sino-Japanese War in 1945. The book's most novel contribution is Muscolino's demonstration that the reduction of conflict can lead to intensified exploitation of resources. In this way, he intends to challenge an important assumption of the "tragedy of the commons" argument.

Fishing Wars begins with an introduction to the problem of the commons. The idea posits people's tendency to maximize resource-use, and it pinpoints the concurrent free-rider effect that often leads to overexploitation of common-pool resources. Although some scholars hold that Hardin's assumption of universal greed is incontrovertible, others diverge from his perspective in their attempts to conceptualize solutions to problems that revolve around the use of resources held in common. Elinor Ostrom (1990) and Feeny et al. (1990), for example, problematize the notion of the commons by emphasizing community strategies for checking unrestrained competition. Muscolino explores these "coordinating strategies" that native-place groups, the modern state, and colonial imperialism deployed in the battleground of the Zhoushan Archipelago in southeast China.

Muscolino shares Hardin's concern with overpopulation as he traces how national growth led to environmental degradation as China's population doubled from 150 million to 300 million during the late Qing Dynasty, in the eighteenth century. The expansion led to an exhaustion of inland resources, and it promoted intense human migration to the lucrative marine fishing grounds. In Zhoushan, however, the ocean was not simply an unregulated commons. People deployed "unofficial strategies" to avert violent conflicts. Native-place networks worked to divide fishing grounds and settle disputes. In addition, local religion postulated the existence of a divine authority with the power to establish fishing prohibitions.

As China became "modernized" during the Republican Period (1911–1949), these local and traditional strategies gradually lost influence. Observing the decrease of fishing output in the early twentieth century, the development-guided state pushed for scientific research that could optimize harvests. They suggested prohibiting the capture of young fish and identifying new fishing grounds, but their plans failed due to funding shortages. Increasing state control over local management, however, was well underway. Nevertheless, the government did not simply displace traditional mechanisms of native organizations. Rather, it relied on the latter to achieve control via tax collection, thereby augmenting the power of local elites.

Whereas many scholars argue that human competition leads to the decline of resources, Muscolino suggests the opposite: aversion of conflict can achieve the same effect and, in some cases, it can make matters worse. He describes three fishing wars that involved international and domestic contests for control over the Zhoushan's fisheries. They included the Japanese drive for colonial expansion; Chinese native-place groups' competition for fishing grounds; and the Republican government's quest to increase tax collection. In each instance, the ebbing of conflicts among colonial forces, state bureaucracies, and local organizations led to a more intense form of fish harvesting.

Although the book presents valuable historical information on how political conflicts intersect with environmental change, the author's binary view of society and environment—as well as his failure to problematize the "tragedy" discourse's assumptions about the inherent destructiveness of human nature—threaten the book's potential contributions. Muscolino crafts a picture of human beings' inevitable and inescapable domination of nature. His conservation ethic, which advocates the preservation of nature "for its own sake" (188), reflects a simplistic idealization of wilderness. In one instance, he deplores the devastating effects of Japan's occupation of the region, especially its destruction of boats and conscripted fishermen. Immediately, however, he stresses the unexpected benefit of the situation: "War had devastating consequences for China's natural landscape, but it was an ecological respite for fish populations" (179).

Muscolino's essentialist approach to human-environment relations reflects his inclination toward environmental determinism and human exceptionalism. In recent years, scholars of science and technology studies have reconceptualized the boundaries between humanity and nature in order to rethink the relations between people and the places in which they live. Donald Moore identifies the potential of "assemblages" to "displace humans as the sovereign makers of history" (2005: 23–24). Hugh Raffles (2002) proposes the notion of "intimacies" to describe the affective relationships between humans and nonhumans in the Amazon, and he criticizes a linear model of "carrying capacity" that positions dwellers as inevitable degraders of land. In her ethnography of the H5N1 virus, Celia Lowe (2010) proposes the term "multispecies clouds" to incorporate the collections of species from viruses, to poultry, to humans, all of which are embroiled in the avian flu pandemic.

The above approaches historicize and problematize notions of "nature" and "population." Moreover, they help us to imagine more dynamic relationships between humans and the nonhuman environment—something that Muscolino fails to do.

Yu Huang
Department of Anthropology
University of Washington

References

Feeny, David, Fikret Berkes, Bonnie J. McCay, James M. Acheson. 1990. "The Tragedy of the Commons: Twenty-Two Years Later." *Human Ecology* 18 (1): 1–19.

Hardin, Garrett. 1968. "The Tragedy of the Commons." *Science* 162 (3859): 1243–1248.

Lowe, Celia. 2010. "Viral Clouds: Becoming H5N1 in Indonesia." *Cultural Anthropology* 25 (4): 625–649.

Moore, Donald. 2005. *Suffering for Territory: Race, Place, and Power in Zimbabwe.* Durham: Duke University Press.

Ostrom, Elinor. 1990. *Government the Commons: The Evolution of Institutions for Collective Action.* New York: Cambridge University Press.

Raffles, Hugh. 2002. *In Amazonia: A Natural History.* Princeton, NJ: Princeton University Press.

PERRAMOND, Eric P. 2010. ***Political Ecologies of Cattle Ranching in Northern Mexico: Private Revolutions,*** 259 pp. Tucson: University of Arizona Press. ISBN 978-0-81652-721-2.

Political Ecologies of Cattle Ranching in Northern Mexico explores the cultural geography of an iconic industry with deep historic roots. This study focuses on ranching in the Río Sonora region located in the north central part of state of Sonora, roughly equidistant between Nogales and Hermosillo. The economic mainstay is beef cattle production. Though the author is cautious about generalizing too quickly from his case, it is important to remember that *ranchero* culture is spread widely and diversely across the Mexican landscape. Depending on where it is found, it may center on dairying, beef cattle, or dual purpose beef and dairy production employing more intensive or extensive production methods. The development of a dominant *ranchero* culture in Mexico's north evolved out of what Marvin Harris (1985) described as the "New World cattle complex" that started with colonial Spaniards in Argentina, parts of Central America, and in western and northern Mexico. It is an industry well suited to certain areas, but was often undertaken because other forms of agriculture were simply impractical. Thus, *rancheros* have often found themselves on Mexico's environmental and political margins.

Perramond makes a number of well-argued points about the misreading of *ranchero* culture both within and outside of Mexico. And so the author sets out to challenge a number of stereotypes. First, he contends that *rancheros* should not be seen through the lens of a simplistic binary of private rancher versus communal *ejidatario.* He points out that many private ranchers are also *ejidatarios,* and that there are diverse forms of private land tenure (e.g., co-owned operations). He also notes with frustration that *rancheros* are often characterized as large-scale oppressors of small-scale *ejidatarios.* He suggests that the lack of attention from the social sciences to private ranching is a by-product of the myth of *rancheros* as hegemonic capitalist bully. I would come at this from a different, but complementary, direction. Following Rosaldo (1989), our approaches are shaped by an ethnographic mapping of the world that draws us to the exotic (e.g., peasants, *ejidatarios,* or *indios* who are not necessarily committed capitalists), and eschews ethnographic subjects that are too much like us (e.g., private ranchers and farmers who are clearly capitalist in orientation).

Second, Perramond contests the notion that *rancheros* form a homogeneous group of large-scale entrepreneurs, which is addressed through a fine-grained examination of the diversity of ranching operations. *Ranchero* enterprises vary in size, tenure system, environmental conditions, and management styles. Indeed, the author argues that smaller operations are better managed and more efficient than their larger (and more locally prestigious) counterparts. This important conclusion counters common wisdom from both neoclassical and Marxist economics, which would contend that the key to success is dominating control of the means of production. Numerous studies suggest that increased landholding is a good predictor of farming/ranching success because unequal access to land and other critical productive resources drive rural inequality. This study reminds us that axiomatic bromides should be approached with caution and skepticism.

Third, Perramond challenges another common binary: human versus environment. He notes, "Class, ethnicity, family roots and relationships, extended kin disputes, and localized negotiations at all levels of governmental power play a role in the cultural geography and political ecologies of natural resource use and abuse" (190). Add free-ranging cattle to the mix and matters are further complicated. His approach is, thus, richly conceived and a welcome addition to the literature on *ranchero* culture.

The book's strength is its ability to engage this complexity and sort it out in a way that is accessible and understandable. Perramond skillfully combines a quantitative approach with an ethnographer's sensibility. One of the most engaging aspects of this work is its interweaving of ethnographic scenes that support and reinforce the analysis. This is also a work about the intense forces of change afoot in Mexico. The opening of the Mexican economy and potencies of globalization have had a devastating effect on many agricultural sectors in Mexico. The beef cattle industry can count itself among them.

In what is otherwise a solid contribution to the literature on rural Mexico, I do have a few quibbles concerning areas that could have received some expanded attention. First, not only are these ranchers struggling with a faltering agrarian economy, they are also threatened by other pernicious forces of globalization. For example, the narcoeconomy's infiltration into the Río Sonora area is examined only briefly. We know that the northern tier states are hotly contested because they are gateways to the lucrative US consumer market. That the area sits just off the corridor that links the coast with Hermosillo and the border town of Agua Prieta should make it prime turf for drug smugglers. The fact that Perramond's identity was questioned (13)—was he a DEA, FBI, or CIA operative?—even in the mid-to-late 1990s suggests this area was already integrating into the growing narcoeconomy. I suspect, though, that this relatively brief treatment may be a result of the majority of the fieldwork, upon which the book is based, being conducted in the mid-to-late 1990s. My own experience working in Michoacán saw the tipping point occur in the early 2000s when the audacious presence of the narcoeconomy became startlingly visible and public. Today I suspect that the presence of the narcoeconomy in the Río Sonora Valley is equally much more directly and dangerously present. Second, though the book explores local political dynamics in interesting ways, it does not deal much with the relationship between ranchers and the state. Specifically, it would be interesting to know how Sonoran development culture operates through agencies like the state's agricultural ministry (Secretaría de Agricultura, Ganadería, Desarrollo Rural, Pesca y Alimentación), which oversees the administration of federal development funds. For example, are most development funds diverted to more lucrative agricultural areas of the state (e.g., irrigated districts with intensive export cash crop production)? For those funds that do make it to the Río Sonora area, what political dynamics shape their distribution? At the local level, it would also be interesting to know more about the local cattlemen's associations and how they operate because they too receive funding and resources from their state-level organization. Again, trading on my Michoacán experience these were hardly transparent organizations. Finally, while the book does explore the historical roots of *ranchero* culture in Mexico (Appendix A), there is little reference to the larger literature on *ranchero* culture outside of Sonora and northern Mexico (e.g., Esteban Barragán on Michoacán, Claudio Lomnitz-Adler on Morelos and the Huasteca Potosina, or Frans Schryer on Hidalgo). I find this surprising since there is some attempt in the book to generalize about the Mexican ranching industry and *ranchero* culture. It is also surprising because the author is very sensitive to the diversity within his microregion. Reflecting his case against those from across Mexico would make for a very interesting comparison.

In sum, quibbles aside, this is an excellent book about arid-lands ranching in the state of Sonora. It should be read by anyone interested in contemporary agrarian struggles in Mexico. By challenging a number of stereotypes surrounding *rancheros*, Perramond has made an important contribution that has value in both academic and applied circles, and will hopefully find a wide audience.

James H. McDonald
College of Humanities and Social Sciences
Southern Utah University

References

Harris, Marvin. 1985. *Good to Eat: Riddles of Food and Culture*. Long Grove, IL: Waveland Press.

Rosaldo, Renato. 1989. *Culture and Truth: The Remaking of Social Analysis*. Boston: Beacon Press.

RINGHOFER, Lisa. 2009. *Fishing, Foraging and Farming in the Bolivian Amazon: On a Local Society in Transition*, 249 pp. London and New York: Springer. ISBN 978-9-04813-486-1.

Fishing, Foraging and Farming in the Bolivian Amazon: On a Local Society in Transition presents an academic description of the materials and energy flow (MEFA) within a Tsimane' village territory. The MEFA methodology seeks to describe the social and environmental interface and the use of the resources (energy) that are available to the social system (internal or external). It describes stocks and flows of human and livestock populations, infrastructure and artifacts (social metabolism) as well as the territory size and the local net primary production (NPP) used to analyze the proportion that is harvested or the human appropriation of net primary production (HANPP). By comparing the Bolivian results with other MEFA studies in Asia, Ringhofer seeks to identify possible external and internal drivers of transitions in the energetic relationship of communities with their environment. She concludes from her findings that Boserup's "intensification theory, based on the analysis of pre industrialized societies, still proves a useful frame for understanding the dynamics of contemporary agrarian societies" (231).

The book includes an extensive background review of the MEFA framework and Social Metabolism theory updating the reader on the theories and methods for tracking energy flows through a social system. In the field, the author measured a Tsimane' community's resources, tracked their sources and estimated the territory's accessible biodiversity. She combined these results with measurements of the time allocated to the different activities that were organized in four economic regimes—individual, family, economic and community by gender and age groups. The results are expressed in energy, joules (TJ, GJ, MJ) per capita per year, permitting comparisons between other communities and habitats.

Ringhofer promotes the use of the MEFA toolbox in the evaluation of development plans and options because of the valuable insights gained through these detailed analysis. For example, the comparison of her results with three other agrarian communities in Asia highlighted the following conventional wisdom that could be very important to decision makers:

> "1. Exposure to markets even over a long period of time, does not automatically lead to a change in local production patterns, or as a consequence, an intensification of land use.
>
> 2. Impacts of state interventions, do not necessarily trigger the move forward in their transition life cycle, but rather had punctual impacts." (231)

There is no doubt that these insights are valuable, but does it take a MEFA evaluation to have understood these points? The author spent eight months collecting data for her analysis, and probably at least a year analyzing to reach the conclusions she has drawn in this book. Her data were supplemented by available anthropological and environmental knowledge from many sources, local and nonlocal, published or verbal, cited or not. Most underdeveloped areas do not have this wealth of information available, nor do the development organizations have the skills to estimate the large factors used Dr Ringhofer, such as NPP, or the harvested portion, the HANPP.

Apart from that limitation, eight months of data on harvest, time allocation, and other economic movements within a community is not an annual cycle. I would have liked to see an inclusive discussion of the extrapolation methods used. The original data was not presented and tables and graphs were labeled as 2004, but data was collected for only the latter half of that year. A straight multiplication as an extrapolation could be debilitating, given the cyclical nature (yearly and seasonally) of resource availability in the Bolivian lowlands. Could any extrapolation insufficiencies have been aggravated by being expanded to landscape level and also interpreted at an individual yearly value? To clear up this doubt, a discussion about the elasticity of these numbers would have been useful because it could show how much conclusions might change if the real value is actually 10 percent of the estimator used in the calculations. This is especially important because the author is also using literature to derive estimators to represent the Tsimane' territory, and the NPP is a particularly critical component in the MEFA framework.

Perhaps in an effort to lighten the heavy intellectual discussion of the academic tools of MEFA, the author's style shifted on occasion into more colloquial descriptions of the Tsimane' people, culture, activities, and environment. But this probably did not help advance Ringhofer's desire for "science to step down from the world of the abstract and feed these insights back to those actors, whose future may be directly at stake" (241). The attempt at popularizing this highly academic work left us with awkward sentence structure, dangling participles, and typographic errors that were numerous and important enough to mention (e.g., page 144 reads, "47 of the entire village area"). "In my mind" (a statement used often by the author) and the continual use of strong adjectives biased the attempts at ethnographic description and sometimes diverted from the point being made. Sometimes the book took a stance on certain topics without knowing enough about them, such as "The humid savannah of the region is seasonally flooded, and combined with its loamy texture, makes it unsuitable for agriculture" (145), whereas some crops actually thrive in these soils, and pre-Columbian populations in the same area used intensive raised field agriculture. Another example closer to my wildlife manager heart appears on page 113: "[Due to] game exhaustion in and around Campo Bello, hunters are increasingly forced to widen their hunting radius to the surrounding savannah region or deeper into the forest. That is why local hunters prey on small animals such as peccaries, rodents, primates and birds. Large mammals like deer, tapir or peccary, while desirable, are becoming increasingly rare." The author does not indicate the source of these statements that she presents as truths, not opinions. Even given her limited understanding of the lowland Bolivian hunting systems, she did not suggest that the size of the wildlife capture basin or the area needed for the production of the harvested wildlife were considered in the estimation of the biodiversity harvested or the HANPP.

The title of this book falls short of describing the breadth of the information contained within. Although the book may not meet its goal to convince development agencies and government planning officials to use the MEFA framework and toolkits described, it does provide students and professionals of sustainability science a unique view into the application of theoretical models to real-life scenarios. The book explains the theoretical background for the MEFA methods and compares the results to other locations, but it does not really answer the development agency's pressing questions such as which crops to develop or how to prevent unsustainable practices. But then probably no book could contain the required local knowledge, skills, and relationships to strengthen critical thinking by local partners and improve decision-making processes for selecting the best options for each unique situation.

Wendy R. Townsend
Noel Kempff Mercado, Natural History Museum
Santa Cruz, Bolivia

SCHELHAS, John, and Max J. PFEFFER. 2008. *Saving Forests, Protecting People? Environmental Conservation in Central America,* 330 pp. Lanham, MD: AltaMira Press. ISBN 978-0-7591-0947-6.

In this timely book, John Schelhas and Max J. Pfeffer offer a valuable addition to the burgeoning literature addressing biodiversity conservation from a social science perspective by exploring the intersection between global environmental discourses and local beliefs and values within communities subject to conservation interventions. As the authors note, a substantial body of work has analyzed conservation practices as an expression of particular environmental discourses; however, relatively little has taken the next step to explore how these discourses manifest within the minds/bodies of discrete individuals in conservation-affected communities. Rejecting the common assumption that environmentalism is solely the elite preoccupation of affluent Westerners, Schelhas and Pfeffer explore how environmentalisms are glocalized in particular contexts through syncretism between global and local ideoscapes, and, ultimately, how this dynamic influences the potential to realize effective conservation within less-developed societies subject to interventions commonly informed by discourses originating in the Global North.

To accomplish this, the authors employ schema theory, a perspective developed within cognitive psychology and anthropology that treats subjects' beliefs and values as inscribed within discrete, bounded units (called schemas) that prescribe, often at a less-than-conscious level, appropriate thought and behavior within a given situation. Through in-depth interviews, the researchers seek to elicit the environmental schemas implicit in informants' explicit statements, coding interview transcripts to identify recurrent words or phrases (termed verbal molecules) that signal the common patterns ordering informants' rhetoric. Their study encompasses two sites in Central America, each comprising several communities of rural farmers adjacent to national parks: La Amistad in Costa Rica (part of a larger transboundary conservation area shared with Panama), and Cerro Azul Meambar in Honduras. These sites were chosen, in part, because they are seen to represent distinct park management strategies: a traditional "fortress" model in la Amistad; and a more inclusive integrated conservation and development (ICD) approach in Cerro Azul Meamber, allowing for a cross-context assessment of the relative influence of each approach. Within each site, extensive interviews guided by schema theory were conducted with select informants, after which a formal survey was administered to a representative sample to test the frequency of common responses documented in initial interviews.

The data reveal that, within both sites, local residents tended to spontaneously avow the importance of preserving intact forests, citing a variety of popular environmentalism concepts to support this position, which the authors summarize as follows: "(1) their importance for purifying air and producing oxygen (often comparing the forest or park to a lung), (2) their role in maintaining rainfall and water for human use … (3) the importance of the forest as a source of food for wildlife, and (4) the importance of forests for future generations" (57). Such statements, indeed, were so commonly repeated that the authors call them "canned responses," the frequency of which "sometimes frustrated us" (207). As a result, Schelhas and Pfeffer contend that these ideas likely held only "lip service" motivation for most informants, meaning that "people can state beliefs and values from dominant (global) social discourses about the environment but that these have little motivating force" (222). At the same time, most locals appeared to be more strongly motivated by utilitarian livelihood concerns, while their actions either pro- or contra-conservation were clearly constrained by regulatory structures enforced by state agents as well.

One wonders, however, to what extent these findings are an artifact of the methods used to generate them. The authors acknowledge that their own status as "expatriate researchers who were on a familiar basis with park and forest conservation staff and were asking questions about forests and values almost certainly led people to put forward the most positive conservation beliefs and values that they had" (207). This is particularly significant given the reality that locals in both sites had at times come into conflict with park officials. Yet Schelhas and Pfeffer display a certain ambivalence concerning the extent to which admittedly canned statements in fact reflected a deeper ecological commitment, suggesting that while their informants appeared to offer "a coherent view of what they think is common opinion with reference to what they think they should (according to outside norms and pressures) be thinking about something," such ideas "may in fact be accepted by them as appropriate belief and value" (209).

The problem is that there is no decisive means, given the available data, to determine whether this is so. This leads to some perplexing equivocation, wherein, for instance, the authors assert "a strong local sentiment for forest conservation" (201) in their Honduran case, while elsewhere qualifying that "it is not clear that strong social norms had developed in local communities" (221). One wonders what other rich and compelling data might have emerged had the researchers been able to employ, say, long-term participant observation, establishing sufficient rapport to accompany informants during their daily activities and gain access to backstage discussions. The research sites seem ripe, in particular, for analysis in terms of covert everyday forms of resistance, a dynamic that the authors acknowledge may in fact be occurring (227) but which they have no way to assess.

In addition, given that the two study sites were selected due to the distinct strategies (fortress vs. ICD) employed in park management, I would have liked to see more explicit comparison of how these different strategies influenced locals' environmental values and conservation behavior. (In this, however, there may be a question of accuracy in their characterizations; I am not familiar with the Honduran site but do have some direct knowledge of the Costa Rican case, and while, until recently, the park was indeed managed primarily on the fortress model by the state, ICD projects were in fact introduced by prominent NGOs during the 1990s.) In general, more discussion of the actual practice of conservation in both sites would be useful, as what is presented speaks to the likelihood of some complex local politics to which the authors' data only begin to allude.

These qualifications, however, do not detract substantially from the overall value of the work. Conceptually, the book challenges us to pay more attention to the intersection of North/South and global/local in the formation of "environmental subjects." Methodologically, the study offers a novel approach for investigating this process (and, in a series of appendixes, a wealth of materials to guide such inquiry), in addition to providing rich food for thought concerning the relationship between study design and results as well as appropriate strategies for studying on-the-ground conservation practice, particularly within contentious communities.

Robert Fletcher
Department of Environment, Peace, and Security
University for Peace
Ciudad Colón, Costa Rica

TRUBEK, Amy B. 2008. *The Taste of Place: A Cultural Journey into Terroir,* 296 pp. Berkeley: University of California Press. ISBN 978-0-52025-281-3.

The French term *terroir* evokes an almost mythical connection between people and place—a connection resulting in a culturally distinct product. Terroir has no equivalent in English and is often simply glossed as territory. This straightforward interpretation fails to capture any of the term's subtle nuances and downplays terroir's potential to rethink physical, material, affective, and conceptual terrains. This narrow conception of terroir as territory also prevents us from understanding how and why terroir associations appear to be proliferating beyond conventional arenas of wine and other food products to a broad range of cultural products that are the result of "local" practices and resources. Trubek's book is a welcome contribution to a small but growing body of academic literature in the US that actively engages with terroir. On her self-described journey to explore terroir as a "set of values, practices and aspirations" (xv), her work contributes to enduring debates within anthropology and other social sciences, including the relationship between nature and culture, an engagement with global systems (one cognizant of the "beleaguered categories of local and global" (xvi), commodities and hierarchies of value, and issues of authenticity, place making, and cultural identity.

From the initial chapters, Trubek locates her exploration in the sensory experience of taste. Her focus plays on our understanding of taste as having the "right" kind of aesthetic judgment (explored in great detail in Bourdieu's *Distinction* [1984]) but one that is inseparable from physiological taste experiences. How is it, she wonders, that we come to think that something tastes good (as well as being in good taste)? How does taste become intertwined in particular places and then, in turn, how do places become a condition through which taste, particularly at a global scale, is imagined? These questions highlight both the tangible and intangible dimensions of taste summarized in the French phrase *le goût de terroir.* When we invoke terroir (and even the more prosaic English taste), it emphasizes the ephemeral or cultural qualities attached to these terms, interwoven with particular places and practices that are transferrable to, and possibly through, people and products. Trubek's appropriation of the double helix in the text, moreover, is intended to provide that visual map of often nonlinear connections among products, places, and peoples across various scales of time and space. Trubek explores these connections over a diverse terrain—from a cultural history of terroir in France, to California as the epicenter in the development of a taste of place in the US and then back east to examine various efforts to build deep and sustainable "buy local" networks (from restaurateurs in Wisconsin to farmer-chef partnerships in Vermont). One of the strengths of the overall text is that Trubek provides a cultural biography of commodities like wine, cheese, hickory nuts, and maple syrup, providing a rich description of the values associated with these objects, how these objects became associated with these values, and some sense of how these values travel through these objects—all without losing a sense that these are material objects produced in and through specific places (Appadurai 1986; Kopytoff 1986).

Trubek's journey into taste, unsurprisingly, begins in France. The first chapter presents a brief history of terroir and explication of the various authorities (from formal state apparatuses to the informal role of France's "tastemakers") involved in creating the French sensibility of taste. Taste in France, she argues, is a form of local, situated knowledge. Chapter 2 uses winemaking and the failure of the American Mondavi family to acquire a French wine domaine to highlight one of the book's recurring themes—the relationships among terroir, nature, culture, and science and the manner in which their intersection varies cross-culturally. The subsequent chapters focus on the development of an American taste of place—one spurred on by changing values around taste, agriculture, and identity in the United States. She argues that the global spread of terroir may both "reveal a food culture and build one" (94). The final substantive chapter explores how food and food practices are made knowable to consumers, arguing against understanding this process solely as a matter of savvy branding. A brief epilogue enumerates major points of the book, arguing that a vibrant taste of place—one resting on values that support sustainable agricultural practices and inform our tastes—is possible within a global food system although this prospect depends on our ability to see food as more than a simple commodity.

The Taste of Place serves as an important contribution to the academic work on terroir. The book represents one of the few monograph-length anthropological explorations of terroir in the United States. Trubek's journey across diffuse settings not only reflects the breadth of terrain where terroir has developed and where it is actively being built, but also captures the sense of terroir as a complex whole encompassing geology, environment, agriculture, and tradition. In this respect, it demonstrates the recursiveness of terroir—where place informs practices and values and where practices and values also shape places. It is here where Trubek argues that a dynamic sense of place (rooted in technological advancement *and* a commitment to local tastes) will save terroir products from simply becoming food commodities. In this respect, Trubek reminds us that these debates are more than economic and technological choices but also reflect broader ethical challenges that we face. Are we committed to local foods, community, and farmers over the cheap foods brought to us by industrialized food systems? Here, the book intersects with public debates and a spate of popular books about globalizing food systems, industrialized agriculture, and their implications. Similarly, Trubek's text is accessible yet remains conceptually rich. Thus, the book can and should be used not only by researchers interested in the politics of terroir, agricultural networks, commoditization, aesthetics, and globalization but also in coursework on culture and consumption, food studies, and ethnography and the sensory experiences.

Megan Tracy
Department of Sociology and Anthropology
James Madison University

References

Appadurai, Arjun. 1986. "Introduction: Commodities and the Politics of Value." Pp. 3–63 in *The Social Life of Things: Commodities in Cultural Perspective*, ed. A. Appadurai. Cambridge: Cambridge University Press.

Bourdieu, Pierre. 1984. *Distinction: A Social Critique of the Judgment of Taste.* Trans. Richard Nice. Cambridge, MA: Harvard University Press.

Kopytoff, Igor. 1986. "The Cultural Biography of Things: Commoditization as Process." Pp. 64–91 in *The Social Life of Things: Commodities in Cultural Perspective*, ed. A. Appadurai. Cambridge: Cambridge University Press.

VAYDA, Andrew P. 2009. *Explaining Human Actions and Environmental Changes*, 303pp. Lanham, MD: AltaMira Press. ISBN 978-0-7591-0323-8.

This book can be highly recommended to scholars and students, not only anthropologists and ecologists but also philosophers. It consists of essays old and new (the old ones updated) by one of the great figures in human ecology. Vayda was a student of Julian Steward, the founder of cultural ecology, and thus got in on the ground floor of anthropological studies of the environment. He went on to a sterling career as a teacher—his students in ecological anthropology range from Roy Rappaport to such current rising leaders as Bradley Walters. Throughout, he has stayed on the cutting edge of the field, never resting on his laurels or falling into repetitiveness. He remains active today.

The essays in this book center on his rigorous ideas about anthropological explanation. Throughout his career, he has grown more and more critical of vast, vague generalizations, and also of simplistic or monocausal explanations in social science. His view, as developed in these essays, is that a particular "event"—any specific occurrence on the ground that one wishes to explain—should be the unit of analysis. Then one should work back to construct a causal chain. In doing so, one needs to keep in mind T. C. Chamberlin's ([1890] 1962) famous "method of multiple working hypotheses," an old idea that never goes out of date. Biological, social, economic, cultural, and any and all other possible causal factors should be considered. For example, in explaining wildfires in Indonesia, it is not enough to consider greedy illegal burners, or drought, or government neglect of forests; one must consider all three of these and more. Vayda points out that one must explain why many forests did not burn, as well as why so many did.

Doing so involves a process of "abduction," a term and concept taken from Charles S. Peirce. Abduction is rather like induction: one looks at what is here and now, tries to figure out how it got that way, and tests the hypotheses one develops. It is more or less like a detective working back from a crime.

Vayda is most scathing—and perhaps at his best—in critiquing simplistic explanations. Current fashions in ecological anthropology include pseudo-Darwinian explanations on the scientistic end of the field and political ecology on the humanistic end. Vayda is merciless to both. Much of the Darwinian work conspicuously lacks Darwin's meticulous experimentation and proof, and thus becomes very close to just-so stories. Simplistic ecosystem explanations, including some of Vayda's own in his earliest work, are similarly critiqued as shaky biology. I am sure Vayda would say the same of the resilience discourse that has peaked since his essays appeared.

Political ecology has tended to blame (note: blame, not explain) global political forces for much of the environmental problems of small-scale communities. Without exonerating the politicians, one may certainly ask what else is going on—how much of the problem is due to climate change, population growth, local migration, or any of a myriad of other factors. Political ecologists have responded that many of them do take account of such matters; only the naive, especially those outside of anthropology (political ecology being an interdisciplinary field), resort to one-factor explanation. Indeed; but we are warned.

Vayda is also merciless to overblown jargon. One footnote captures the spirit: "Extreme current examples of claims of the latter kind [that vast, vague entities can "cause" things] are the many claims involving 'globalization,' which … has transmogrified from being a label for certain modern-world changes that call for explanation to being freely invoked as the process to which the changes are attributed" (24). I would add neoliberalism, governmentality, and resilience as other examples of this depressing tendency, so ancient and familiar among scholars.

My one criticism is that Vayda sometimes ignores his own strictures. For example, in discoursing on sacred groves (35), he cites some anecdotes to argue that sacred groves are protected by distance from settlements rather than by sacredness. This flies in the face of many studies (including mine) that describe sacred groves adjacent to large villages, towns, and temples, and preserving major biodiversity in spite of it. There are thousands of such groves in the world, and they deserve the kind of abductive explanation that Vayda advocates.

Such minor criticisms merely make Vayda's case even stronger. This book is a highly important cautionary note for those who would explain human actions.

E. N. Anderson
Deptartment of Anthropology
University of California, Riverside

Reference

Chamberlin, T. C. [1890] 1965. "The Method of Multiple Working Hypotheses." *Science* 148: 748–759.

NOTES TO CONTRIBUTORS

The *Environment and Society* Editorial Board welcomes contributions. Authors should submit articles as attachments by e-mail, formatted as Microsoft Word or Rich Text Format files. Please e-mail articles for consideration or questions regarding submissions to the Editors:

Paige West, Dan Brockington, ares.journal@gmail.com

Please direct questions regarding books for review or expressions of interest in reviewing books to the Reviews Editors:

Michael Cepek, michael.cepek@utsa.edu
Jamon Halvaksz, Jamon.Halvaksz@utsa.edu

All submitted articles should be original works and not concurrently under consideration by any other publication. Articles should total approximately 7,000 words (including notes and references), although longer articles might be considered.

Each article must include an abstract of not more than 150 words and 6 to 8 keywords. The abstract should summarize or paraphrase the text and should avoid duplicating it verbatim. Complete contact information for each author is required, and biographical data of approximately 100 words must be sent with the article.

The *Environment and Society* style guide is based on the *Chicago Manual of Style*, 14th and 15th editions, with occasional deviations based on the publisher's preferences. The journal uses American English punctuation and spelling, according to *Merriam-Webster's Collegiate Dictionary* or the *American Heritage College Dictionary*.

Please consult the journal's style guide for the following conventions, among others: the presentation of numbers and dates; the proper use of notes and the citing of references; the requirements for the submission of graphic materials (illustrations, maps, photographs, figures, tables).

Manuscripts that have been accepted for publication but do not conform to the *Environment and Society* style guide may be returned to the author for amendment. Upon acceptance, authors are required to submit copyright agreements and all necessary permission letters for reprinting or modifying copyrighted materials. The author is fully responsible for obtaining all permissions and resolving any associated fees.

www.ingramcontent.com/pod-product-compliance
Lightning Source LLC
LaVergne TN
LVHW060035210926
848481LV00013B/3869